W9-CMP-983

DEMCO

ENVIRONMENT, DEVELOPMENT, AND PUBLIC POLICY

A series of volumes under the general editorship of
Lawrence Susskind, *Massachusetts Institute of Technology*
Cambridge, Massachusetts

CITIES AND DEVELOPMENT

Series Editor: Lloyd Rodwin, *Massachusetts Institute of Technology*
Cambridge, Massachusetts

CITIES AND CITY PLANNING
Lloyd Rodwin

THINKING ABOUT DEVELOPMENT
Lisa Peattie

CONSERVING AMERICA'S NEIGHBORHOODS
Robert K. Yin

MAKING-WORK
Self-Created Jobs in Participatory Organizations
William Ronco and Lisa Peattie

CITIES OF THE MIND
Images and Themes of the City in the Social Sciences
Lloyd Rodwin and Robert M. Hollister

NEIGHBORHOODS, PEOPLE, AND COMMUNITY
Roger S. Ahlbrandt, Jr.

HERE THE PEOPLE RULE
Selected Essays
Edward C. Banfield

THE ART OF PLANNING
Selected Essays of Harvey S. Perloff
Edited by Leland S. Burns and John Friedmann

Other subseries:

ENVIRONMENTAL POLICY AND PLANNING
Series Editor: Lawrence Susskind, *Massachusetts Institute of Technology*
Cambridge, Massachusetts

PUBLIC POLICY AND SOCIAL SERVICES
Series Editor: Gary Marx, *Massachusetts Institute of Technolgoy*
Cambridge, Massachusetts

Here the People Rule
Selected Essays

Edward C. Banfield
Harvard University
Cambridge, Massachusetts

Plenum Press • *New York and London*

Library of Congress Cataloging in Publication Data

Banfield, Edward C.
  Here the people rule.

  Includes bibliographical references and index.
  1. United States — Politics and government — Addresses, essays, lectures. 2. Political
participation — United States — Addresses, essays, lectures. I. Title.
JK21.B24   1985   320.973   85-9504
ISBN 0-306-41969-6

© 1985 Plenum Press, New York
A Division of Plenum Publishing Corporation
233 Spring Street, New York, N.Y. 10013

To Laura

Acknowledgments

I wish to acknowledge debts to several friends: Lloyd Rodwin, who—although he does not share many of my views—put a lot of effort into getting this project under way; Martin Meyerson, who helped me find an organizing principle that suited me; Christopher C. DeMuth, who made illuminating comments on some of the introductory notes; and Michael O'Hare, who gave good advice on what to include in the final section.

No less than five chapters are reprinted from books edited by Robert A. Goldwin. These would not have been written were it not for the very special flavor of the public policy conferences that he invented and over which he still presides. For more than thirty years Bob has charmed, stimulated, and—within the limits of the possible—educated me. Need I say that I am grateful?

Barbara M. Verdi, production editor at Plenum Press at the time the book was made, has my appreciation for a first-rate job.

Contents

PART I.   COMPROMISE AND REFORM   1

1.   Federalism and the Dilemma of Popular Government   5

2.   In Defense of the American Party System   20

3.   Party "Reform" in Retrospect   38

PART II.   DEMOCRATIC POLICYMAKERS   53

4.   Making a New Federal Program   57

5.   Evaluating a Federal Program   97

6.   Revenue Sharing in Theory and Practice   110

PART III.   POLITICS VERSUS ADMINISTRATION   123

7.   Policy Science as Metaphysical Madness   127

8.   Corruption as a Feature of Governmental Organization   147

9.   Ends and Means in Planning   171

10.   The Training of the Executive   182

PART IV.   THE CAPACITIES OF LOCAL GOVERNMENT   209

11.   The Political Implications of Metropolitan Growth   211

12.   The City and the Revolutionary Tradition   230

13.   A Critical View of the Urban Crisis   245

14.   The Zoning of Enterprise   256

PART V.   THREE PROBLEMS FOR DEMOCRACY   269

15.   Present Orientedness and Crime   271

16.   How Many, and Who, Should Be Set at Liberty?   282

17.   The Dangerous Goodness of Democracy   299

PART VI.   POLITICAL ECONOMIZING   303

18.   Some Alternatives for the Public Library   305

19.   Economic Analysis of Political Problems   317

20.   Are *Homo Economicus* and *Homo Politicus* Kin?   328

          Index   341

Introduction

Most of the essays in this volume have appeared in scholarly journals or in books edited by others. A few are published here for the first time. None has been taken from one of my books. A would-be reader would have to go to much trouble to find them; that is the reason for bringing them together.

Collections of essays are frequently miscellanies. This one is not. Except for the final two chapters, all deal with some aspect of the American political system. Some have to do with the structure and functioning of the federal system, others with the nature of public—and incidentally other—organization, and still others with the causes and supposed cures of the social problems that government is nowadays expected to solve or cope with. The two final chapters are about the relationship between economics and political science; for lack of a better term they may be called *methodological*.

The essays are at once descriptive and analytical. Concretely, their subject matter is the American political system. In my view, the feature of that system that is analytically most interesting—that is, the one that is most useful in explaining it or (if there is one) any other system having the same feature—is the extreme fragmentation of formal authority. Many consequences follow this fragmentation, the most obvious perhaps being widespread participation—interference, some might call it—in the day-to-day conduct of government. It is to stress this analytical focus that the book is titled *Here the People Rule*, and not, as it might have been if the focus were on the concrete and descriptive, *American Government and Politics*. The title, incidentally, is a sentence taken from the statement made by Gerald Ford when, after the resignation of Richard Nixon, he became president of the United States. One may profitably contrast it with the famous phrase of Jefferson's favorite philosopher, John Locke: "the people shall judge" (also now the title of a book).

Several analytical themes recur in the essays, most of them related to what a few lines before I called the analytical focus of the book. Each section of the book is preceded by a short introductory note, the purpose of which is to call attention to the themes that unify that section and to relate them to the most general theme of the book. The essays have been reprinted without change under their original titles.

It may be useful to give some account of the mode of thought that is reflected in these essays. For space and other reasons, it will have to be a very brief account; nevertheless it may help the reader to place what follows in a framework that will make it more intelligible.

Let us begin with the nature of man. In the eighteenth century, it was widely believed that man is by nature mainly a creature of passions and only incidentally one of reason. His overriding passion is to be admired (no man ever got enough admiration to satisfy him, John Adams wrote, and the neglect or contempt of the world "is as severe a pain as the gout or stone"). This and other passions lead men to be good members of society, but it also leads them to violate laws and to tyrannize when doing so will bring them admiration. The great problem of government is to place limits on the lust for power, and this, as everyone knows, was a principal reason for the Founders' preoccupation with devising a system of checks and balances and a division of powers; as Publius wrote in *Federalist* 51, "ambition must be made to counteract ambition."

Whatever limitations it may have from other points of view, this conception of the nature of man is, in my opinion, the only appropriate one for the political thinker. It makes conflict, actual or potential, the central, ineradicable fact of life. Reasonable discussion does of course occur, but the nature of man is such that it can never be the principal basis of order in society. Order depends ultimately upon the exercise, actual or potential, of political power.

The emulative or, if one prefers, selfish nature of man precludes in general the possibility of voluntary action to secure a good in which all will share. This point is developed somewhat in the introductory note to Part I.

The system of checks and balances that arises from a recognition of the emulative nature of man results in a wide distribution of power and in a public opinion that holds (the Founders to the contrary notwithstanding) that everyone has a right, indeed an obligation, to participate—or interfere—in the day-to-day conduct of government. As I said before, this is the defining characteristic of rule by the people. In other democracies it is taken for granted that the people will give or withhold consent at intervals and that in the meantime the government, so long as it has a working majority in parliament, has the exclusive power to make and implement plans. Rule by the people, by contrast, necessitates creating a sufficient coalition of powerholders issue by issue.

In such a system, the function of the politician is to find the terms on which the many holders of bits and pieces of power will collaborate: he is a broker who arranges deals in this way. Obviously his talents and skills are very different from those of the statesman, whose function is to discern the content of the common good and to devise strategies for securing it. Although they are analytically unlike, the politician and the statesman may be concretely the same: no one will doubt that Lincoln, for example, was both a politician and a statesman. It will be seen, however, that it is in the nature of rule by the people (as opposed, say, to parliamentary democracy) to make more use of politicians than of statesmen.

The intellectual, meaning here one whose talent is for the manipulation of abstract ideas, has nothing in common with the politician and, except as he may sometimes be used by the politician, plays no part in American politics. As in the previous paragraph, I speak here of analytical types. It sometimes happens that an intellectual is also a politician: Lincoln is a good example here too. The combination does not occur very often, however, and when it does the individual finds it expedient to keep his intellectuality out of public view.

Rule by the people allows no place for the making and carrying out of comprehensive, internally consistent plans. Planning in this sense is incompatible with the character of a highly fragmented political system. To be sure, the universities give degrees in planning, and there are countless governmental planning bodies, most of which produce documents called plans. When government acts, however, it is not on the basis of these plans; rather it is on the basis of compromises that have been reached among competing interests. For further testimony on this, see the words of Woodrow Wilson quoted in the introductory note to Part III.

Government by compromises often leads either to stalemate or to outcomes that are indefensible from a public standpoint. The reader will find examples in the accounts given here of the Model Cities Program (Chapter 4) and of site selection by the Chicago Housing Authority

(Chapter 9). Such outcomes often impose very high costs upon those most affected, and, what is worse, they raise doubts about the ability of the government to meet emergencies of a kind that require an internally consistent line of action. Government by pressure groups, as political scientists are fond of pointing out, has its good—even indispensable— side in that pressure groups not only bring information that would other- wise be overlooked to the attention of officials but also oblige them to take it into account. Moreover, not all such groups are "selfish." As Jeffrey M. Berry writes in *Lobbying for the People*,[1] there are many "public interest" groups, the defining characteristic of which is that the benefits they seek are not for members, supporters, or activists but for "the public." It goes without saying that such groups sometimes have narrow or mis- taken notions of what constitutes the public interest; nevertheless the effort to distinguish public from private interest is significant.

Taking the pluses and minuses into account as best I can, it seems to me that government by compromise is a luxury that democracy can and should afford except possibly in the gravest emergencies of war.

The citizen is rarely a very competent ruler. As Joseph Schumpeter remarks in *Capitalism, Socialism, and Democracy*,[2] when he deals with public as opposed to private affairs (himself, his family, his business, and so on), the individual's sense of familiarity and responsibility is usually much reduced, in which case "ignorance will persist in the face of masses of information however complete and correct."

Acting without any immediate responsibility for the outcome, the citizen, it seems to me, is apt to show naiveté in three contexts that are of particular interest here.

First, he is likely to favor any changes ("reforms") that are likely to make the political system more democratic. The democratic ideal allows for the exercise of only such power as arises from reasonable discussion about the common good in which all participate; all other sources of power—for example, "machines" and "bosses"—are incompatible with the ideal and should therefore (the citizen is apt to think) be eliminated. The ideal, however, is unrealizable: government cannot operate *solely* on the basis of reasonable discussion, and the elimination of illicit sources of

---

[1] Jeffrey M. Berry, *Lobbying for the People* (Princeton: Princeton University Press, 1977).

[2] Joseph Schumpeter, *Capitalism, Socialism, and Democracy* (New York: Harper and Brothers, 1942).

power will, if carried far enough, render it ineffective. For a discussion of these matters in concrete terms, the reader is referred to Chapters 2 and 3 on reform of the American party system.

Second, the citizen is likely to overlook the systemic character of institutions and to suppose that an intended change can be effected without thereby giving rise to consequences that are neither intended nor wanted. Here again the chapters on reform of the party system are illustrative. A (political) system is a set of relationships such that a disturbance at one point in the system is felt throughout it and may change its entire character. It is naive for a reformer to think that he can change one feature of a system without affecting others or to ignore the real possibility that in improving matters in one respect he may unwittingly make them worse in other, perhaps more important, respects.

Third, the citizen is prone to assume that the morality appropriate to personal relations (with family, friends, neighbors, business associates, and so on) is also appropriate to affairs of state. Chapter 17, "The Dangerous Goodness of Democracy," contends that this is mistaken simplicity.

An explanatory principle put forward in several of these essays is the special outlook, or ethos, of the growing American middle class. For present purposes, the middle class consists of those persons who expect to send their children to college or who have attended college themselves. Whether because of their interest in formal education or for other reasons, middle-class people tend to take more account of the future than do working-class people, to have more interest in public affairs, to be more willing to make small sacrifices of private for public interest, and to participate to a greater extent in public affairs. One might infer from this that continuing "middle classification" will eventually immunize public opinion against the kinds of naiveté that I have described. On the contrary. In my judgment, the middle class, especially its higher reaches, has a stronger attachment to the democratic ideal than does the working class and a more unquestioning faith that good intentions will produce good consequences. Confident that goodwill and expertise can solve, or at least alleviate, all social problems, a middle-class person is very likely to feel a positive obligation to undertake reform no matter what the costs.

The typical middle-class reformer has undue confidence in the use of social science and of organized knowledge generally. He fails to see that the problems of society are at bottom moral and political, arising from differences of opinion as to what is and, especially, as to what ought to be. Insofar as such problems are solvable, it is by discussion, bargaining,

and struggle; to try, as the reformer is prone to do, to substitute information for wisdom is likely to exacerbate rather than to alleviate these
problems, which arise from the nature of man and in one form or another
are ineradicable.

While failing to recognize the essentially moral nature of social
problems, the reformer is given to moralizing about them, and this is
worse than futile. The moralizer urges as a basis of action moral principles
or standards that cannot be applied in the circumstances in which action
must be taken: he averts his gaze from the facts that constitute the problem
while advocating what, if the situation were different, would be a right
course of action.

It should be emphasized that the reformist impulse exists among
conservatives as well as liberals. Even among conservatives of the lib-
ertarian variety—perhaps especially among them—there is a distaste for
politics and compromise, a persistent effort to substitute for it a system
of rules adherence to which will produce "welfare" ("the public interest"),
and a characteristic confidence in the efficacy of information as opposed
to wisdom. Like the liberal, the conservative reformer wants to break up
concentrations of government power. Ironically, the efforts of both lead,
albeit in somewhat different ways, to increasing the concentrations. Neither the liberal nor the conservative seems able to accept the central fact,
which is that the nature of the American political regime does not allow
of more than a dash of consistency in policy. A reformer who wants to be
effective must reconcile himself to recycling old compromises. That rule
by the people is incompatible with rule by general principles is a painful
reality that reformers cannot bring themselves to accept.

In my view, the reformer-moralizer constitutes a serious problem
for rule by the people. For one thing, he recommends measures that
cannot work and are apt to make matters worse. For another, he blames
"society" for its failure to do the undoable and by so doing he creates
disenchantment, cynicism, and mutual distrust: in short, he undermines
the consensual basis of the society. Thus Ramsey Clark, quoted in Chapter
1, asked: "What is to be said of the character of a people who, having the
power to end all this [slums, racism, crime, etc.], permit it to continue?"

To say (as I said previously) that the essays in this book are descriptive and analytical implies that they are free, or almost free, of value
judgments. I do not accept the categorical distinction that is often made
between *facts* and *values*: each, it seems to me, partakes in some degree
of the nature of the other and all judgments, including of course those of

probability, have a subjective element. To say what the world *ought* to be like is a task that challenges the highest faculties of intellect as well as imagination. It is not, however, the one to which I have addressed myself. Although this book is by no means free of opinions, those that it contains are, I think, opinions about the nature of what is, rather than of what ought to be.

I hasten to add that I do not regard these as "mere" opinions. The assertions on which most of the weight of argument rests—that the American political system is highly fragmented, that interest groups play an important part in the day-to-day conduct of affairs, that comprehensive plans are rarely if ever made the basis of action, that the middle class is the main support of reform efforts—these and other assertions of the same generality are supported not only by data presented or referred to in the essays but are in the nature of common knowledge among students of the American political system. Where there is difference of opinion it is mostly not about the facts but about their causes and, especially, their probable consequences.

Although they all discuss problems of one sort or another, none of the essays ends with a set of policy recommendations. (A partial exception is Chapter 18, where some "minor changes" in library practices are put forward as "illustrative ideas.") The reason for not making recommendations is that I do not see how, after explaining why things are as they are, one can go on to say how they could be different. The point is made by Frank H. Knight in words that I quote in Chapter 10: "scientific explanation of what is demonstrates that it is inevitable under the given conditions."[3] To this one may of course reply that a recommendation need only consist of showing how the given circumstances may be changed so as to make possible a different and better outcome. No doubt it is practicable to make such a showing in some matters. With regard to those discussed here, however, I do not think that it is at all practicable. The "givens" here are altogether too immovably given: to propose a change in the structure of the American government or in the ethos of the middleclass, for example, would be a waste of words at best and a misrepresentation of reality at worst.

This view cannot properly be called *Panglossian*. I do not claim that all is for the best in this best of all possible worlds: only that I am not

[3]Frank H. Knight, *On the History and Method of Economics* (Chicago: University of Chicago Press, 1956), 143.

able to specify the means by which it may be made better in important respects.

It is, of course, a conservative view. It is not the kind of conservatism that says most people are not worth bothering with or the kind that says the devil take the hindmost. What it *does* say is that the nature of man makes conflict inevitable, that social institutions are the outcome of a long process of trial and error, that they cannot be reformed by central planning and direction, and that the goal of public life, which is to afford every person a real opportunity to live a life that is fully human, requires above all the exercise of prudent judgment, a faculty given to relatively few.

One may well wish that the world were different—that men were capable of governing themselves and of realizing their full human potential by means of reasonable discussion. But as these essays show, the world is not what one would wish, and government, even government by the people, must take it as it is.

PART I

Compromise and Reform

The essays in this first section deal with the structure of the American political system and how the unfolding of two different and opposed logics affects the most essential and characteristic feature of the system, its extreme formal decentralization.

The intention of the principal figures among the Founders was to establish a strong and stable government based upon the consent of the people. This was impossible if, as most of the Founders believed, men were both rational and self-interested. As the philosopher Thomas Hobbes had pointed out more than a century before, unless such men stood in fear of coercion they would renege upon their agreement (a constitution, for example) whenever they thought it to their advantage to do so. Although none seems to have had Hobbes's warning in mind, some of the Founders—certainly Madison—saw very clearly that it would be difficult, perhaps impossible, to construct a government that would guarantee freedom while at the same time allowing its exercise.

Ingenious as were the efforts of Madison and others to cope with the Hobbesian dilemma, they failed. "Federalism and the Dilemma of Popular Government" shows that the Founders themselves reneged on the terms of their agreement almost at once, and it traces the subsequent ending of the Great Compromise by the terms of which the powers of the national government were to be "few and defined." The paper concludes that however much one may deplore it, there is indeed a tendency in popular government to exceed whatever limits are constitutionally set upon its sphere.

It may be objected that the Hobbesian logic applies only insofar as the members of a public are radically self-interested. In fact, modern

1

*theorists believe—as did the Founders—that people acting in large groups
generally* do *act in this way (see the citations in Footnote 17 of the essay),
and some have pointed out that even a society of altruists would be
incapable of acting in the general interest in certain circumstances (see
the discussion of "the problem of assurance" in John Rawls).* [1]

*The other two essays in the section deal with an element of the
political system that the Founders did not contemplate but that has nevertheless long been one of its principal components: the political party. Here
again the argument is that the unfolding of a certain logic tends to undermine and destroy the institutions that support, albeit imperfectly, the
value premise upon which the logic depends. But whereas in the first
essay the problem arose from the prevalence of radical self-interestedness,
here it arises from something almost its opposite: commitment to an ideal
of democracy that deems all power illegitimate except that that arises
from reasonable discussion about the common good in which all participate.*

*"In Defense of the American Party System," written in 1960, con-
tends that the logic of this ideal of democracy requires "reform" to eliminate all forms of power—for example, corruption, log rolling, the exercise
of charm or charisma—that do not arise from reasonable discussion about
the common good in which all participate. The efforts of party reformers,
accordingly, are to eliminate the many features of party organization and
practice that are irreconcilable with the democratic ideal. Unfortunately,
these very features are indispensable to the functioning of the party system
and thus also to the maintenance of a political system that has been a
bulwark of civilization. Government deriving its power solely from reasonable discussion is impossible, but reformers inspired by the ideal of
democracy will be impelled to make reforms nevertheless until the eventual
culmination of their efforts: the destruction of the imperfect democracy
that they are endeavoring to bring into accord with the impossible ideal.*

*The third essay, written some twenty years later, describes changes
that have been made, or have occurred, in the party system in the interim.
While acknowledging that the reforms of these years (mostly in the Democratic party) were motivated in part by the special interests of candidates
and constituencies, the author shows that the most important reforms
were justified on the grounds that they would make the party more democratic. In fact, the essay argues, the reforms seriously weakened the
parties without really making them more democratic.*

[1] John Rawls, *A Theory of Justice* (Cambridge: Harvard University Press, 1971).

In essence, the retrospective look at the parties says, "I told you so." In further retrospect—that of the summer of 1984—some third thoughts are called for. The reforms that enabled George McGovern and Jimmy Carter to be nominated were reversed by the Democratic party after 1976 sufficiently to enable Walter Mondale, a product of the old-style "back room" professionals, to defeat Senator Gary Hart, who won more primaries than Mondale. Obviously the logic of the democratic ideal does not have quite the imperative force that the essays claim. But the end is not yet. (For another and up-to-date account of party reform, the reader may wish to consult Nelson Polsby.[2])

---

[2]Nelson Polsby, *Consequences of Party Reform* (New York: Oxford University Press, 1983).

CHAPTER 1

Federalism and the Dilemma of Popular Government

Beginning with the Truman administration, there have been persistent and sometimes strenuous efforts to devolve many federal activities to state and local governments.[1] These efforts have almost entirely failed. Meanwhile, the number and variety of federal interventions in what until recently were generally considered local matters has increased at an accelerating rate. One gets an indication of what has been happening from the fact that federal domestic spending more than doubled as a percentage of gross national product in the twenty-five years prior to 1980.

The endeavor to reduce the federal government's domestic role moves along three lines: transfer of programs to the states, relaxation of conditions attached to grants, and imposition of a fixed limit on federal spending. Curiously, there has been almost no effort to find and set forth some general principles that will define and limit the legitimate sphere of government, especially the federal government. (The Advisory

---

From *How Federal is the Constitution?*, eds. R. A. Goldwin and W. Schambra (Washington, D.C.: The American Enterprise Institute, 1985). Copyright 1985 by The American Enterprise Institute for Public Policy Research. Reprinted by permission.

[1] For an account of early efforts to devolve federal programs, see Morton Grodzins, "Centralization and Decentralization in the American Federal System," in Robert A. Goldwin, ed., *A Nation of States* (Chicago: Rand McNally & Co., 1961), especially pp. 4–15. For comprehensive treatment of a wide variety of issues relating to federalism, see the eleven volumes published by the Advisory Commission on Intergovernmental Relations under the series title *The Federal Role in the Federal System: The Dynamics of Growth*. The final volume was published in 1981.

Commission for Intergovernmental Relations proposed in 1980 that the president convene key federal, state, and local officials to discuss—among other things—"what the term 'national purpose' now means in a regulatory and programmatic sense," but nothing came of this.)[2]

The reason for this inattention to principles may be that practical people sense that an effort to agree upon principles sufficiently definite to serve as criteria, or to implement them if they are agreed upon, would be both futile and divisive. Americans are accustomed to think of their government as a limited one; the doctrine of limited government was, as Herbert J. Storing has written, "the great principle of the revolution."[3] But they are also accustomed to thinking that government ought to serve the people in whatever ways they want, and—as the vastness of the federal establishment testifies—they approve, tacitly at least, of political arrangements that bring government into almost every aspect of life.

The import of this paper is that it is indeed futile to try to limit government to some defined sphere. Nothing of importance can be done to stop the spread of federal power, let alone to restore something like the division of powers agreed upon by the framers of the Constitution. The reason lies in human nature: men cannot be relied upon voluntarily to abide by their agreements, including those upon which their political order depends. There is an antagonism, amounting to an incompatibility, between popular government—meaning government in accordance with the will of the people—and the maintenance of limits on the sphere of government. This fact, clearly recognized by Thomas Hobbes, John Locke, and many later writers, constitutes "the dilemma of popular government."[4]

In brief, the argument is that leading figures at the Convention of 1787 saw the failure of state governments and of the government (if it

[2]The commission's recommendations appear in *Conference on the Future of Federalism: Report and Papers* (Washington, D.C.: ACIR). The quoted words are on page 132.

[3]Herbert J. Storing, *What the Anti-Federalists Were For* (Chicago: University of Chicago Press, 1981), 53.

[4]Hobbes wrote, "There must be some coercive power to compel men equally to the performance of their covenants, by the terror of some punishment, greater than the benefits they expect from the breach of their covenant." *Leviathan*, Part 1, Chapter 15. For Locke's formulation of the dilemma, see Robert Horwitz, "John Locke and the Preservation of Liberty," *The Political Science Reviewer* 6 (Fall 1976): 325–353. See also Robert A. Goldwin's discussion in his chapter on Locke in Leo Strauss and Joseph Cropsey, eds., *History of Political Philosophy*, 2nd ed. (Chicago: University of Chicago Press, 1981).

can be called that) existing under the Articles of Confederation as resulting from the unwillingness of individuals and of state governments to subordinate their special interests to the general interest; that—judged by the intentions of these leading figures—the convention itself failed from this same cause; that Madison thought (or affected to think) that the structure of the new government might enable it to escape the dilemma; that the agreement the framers reached on the division of powers was immediately breeched by the leaders among them; that during fifty years of "elite democracy" the powers of the federal government were much enlarged; and, finally, that the pressures of pluralistic democracy subsequently eroded most of the remaining limits on federal power.

It was the dilemma of popular government that brought the framers together in Philadelphia on May 18, 1787. The great principle proclaimed in the Declaration of Independence—that governments derive their just powers from the consent of the governed—had been made the basis of state constitutions during the Revolution. Now, in the 1780s, it appeared that the revolutionary principle had not worked and perhaps could not be made to work. The federation of states organized under the Articles of Confederation had driven George Washington almost to distraction during the war; it was manifestly incapable of coping with the great problems of the new nation—stabilization of money and credit, foreign affairs, the disposal of public lands, and the rest. Washington, Alexander Hamilton, James Madison, Benjamin Franklin, James Wilson, Gouverneur Morris, and others thought it obvious that a national government exercising its power upon people, rather than upon states, was urgently and indispensably needed. "The primary source of all our disorders," Washington wrote, "lies in the different State Governments, and in the tenacity of that power which pervades the whole of their systems."[5]

The power to which he referred was popular government or, as some saw it, democracy run riot. To Madison and some others the need to protect people from capricious and unjust state governments was as compelling as that to create a power capable of dealing with national interests. The failures and the injustices of state governments, Madison said, were "so frequent and so flagrant as to alarm the most steadfast friends of Republicanism." These evils, more than anything else, had produced the uneasiness that led to the convention and prepared the

[5]Max Farrand, ed., vol. 3, *The Records of the Federal Convention of 1787* (New Haven: Yale University Press, 1937), 51.

public mind for a general reform.[6] Under the sway of popular leaders
(*local demagogues*, the nationalists called them), the state legislatures
passed paper money, legal tender, debt installment, and other laws serving dominant interests—debtors or creditors, merchants, or members of
a religious sect—without regard to people's rights or to the general interest of the society.[7] The powers of government exercised by the people
"swallowed up the other branches," Edmund Randolph told the convention; none of the state constitutions had provided sufficient checks against
democracy. In tracing to their origins the evils under which the United
States labored, he said, every man had found there "the turbulence and
follies of democracy."[8]

Some of the nationalists wanted to abolish the states entirely. That,
however, was politically impossible. In presenting to the convention the
Virginia Plan (in which, incidentally, the word *national* appeared no less
than nineteen times), Randolph proposed that the idea of states be "nearly
annihilated."[9] Madison, the principal drafter of the plan that Randolph
presented, had in mind a way to end state influence. It was to give the
national government a veto over state laws. The necessity of a general
government, he said, arose from the propensity of the states to pursue
their particular interests in opposition to the general interest. Nothing
short of a veto on their laws could stop this.[10]

Citizens and governments, it appeared, each presented the dilemma
of popular government. Citizens used their freedom in ways that infringed
upon the rights of fellow citizens: debtors, for example, used government
to cheat their creditors. States used their freedom to serve local interests
at the expense of national ones; as Wilson put it, "Each endeavored to
cut a slice from the common loaf, to add to his own morsel, till at length
the confederation became frittered down."[11]

In July, the small states asserted their special interest over the
national one by demanding equal weight with the large states in the

[6]Quoted by Gordon S. Wood, *The Creation of the American Republic, 1776–
1787* (New York: W. W. Norton, 1969), 467.
[7]Charles F. Hobson, "The Negative on State Laws: James Madison, the Constitution, and the Crisis of Republican Government," *William and Mary Quarterly*,
third series, 36:2 (April 1979): 222.
[8]Farrand, *Records* 1:27, 51.
[9]Ibid., 1:24.
[10]Ibid., 2:27.
[11]Ibid., 1:166–67.

general government. The result was the Great Compromise. Each state legislature would send two representatives to the Senate; in the House representation would be by popular vote. The new government would thus be "partly federal, and partly national"—"a composition of both."[12]

To the nationalists, this was failure. After the compromise committee had made its report, Washington wrote Hamilton: "I *almost* despair of seeing a favorable issue . . . and do therefore repent having had any agency in the business."[13] Madison thought that the union was now "a feudal system of republics," which was hardly better than a "Confederation of independent States." The Constitution, he wrote Jefferson shortly before the convention adjourned, would "neither effectually *answer* its *national object* nor prevent the local *mischiefs* which everywhere *excites disgusts* ag[ain]st the state governments."[14] Wilson was sensible that perfection was unattainable in any plan; that (an equal voice for the states in the second branch), however, was "a fundamental and a perpetual error."[15]

Popular government in the states had prevented the making of a popular government in the nation. So at least it seemed to Washington. "The men who oppose a strong & energetic government," he wrote Hamilton, "are, in my opinion, narrow minded politicians, or are under the influence of local views."[16]

Madison saw clearly the dilemma of popular government (which, of course, would have existed even if the states had been annihilated). In the convention debates and later in the first paper that he wrote for *The Federalist,* he grappled with it. The problem for a friend of popular governments, he wrote in *Federalist* 10, is to find a proper cure for the violence of faction "without violating the principle to which he is attached." In other words, it is "to secure both the public good and the rights of other citizens" while preserving "the spirit and the form of popular government."

Faction, which he defined as any combination of citizens acting contrary to the rights of others "or to the permanent and aggregate interests of the community" arose from the natural diversity in the faculties of man,

---

[12]*Federalist* No. 39.
[13]Farrand, *Records* 3:56.
[14]Quoted by Hobson, *Creation of American Republic*, 230–31.
[15]Farrand, *Records*, 2:10.
[16]Ibid., 3:56

the protection of which (the dilemma again) "is the first object of government."

The problem was not that men were radically selfish (if that were the case, Madison wrote in *Federalist* 55, "nothing less than the chains of despotism can restrain them from destroying and devouring one another"); rather it arose from the diversity of faculties, which produced motives by no means all of which were blameworthy—zeal for opinions and attachments to leaders, for example—but which led to differences and animosities and to the formation of factions. The principal task of legislation was to regulate faction. "But [the dilemma again] what are the different classes of legislators themselves but advocates and parties to the causes which they determine?"

The danger of *majority* faction troubled Madison most. In *Federalist* 10 he acknowledged in two sentences that minority faction might clog administration or even convulse society; it could be defeated, however, by regular vote of the majority. Majority faction, on the other hand, was capable of destroying popular government. Here he saw but did not fully integrate into his argument a phenomenon which has only recently been rationalized by decision theorists—namely that men are motivated differently in large groups than in small.[17] Religious and moral motives, which are not adequate to control injustice and violence on the part of individuals, he wrote in *Federalist* 10, "lose their efficacy in proportion to the number combined together, that is [another dilemma] in proportion as their efficacy becomes needful." The point was one that he had elaborated in the convention:

> In all cases where a majority are united by a common interest or passion, the rights of the minority are in danger. What motives are to restrain them? A prudent regard to the maxim that honesty is the best policy is found by experience to be as little regarded by bodies of men as by individuals. Respect for character is always diminished in proportion to the number among whom the blame or praise is to be divided. Conscience, the only remaining tie is known to be inadequate in individuals; in large numbers, little is to be expected from it.[18]

[17] See James M. Buchanan, "Ethical Rules, Expected Values, and Large Numbers," *Ethics* LXXXVI:1 (October 1965): 1–13, and Mancur Olson, *The Logic of Collective Action* (Cambridge: Harvard University Press, 1965).

[18] Farrand, *Records* 1:135.

If moral motives have less effect on men in large bodies than small, one might infer that governments should be small. (Madison acknowledged the logic of this in a later writing; the "more extensive a country, the more insignificant is each individual in his own eyes," a condition that "may be unfavorable to liberty.")[19] Other considerations, however, were of more importance. There being no motives that could be relied upon to restrain a majority, the need was to prevent a majority from having the same passion or interest at the same time or, failing that, to prevent it from uniting to carry out its schemes. This would be impossible in a "pure democracy" (a society consisting of a small number of citizens who assemble and administer the government in person); in such a society there could be no cure for the mischiefs of faction. By contrast, where the sphere of government was extensive, the community would be divided into so great a number of interests and parties that they might be prevented from cohering and acting as a majority.

In the convention, where he had supported his argument for an extensive government with the claim that there was less faction and oppression in the large states than in the small (delegates from Connecticut and Maryland asserted the opposite),[20] Madison said that an extensive government "was the only defense agst. the inconveniences of democracy consistent with the democratic form of Govt."[21]

In *Federalist* 10 he did not repeat this claim. Here he proposed (as he had in the convention) an additional device that did not altogether escape the dilemma. This was creation of a power capable of coercing a faction, including, of course, a majority one. This was the way he had hoped to end faction in the state governments: by virtue of its veto power, the central government would act as a "disinterested and dispassionate umpire" without becoming a tyrant.[22] This, he had told the convention, "is at once the most mild & certain means of preserving the harmony of the system." Its usefulness could be seen in the example of the British system where the prerogative of the Crown "stifles in the birth every Act of every part tending to discord or encroachment." The prerogative had

---

[19]Gaillard Hunt, ed., *The Writings of James Madison* vol. 6 (New York: G. P. Putnam's Sons, 1906), 70.
[20]Farrand, *Records* 1:406, 2:4.
[21]Ibid., 1:134–35.
[22]Hobson, *Creation of American Republic*, 231.

sometimes been misapplied, to be sure, but Americans did not have the same reason to fear misapplications in their own system.[23]

In *Federalist* 10, the function of the disinterested and dispassionate umpire was to be performed by representatives "whose enlightened and virtuous sentiments render them superior to local prejudices and to schemes of injustice." The public views would be refined and enlarged

> by passing them through the medium of a chosen body of citizens,
> whose wisdom may best discern the true interest of their country,
> and whose patriotism and love of justice will be least likely to sacrifice
> it to temporary or partial considerations.

The public voice, pronounced by such representatives, might be more consonant with the public good than if pronounced by the people themselves. But there was a danger that the wrong sort of people might be elected.

> Men of factious tempers, of local prejudices, or of sinister designs,
> may, by intrigue, by corruption, or by other means, first obtain the
> suffrages, and then betray the interests, of the people.

The more extensive the government, Madison went on to argue, the greater the probability that "proper guardians of the public weal" would be elected. Obviously his second device—the exercise of a prerogative of enlightened and virtuous representatives—was not completely reliable.

Madison, like most of the other nationalists, devoted himself to devising institutional arrangements. He was well aware that republican government presupposed a certain degree of public spiritedness and of reverence for authority:[24] "I go," he said, "on this great republican principle, that the people will have virtue and intelligence to select men of virtue and wisdom."[25] For purposes of constitution making, men had to be taken as they were or, rather, as they would have to be if popular government was to have a chance of succeeding.

In the discussions preceding and during the convention there was general agreement that the government should have limited powers. The only question was where and how the limits were to be drawn. At the start of the convention, leading nationalists, among them Randolph and

---

[23]Farrand, *Records* 2:28.

[24]*Federalist* 55 and Hunt, *Writings of James Madison* 6:85.

[25]Quoted by Alexander Landi, "Madison's Political Theory," in *The Political Science Reviewer* VI (Fall 1976): 87.

Wilson, thought an enumeration of powers impossible. Madison came to Philadephia with a "strong bias" in favor of one, but, while still favoring one, he very soon began to doubt that an enumeration was possible.[26]

In July, after the Great Compromise had been reached, it was clear that there would have to be an enumeration. Now that it had been agreed that the national government would be sovereign in some matters and the states in others, the national government's powers would have to be specified.

This was done in Article One, Section 8. The powers of the national government, Madison said in *Federalist* 45, were "few and defined." In fact there were eighteen powers, very few of which could properly be called *defined*. Some terms, like *necessary* and *proper* ("to make all Laws which shall be necessary and proper for carrying into Execution the foregoing Powers") could hardly have been more ambiguous. It is impossible to believe that the delegates were not well aware that such language would bear different constructions. But it is also impossible to believe that they did not expect it to limit the powers of the central government.

Certainly the voters in the ratifying conventions were told that it would do so. Publius gave the voters of New York very positive assurances. "The powers of the government," said *Federalist* 39, "extend to certain enumerated objects only, and leave to the several states a residuary and inviolable sovereignty over all other objects." *Federalist* 41 considered the charge of some opponents of the Constitution that

> the power to lay and collect taxes, duties, imposts and excises, to pay the debts, and provide for the common defense and general welfare of the United States amounts to an unlimited commission to exercise every power which may be alleged to be necessary for the common defense or general welfare.

This interpretation, Publius said, was absurd. "For what purpose would the enumeration of particular powers be inserted, if these and all others were meant to be included in the preceding general power?" In *Federalist* 17 Publius confessed that, allowing for the utmost latitude to the lover of power which any reasonable man can require, he could not see how an administrator of the general government could be tempted to usurp authorities belonging to the states. It would, he said, always be far easier for state governments to encroach upon the national one.

[26]Farrand, *Records* 1:60, 65.

Charles Cotesworth Pinckney, addressing the legislature of South Carolina, gave slave owners assurances as solid as those that Publius gave the voters of New York. By this settlement, he said,

> we have a security that the general government can never emancipate them [the slaves], for no such authority is granted; and it is admitted, on all hands, that the general government has no powers but what are expressly granted by the Constitution, and all rights not expressed were reserved by the several states.[27]

That the ratifying conventions might have still further assurance, the delegates let it be known that they would support amendments to the Constitution, one of which (it turned out to be the Tenth) would state explicitly that powers not delegated to the United States were reserved to the states.

The ink was not yet dry on the Constitution when its revision began. Ironically, it was not men of factious tempers, local prejudices, or sinister designs who did the revising. Rather it was done by the men of "enlightened views and virtuous sentiments"—those who had, albeit reluctantly, committed themselves to the Great Compromise and urged it upon the voters.

In his first message to Congress, President Washington proposed giving assistance to agriculture, commerce, and manufacturing. The First Congress, sixteen of whose thirty-nine members had been delegates to the convention, made tariff legislation its first order of business. The declared purpose of the Tariff Act was certainly constitutional—Congress had power to lay duties and collect revenues—but this purpose concealed another, the subsidization of particular occupations and interests, which the Constitution did not authorize. "Legislative procedure, as contrasted with economic or political theory," the historian of that Congress, E. A. J. Johnson, has written, "simply assumed that the machinery of government ought to be employed to aid importuning interests."[28]

In 1791, Hamilton, as secretary of the treasury, proposed that Congress charter a national bank. Thomas Jefferson, asked by Washington for his opinion on the constitutionality of the bill, said that the convention had voted against giving Congress the very power now claimed, that the word *necessary* (in the "necessary and proper" clause) did not mean "convenient" and that to give it that meaning "would swallow up all the

---

[27]Ibid. 3: 254.
[28]E. A. J. Johnson, *The Foundations of American Economic Freedom* (Minneapolis: University of Minnesota Press, 1973), 260.

delegated powers, and reduce the whole to one power."[29] Hamilton, in his opinion, said that the convention had probably not meant to deny Congress the power of incorporation (the records of the convention, made public many years later, proved Hamilton wrong and Jefferson right),[30] that "necessary" often means "no more than needful, requisite, incidental, useful, or conducive to," and—most important—that the general legislative authority of Congress *implied* the power to create corporations.[31]

Washington accepted Hamilton's opinion, and the federal sphere was thereby enlarged. It was further enlarged by Hamilton's interpretation of the "general welfare" clause. There was no doubt, he wrote in his *Report on Manufactures*, that the clause necessarily embraced "a vast variety of particulars, which are susceptible neither of specification or of definition." Whatever concerns these matters, he said—he mentioned learning, agriculture, manufactures, and commerce—is within the power of Congress to legislate *as far as regards an application of money* and so long as the appropriation be general rather than local in application.[32]

This interpretation prevailed, despite the strenuous objections of Madison, who was now anti-Hamilton and antinationalist. (In *Federalist* 41, it will be recalled, he had called such an interpretation absurd.) In his *Report on the Virginia Resolutions*, written ten years later, he warned that it would enlarge the powers of the presidency, thus increasing the amount of presidential patronage, which in turn would render elections more important, making them so violent and corrupt that the public might call for an hereditary succession.[33] Much later (1830), he wrote that the general welfare clause had got into the Constitution more or less by accident, or, as he put it, "inattention to the phraseology occasioned doubtless by its identity with the harmless character attached to it in the Instrument [the Articles of Confederation] from which it was borrowed."[34] ("Inattention" seems unlikely in light of the fact that when Morris, a

---

[29]Merrill D. Peterson, ed., *The Portable Thomas Jefferson* (New York: Penguin Books), 264, 265.

[30]Farrand, *Records* 2:321–22.

[31]Jacob E. Cooke, ed., *The Reports of Alexander Hamilton* (New York: Harper Torchbooks, 1964), 88.

[32]Ibid., 172.

[33]Paul C. Peterson, "The Statesmanship of James Madison," eds., Ralph A. Rossum and Gary L. McDowell, *The American Founding* (Port Washington, N.Y.: Kennikat Press, 1981), 128.

[34]Farrand, *Records* 3:486.

member of the drafting committee, slyly inserted a semicolon after the words "to lay and collect taxes, duties, imposts and excises"—thus transforming the words that followed, "to pay the debts and provide for the common defence and general welfare of the United States," into a broad grant of power—the semicolon was replaced with a comma after the trick was discovered.)[35]

The revisers of the Great Compromise were not all nationalists. Jefferson, as Washington's secretary of state, wrote a *Report on the Fisheries* which suggested that Congress do something, perhaps remit taxes, to aid the fishing industry.[36] As president, he acknowledged making "a blank paper" of the Constitution by purchasing Louisiana, and his secretary of the treasury, Gallatin, originated vast plans for internal improvements. When Gallatin sought to establish a branch bank of the United States, Jefferson, after protesting that the bank is "of the most deadly hostility existing against the principles and form of our Constitution," gave his approval.[37]

When the election of Andrew Jackson ended the line of presidents "of enlightened views and virtuous sentiments," the powers of the United States had been very substantially enlarged without recourse to the amending process that the Constitution provided. (The Eleventh and Twelfth Amendments were not for the purpose of extending federal powers.) One may say that the revisions were necessary, that the United States could not have prospered—perhaps even survived—without them, and that even if they could have been made by the amending process (which is doubtful), it was best that they were made without it because, as Publius said in *Federalist* 62, people feel some loss of attachment and reverence toward a political system when it is changed. But whatever judgments are made in these matters, there is no denying that the experience of the first fifty years supports the proposition that popular government cannot be relied upon to abide by the principles it has established itself upon.

The leading nationalists held a common conception of popular government: that there existed a common interest of the society; that most men, if they recognized the common interest (which was unlikely), would,

---

[35]Forrest McDonald, *E Pluribus Unum* (Indianapolis: Liberty Press, 1965), 306.
[36]E. A. J. Johnson, *Foundations of American Economic Freedom*, 278.
[37]Walter Berns, "The Meaning of the Tenth Amendment," ed. Robert A. Goldwin, *A Nation of States* (Chicago: Rand-McNally & Co., 1974), 148.

out of selfishness or other weakness, subordinate it to their special interests; that therefore, it was incumbent upon the wise and disinterested
few to guide and direct the many, knowing that the cool and deliberate
sense of the community, even if wrong, must ultimately prevail. This
conception, on the basis of which the American government was formed
and administered for half a century, was replaced by an altogether different one when John Quincy Adams left the White House. Tocqueville,
who came to America just then, found Americans devoted ("beyond reason") to the notion that man is endowed with an infinite capacity for
improvement and committed also to the principle of self-interest rightly
understood—the Americans' "chief security against themselves."[38] To the
Founders (with the notable exception of Wilson), the notion of the goodness, let alone the perfectability, of man was absurd: it was because men
were not angels that government was necessary. As for self-interest, however understood, the Founders believed that if (as Tocqueville predicted)
it became the sole spring of men's actions, it would be fatal to popular
government. Republican government, Publius observed in *Federalist* 55,
presupposes the existence of certain qualities—presumably wisdom and
virtue—in a higher degree than any other form. Madison, writing while
a member of the First Congress, described a type of government which
he said did not exist on the west side of the Atlantic:

> A government operating by corrupt influence; substituting the motive
> of private interest in place of public duty; converting its pecuniary
> dispensations into bounties to favorites, or bribes to opponents;
> accommodating its measure to the avidity of a part of the nation
> instead of the benefit of the whole; in a word, enlisting an army of
> interested partizans, whose tongues, whose pens, whose intrigues,
> and whose active combinations, by supplying the terror of the sword,
> may support a real domination of the few, under an apparent liberty
> of the many.[39]

The changed conception of human nature implied that there was
no need to enlarge the views of the ordinary man by passing them through
the medium of a chosen body of citizens. And the legitimation of self-
interest (whether rightly understood or not) produced the politician, whose
function was to bring interests, however selfish, into the making of policy,
in place of the statesman, whose function had been to discern the common

---

[38] Alexis de Tocqueville, *Democracy in America*, vol. 2 (New York: Knopf, 1948),
34, 123.
[39] Hunt, *Writings of James Madison*, 94.

good and lead men towards it. ("The race of American statesmen," Tocqueville wrote, "has evidently dwindled remarkably in the course of the last fifty years.")[40]

Political parties, which at first had represented opposed conceptions of the public interest, came more and more to represent private interests. ("America has had great parties, but has them no longer" Tocqueville remarked.)[41] For several decades, parties and politicians, especially those who saw national power as a threat to slavery, checked its growth. Lincoln, of course, asserted national power without limit; he violated the Constitution in order to preserve it.

Republican institutions were transformed into democratic ones. The electoral college remained, but presidential elections came to turn on the popular vote. In some states, senators were chosen de facto by straw ballot long before the passage in 1913 of the Seventeenth Amendment. The passage of the Sixteenth Amendment in that same year gave the federal government an ample revenue base and, incidentally, as George F. Break has observed, created a direct link between each taxpayer and Washington, one that invited "a whole new way of thinking about Washington's responsibility."[42] As more and more people left the relative security of the farm for the uncertainties of wage work in the cities, dependence upon government in times of distress grew.

Early in the New Deal the Supreme Court undid much of what remained of the Great Compromise by adopting Hamilton's interpretation of the "general welfare" clause. In *United States* v. *Butler* (297 U.S. 1, 66, 1936), in the course of finding the Agricultural Adjustment Act unconstitutional (as an invasion of the reserved rights of the states!), the Court declared that "the power of Congress to authorize expenditures of public moneys for public purposes is not limited by the direct grants of legislative power found in the Constitution." Later, in *Buckley* v. *Valeo* (U.S. 1, 424, 1976), it found that Congress' power to provide for the general welfare extends to regulation of the financing of political campaigns.

It was not until after the New Deal that conditional grants became an important means of exercising Federal influence. Beginning about

[40]Tocqueville, *Democracy in America* 1:200.
[41]Ibid., 175.
[42]George F. Break, "Fiscal Federalism in the United States," in Advisory Commission for Intergovernmental Relations, *Conference on the Future of Federalism: Report and Papers* (Washington, D.C., July 1981), p. 44.

1960, the number of grant programs as well as the number and stringency of the conditions attached to them grew year by year until the states and cities depended upon the Federal government for about one-quarter of their revenue. Conditional grants do not give the Federal government the veto power that Madison wanted for it, but their tendency, the editors of the *Yale Law Journal* have written, is to "allow the national government not merely to influence policy, but to make it." If federalism is to be taken seriously, they say, the states must have a zone of power belonging exclusively to them.[43]

This paper has tried to show that there is a tendency inherent in popular government to exceed whatever limits are constitutionally set upon its sphere. Insofar as men are free, they must be expected to use their freedom to renege upon agreements they deem contrary to the public interest or to their private interests. It is in the nature of politics that it cannot be confined to an agreed-upon arena.

The intention has not been to deplore the growth of federal power. It would be pointless to deplore what one maintains is inevitable. Apart from that, it is hard to believe that this nation could have prospered, or even survived, if—as Wilson assured the Pennsylvania ratifying convention would be the case—the congressional power had been collected "not from tacit implication, but from the positive grant expressed in the instrument of the union."[44] The state governments are well on the way to becoming (what many of the Founders wanted them to become!) mere administrative districts, not very different from counties; that may be regrettable, but it does not presage any calamity.

What *is* dismaying is the prospect that eventually—perhaps soon—the American people, having forgotten that the great principle of the revolution was limited government, will demand that government do what cannot be done and the attempting of which will destroy popular government itself.

[43]"Taking Federalism Seriously: Limiting State Acceptance of National Grants," *Yale Law Journal* (June 1981): 1695, 1713.
[44]Quoted by Storing, *What the Anti-Federalists Were For*, 63.

CHAPTER 2

In Defense of the
American Party System

The American party system has been criticized on four main grounds:
(1) the parties do not offer the electorate a choice in terms of fundamental
principles; their platforms are very similar and mean next to nothing;
(2) they cannot discipline those whom they elect, and therefore they
cannot carry their platforms into effect; (3) they are held together and
motivated less by political principle than by desire for personal, often
material, gain, and by sectional and ethnic loyalties; consequently party
politics is personal and parochial; and (4) their structure is such that they
cannot correctly represent the opinion of the electorate; in much of the
country there is in effect only one party, and everywhere large contributors and special interests exercise undue influence within the party.[1]

From *Political Parties, U.S.A.*, ed. Robert A. Goldwin (Chicago: Rand McNally,
1961), 21–39. Copyright 1961 by Kenyon College Public Affairs Conference Center. Reprinted
by permission.

[1]These criticisms are made, for example, by the French political scientist, Maurice
Duverger, in *Political Parties* (New York: Wiley, 1954). For similar criticisms by
Americans, see especially Committee on Political Parties of the American Political
Science Association, *Toward a More Responsible Two-Party System* (New York:
Rinehart, 1950),and E. E. Schattschneider, *Party Government* (New York: Farrar
& Rinehart, 1942). Criticisms of American parties are summarized and analyzed
in Austin Ranney, *The Doctrine of Responsible Party Government* (Urbana:
University of Illinois Press, 1954). Defenses of the American party system include
A. Lawrence Lowell, *Essays on Government* (Boston: Houghton Mifflin, 1889),
chs. I, II; Arthur N. Holcombe, *The Political Parties of Today* (New York: Harper,
1925); and *Our More Perfect Union* (Cambridge: Harvard University Press, 1950);
Pendleton Herring, *The Politics of Democracy* (New York: Norton, 1940); and
Herbert Agar, *The Price of Union* (Boston: Houghton Mifflin, 1950).

These criticisms may be summarized by saying that the structure and operation of the parties do not accord with the theory of democracy or, more precisely, with that theory of it which says that everyone should have a vote, that every vote should be given exactly the same weight, and that the majority should rule.

"It is a serious matter," says Maurice Duverger, a French political scientist who considers American party organizations "archaic" and "undemocratic," "that the greatest nation in the world, which is assuming responsibilities on a world-wide scale, should be based on a party system entirely directed towards very narrow local horizons."[2] He and other critics of the American party system do not, however, base their criticism on the performance of the American government. They are concerned about procedures, not results. They ask whether the structure and operation of the parties is consistent with the logic of democracy, not whether the party system produces—and maintains—a good society, meaning, among other things, one in which desirable human types flourish, the rights of individuals are respected, and matters affecting the common good are decided, as nearly as possible, by reasonable discussion.[3]

If they were to evaluate the party system on the basis of results, they would have to conclude that on the whole it is a good one. It has played an important part (no one can say how important, of course, for innumerable causal forces have been at work along with it) in the production of a society which, despite all its faults, is as near to being a good one as any and nearer by far than most; it has provided governments which, by the standards appropriate to apply to governments, have been humane and, in some crises, bold and enterprising; it has done relatively little to impede economic growth and in some ways has facilitated it; except for the Civil War, when it was, as Henry Jones Ford said, "the last bond of union to give way,"[4] it has tended to check violence, moderate conflict, and narrow the cleavages within the society; it has never produced, or very seriously threatened to produce, either mob rule or tyranny, and it has shown a marvelous ability to adapt to changing circumstances.

[2]Duverger, *Political Parties*, 53.

[3]The report of the Committee on Parties of the American Political Science Association, cited before, discusses the "effectiveness" of parties entirely in terms of procedure. Duverger does the same.

[4]Henry Jones Ford, *The Rise and Growth of American Politics* (New York: Macmillan, 1900), 303.

Not only has the American party system produced good results, it has produced better ones than have been produced almost anywhere else by other systems. Anyone who reflects on recent history must be struck by the following paradox: those party systems that have been more democratic in structure and procedure have proved least able to maintain democracy; those that have been most undemocratic in structure and procedure—conspicuously those of the United States and Britain—have proved to be the bulwarks of democracy and of civilization.

This paper explores this paradox. It maintains that there is an inherent antagonism between "democracy of procedure" and "production of, and maintenance of, a good society"; that some defects of procedure are indispensable conditions of success from the standpoint of results, and that what the critics call the "archaic" character of the American party system is a very small price to pay for government that can be relied upon to balance satisfactorily the several conflicting ends that must be served.

DIFFICULTIES IN PLANNING CHANGE

Before entering into these matters, it may be well to remind the reader how difficult is the problem of planning social change.

Social relationships constitute systems: they are mutually related in such a manner that a change in one tends to produce changes in all of the others. If we change the party system in one respect, even a seemingly trivial one, we are likely to set in motion a succession of changes which will not come to an end until the whole system has been changed. The party system, moreover, is an element of a larger political system and of a social system. A small change in the structure or operation of parties may have important consequences for, say, the family, religion, or the business firm.

The changes that we intend when making a reform, if they occur at all, are always accompanied by others that we do not intend. These others may occur at points in the system far removed from the one where the change was initiated and be apparently unrelated to it. Commonly changes produced indirectly and unintentionally turn out to be much more important than the ones that were sought. This is a fact that is seldom fully taken into account. Those who support a particular reform are often indifferent to its consequences for values that they either do not share or consider subordinate. Even those who feel obliged to take a wide range of values into account do not usually try very hard to anticipate the indirect consequences of reforms—often for a very good reason: the complexity of the social system makes the attempt implausible. Usually we take it on faith that the consequences we get by intention justify the risk we take of incurring others that we do not intend or want. Since these others are seldom recognized as consequences of our action at all (they either go unnoticed or seem to have "just happened"), the basis of our faith is not called into question.

No doubt it is a great help to the practical reformer to have tunnel vision. But those who are concerned with the welfare of society as a whole must take the widest perspective possible. They must try to identify all of the consequences that will follow from a reform—the unintended ones no less than the intended, the remote, contingent, and imponderable no less than the immediate, certain, the specifiable. And they must evaluate all of these consequences in the light of a comprehensive system of values.

Those who devise "improvements" to a social system can rarely hope to attain all of their ends; usually they must be prepared to sacrifice some of them to achieve others. This is so because resources are usually limited and also because there are often incompatibilities among ends such that a gain in terms of some necessarily involves a loss in terms of others. The reformer must therefore economize. He must be able to assign priorities to all ends in such a way that he can tell how much of each to sacrifice for how much of others, on various assumptions as to "supply."

The critics of the party system tend to value democratic procedure for its own sake, that is, apart from the results it produces. There is no reason why they should not do so. But they are in error when they do not recognize that other values of equal or greater importance are often in conflict with democratic procedure, and that when they are, some sacrifice of it is essential in order to serve the other values adequately. If they faced up to the necessity of assigning priorities among all of the relevant ends, they would not, it is safe to say, put "democratic procedure" first. Probably they, and most Americans, would order the ends as follows:

1. The party system must above all else provide governments having the will and capacity to preserve the society and to protect its members. Any sacrifice in other ends ought to be accepted if it is indispensable to securing this end.
2. The party system must insure periodic opportunity to change the government by free elections. Any sacrifice of other ends

(except the one mentioned) ought to be accepted if it is indispensable to securing this one.

3. The party system should promote the welfare of the people. By "welfare" is meant some combination of two kinds of values: "principles," what is thought to be good for the society, described in rather general terms, and "interests," the ends individuals and groups seek to attain for their own good, as distinguished from that of the society. The party system should produce governments that assert the supremacy of principles over interests in some matters; in others it should allow interests to prevail and should facilitate the competitive exercise of influence.

4. The party system should moderate and restrain such conflict as would threaten the good health of the society. Other conflict it should not discourage.

5. The party system should promote and exemplify democracy, meaning reasonable discussion of matters affecting the common good in which every voice is heard.

These ends have been listed in what most Americans would probably consider a descending order of importance. In devising a party system, we ought not to try to serve fully each higher end before serving the one below it at all. The first two ends are exceptions to this rule, however: each of them must be attained even if the others are not served at all. With respect to the remaining three, the problem is to achieve a proper balance—one such that no reallocation from one end to another would add to the sum of value.

Finally, we must realize that we can rarely make important social changes by intention. The most we can do is to make such minor changes as may be consistent with, and more or less implied by, the fixed features of the situation in which we are placed. Even to make minor changes in an institution like a political party requires influence of a kind and amount that no group of reformers is likely to have or to be able to acquire. It is idle to propose reforms that are merely desirable. There must also be some possibility of showing, if only in a rough and conjectural way, that they might be carried into effect.

With respect to the American party system, it seems obvious that the crucial features of the situation are all fixed. The size of our country, the class and cultural heterogeneity of our people, the number and variety of their interests, the constitutionally given fragmentation of formal authority, the wide distribution of power which follows from it, the inveterate taste of Americans for participation in the day-to-day conduct of government when their interests are directly at stake—these are all unalterable features of the situation. Taken together, they mean that the party system can be reformed only within very narrow limits.

A MODEL PARTY SYSTEM

Let us imagine a system free of the alleged defects of ours. In this model system, every citizen is motivated—highly so—by political principles, not subsidiary ones, but ones having to do with the very basis of the society. (In France and Italy, Duverger says approvingly, political warfare "is not concerned with subsidiary principles but with the very foundations of the state and the nature of the regime."[5]) The electoral system, moreover, is such as to give every side on every issue exactly the weight that its numbers in the population warrant; no group or interest is over- or underrepresented. ("One's thoughts turn," Duverger says, "to the possibility of a truly scientific democracy, in which parliament would be made up of a true sample of the citizens reproducing on a reduced scale the exact composition of the nation, made up, that is, according to the very methods that are used as a basis for public opinion surveys like the Gallup polls."[6])

Assuming that the society is divided by the usual number of cleavages (e.g., haves versus have-nots, segregationists versus antisegregationists, isolationists versus internationalists, etc.), the following would result:

1. There would be a great many parties, for no citizen would support a party with which he did not agree fully.
2. The parties would tend to be single-issue ones. If logically unrelated issues (for instance, segregation and isolationism) were linked together in a party program, only those voters would support the party who chanced to be on the same side of all of the linked issues. The number of these voters would decrease as the number of issues so linked increased.

[5]Duverger, *Political Parties*, 419.
[6]Ibid., 158.

3. Parties would be short lived. They would come into and pass out of existence with the single issues they were organized to fight.

4. In their election campaigns and propaganda, parties would emphasize their single defining principles. This would tend to widen the cleavages along which the parties were formed.

5. Ideological issues, not practical problems, would constitute the substance of politics.[7]

6. The number of such issues pressing for settlement at any one time (but being incapable of settlement because of their ideological character) would always be more than the system could accommodate.[8]

7. Coalitions of parties would seldom form, and such as did form would be highly unstable. Party leaders would find compromise almost impossible because it would lead to loss of highly principled supporters.

8. Coalitions of parties being unstable, governments would also be unstable and therefore lacking in power and decision.

9. Those selected for positions of political leadership would tend to be ideologues skilled in party dialectics and symbolizing the party and its positions. Practical men, especially those with a talent for compromise and those symbolizing qualities common to the whole society, would be excluded from politics.

10. Matters having no ideological significance (a category that includes most local issues) would either be endowed with a spurious one or else would be left outside the sphere of politics altogether.[9]

---

[7]In France, according to Siegfried, "every argument becomes a matter of principle; the practical results are relegated to second place." André Siegfried, "Stable Instability in France," *Foreign Affairs*, XXXIV (April 1956): 395.

[8]According to Siegfried: "The difficulty is that too many questions of fundamental importance on which the various parties have cause to disagree have come up for discussion at one time." Ibid., 399.

[9]In France, Luethy says, "politics," which deals with ideological matters, and the "state," i.e., the bureaucracy, which deals with practical ones, function "in watertight compartments" with the consequence that French democracy is an amalgam of absolutist administration on the one hand and of anarchy, tumultuous or latent, on the other. Herbert Luethy, *France against Herself* (New York: Meridian Books, 1957), 61. On this see also Siegfried, "Stable Instability," 399.

These points should suffice to show that a system with a perfectly democratic structure would not produce results acceptable in terms of the criteria already listed.

Now let us introduce into the model system one of the alleged defects which the critics find most objectionable in the American party system. Let us suppose that at least half of the electorate is prevailed upon to exchange its vote in matters of fundamental principle for advantages that have nothing to do with principle, especially private profit, sectional gain, and nationality "recognition."

One effect of this would be to reduce greatly the intensity of ideological conflict and to make political life more stable and conservative. This, in fact, seems to be what happened when American parties first came into being. John Adams tells in his diary how in 1794 "ten thousand people in the streets of Philadelphia, day after day, threatened to drag Washington out of his house and effect a revolution in the government, or compel it to declare war in favor of the French Revolution and against England."[10] After parties had been organized, however, patronage took the place of ideological fervor. "The clubs of the social revolutionists which had sprung up in the cities, blazing with incendiary ideas caught from the French Revolution," Henry Jones Ford says, "were converted into party workers, and their behavior was moderated by considerations of party interest."[11]

Another effect would be to encourage the formation of a few (probably two) stable parties. These might begin as alliances among the profit minded, the sectional minded, and the nationality minded, but to attract support from principled voters the parties would have to seem to stand for something—indeed, for anything and everything. Since no faction of them could hope to win an election by itself, principled voters would attach themselves to those parties that they found least objectionable. The parties would develop corporate identities and mystiques; principled voters would then subordinate their differences out of "loyalty" to the party and in response to its demands for "regularity." Competition for middle-of-the-road support would cause the parties to offer very similar programs. This competition might lead to there being only two parties, but this result would probably be insured by introducing another supposed defect into the system: a principle of representation (single-member

[10]Quoted by Henry Jones Ford, *American Politics*, 125.
[11]Ibid., 144.

districts and plurality voting) which, by letting the winner take all, would force small parties to join large ones in order to have some chance of winning.

In one way or another, the "defects" of the system would tend to produce these consequences—consequences which have in fact been produced in the United States:

1. A strong and stable government would be possible. The country would be governed by the party that won the election, or (given the particular complexities of the American system) by two closely similar parties engaged in give-and-take and, therefore, in a sense constituting one party under two names.
2. There would be a high degree of continuity between administrations elected from different parties. Elections would not shake the nation to its foundations because the competing parties would be fundamentally in agreement. Agreement would be so built in by countless compromises within the parties (each of which would be under the necessity of attracting middle-of-the-road support) that a change of party would seldom entail complete reversal of policy in an important matter.
3. There would exist many substructures of power that would be largely or wholly impervious to the influence of political principle or ideology. "Machines"—party organizations of the profit minded, the sectional minded, and the nationality minded—would not be inclined to offer pie in the sky or to stir the emotions of the masses because they could count upon getting their votes in other ways. These essentially apolitical centers of power would therefore exert a stabilizing and conservative influence throughout the political system. By making businesslike deals with the leaders of the "machines," the president could sometimes buy freedom to do as he thought best in matters of principle.
4. The diversity of the principles and the multiplicity of the interests within the party would be another source of strength to the leader elected from it. He could afford to offend some elements of the party on any particular question because there would be enough other elements unaffected (or even gratified) to assure his position. The more fragmented his party, the less attention he would have to pay to any one fragment of it.

5. The assertion of interests (as distinguished from principles) would be encouraged. The profit minded, the sectional minded, and the nationality minded would in effect give up representation on matters of principle in order to get it on matters involving their interests. Thus two different systems of representation would work simultaneously. The party leader would act as a trustee, disregarding interests in favor of principles. ("Congress represents locality, the President represents the nation," Ford wrote in 1898.[12]) Meanwhile legislators dependent on machines and, in general, on profit minded, sectional minded, and nationality minded voters would act as agents of interests. The trustee of principles (the president) and the agents of interests (congressmen) would of necessity bargain with each other; by allowing the agents of interests some successes—but only in this way— the trustee of principles could win their support in the matters he considered most important. Thus, there would be achieved that balancing of interests and of interests against principles (the most important principles usually being vindicated) that a good party system should produce.

6. The formation of deep cleavages would nevertheless be discouraged. The competition of the parties for the middle-of-the-road vote; their tendency to select practical men of wide popular appeal, rather than ideologues, for positions of leadership; and the definition of the politicians' task as being that of finding the terms on which people who disagree will work together, rather than that of sharpening ideological points—these would all be unifying tendencies.

Some critics of the American party system have attributed its alleged defects to the absence of class consciousness in our society. No doubt there is some truth in this. But causality may run the other way also. We may be lacking in class consciousness because our politicians are prevented by the nature of the party system from popularizing the rhetoric of the class struggle; the party system actually induces the voter to forego the allurements of principle and ideology by offering him things he values

[12]Ibid., 187. For a recent brilliant account of how the two systems of representation work, see Willmoore Kendall, "The Two Majorities," *Midwest Journal of Political Science* IV, no. 4 (November 1960), 317–45.

more: for example, personal profit, sectional advantage, and nationality "recognition."[13]

In those countries where the voter expresses at the polls his ideology rather than his interests, he may do so not from choice but because the party system leaves him no alternative. In such countries, class warfare may be the principal subject matter of politics simply because matters of greater importance to the voters are not at stake.

Experience in the underdeveloped areas seems to bear out the claim that certain "defects" in a party system may be essential to good government. The transplanted "defects" of the American party system are among the factors that have made the Philippines the most democratic country in Southeast Asia. According to Professor Lucian W. Pye:

> The image of leadership that evolved in the Philippines was clearly that of the politician who looked after the particular interests of voters. Elsewhere the pattern of the Western impact under colonialism gave emphasis to the role of the rational administrator who apparently operated according to the principles of efficiency and who was not supposed to be influenced by political pressures within the society. Consequently, when the politicians emerged in these societies, they tended to become the champions of nationalistic ideologies and even the enemies of the rational administrators.[14]

In the Philippines, as at home, our party system has had the defects of its virtues—and the virtues of its defects. On the one hand, Pye says, the Philippines have never had an efficient administrative machinery, and the demand for higher standards of personal integrity among their public officials is reminiscent of the muckraking era of American politics; on the

[13]"In coordinating the various elements of the population for political purposes," Ford says, "party organization tends at the same time to fuse them into one mass of citizenship, pervaded by a common order of ideas and sentiments, and actuated by the same class of motives. This is probably the secret of the powerful solvent influence which American civilization exerts upon the enormous deposits of alien population thrown upon this country by the torrent of emigration. Racial and religious antipathies, which present the most threatening problems to countries governed upon parliamentary principles, melt with amazing rapidity in the warm flow of a party spirit which is constantly demanding, and is able to reward the subordination of local and particular interests to national purposes." (*American Politics*, 306–7.)

[14]Lucian W. Pye, "The Politics of Southeast Asia," eds. G. Almond and J. Coleman, *The Politics of the Developing Areas* (Princeton, N.J.: Princeton University Press, 1969), 97.

other hand, "the Philippine electorate seems to recognize that the most fundamental question in politics is who is going to control the government, and thus, while the parties have not had to expend much effort in trying to distinguish themselves ideologically from each other, the expenditures of money on political campaigns in the Philippines are probably the highest in proportion to per capita income of any country in the world."[15]

MAKING PARTIES "RESPONSIBLE"

Some think that the American party system can be reformed without changing its nature essentially. Several years ago, a Committee on Parties of the American Political Science Association proposed making certain "readjustments" in the structure and operation of the party system to eliminate its "defects." These readjustments, the committee said, would give the electorate "a proper range of choice between alternatives" in the form of programs to which the parties would be committed and which they would have sufficient internal cohesion to carry into effect. Thus, the two-party system would be made more "responsible."[16]

What this means is not at all clear. *Responsibility* here seems to be a synonym for accountability, that is, the condition of being subject to being called to account and made to take corrective action in response to criticism. In the case of a party, this can mean nothing except going before an electorate, and in this sense all parties are by definition responsible. *Responsibility* can have no other meaning in this context; as William Graham Sumner remarked, "a party is an abstraction; it cannot be held responsible or punished; if it is deprived of power it fades into thin air and the men who composed it, especially those who did the mischief and needed discipline, quickly reappear in the new majority."[17]

Leaving aside both the question of what "responsibility" means when applied to a party and the more important one of whether as a matter of practical politics such "readjustments" could be made, let us consider how the political system would probably be affected by the changes proposed.

[15] Ibid., 123, 126.
[16] See the Committee Report, *Two Party System*, 1, 85.
[17] William Graham Sumner, *The Challenge of Facts* (New Haven, Conn.: Yale University Press, 1914), 271–72.

The hope that the two-party system might be made to offer a choice between distinct alternatives is illusory for at least two reasons. One is that a party which does not move to the middle of the road to compete for votes condemns itself to defeat and eventually, if it does not change its ways, to destruction. But even if this were not the case, the parties could not present the electorate with what reformers think of as "a valid choice." The reason is that the issues in our national life are such that there does not exist any one grand principle by which the electorate could be divided into two camps such that every voter in each camp would be on the "same" side of all issues. The idea of "left" and "right" is as close as we come to having such a grand principle, and it has little or no application to many issues.[18] The logic of "left" and "right" does not, for example, imply opposite or even different positions on (for example) foreign policy, civil liberties, or farm subsidies. Without a grand principle which will make unities—opposed unities—of the party programs, the electorate cannot be offered "a valid choice." A choice between two market baskets, each of which contains an assortment of unrelated items, some of which are liked and some of which are disliked, is not a "valid" choice in the same sense that a choice between two market baskets, each of which contains items that "belong together" is a "valid" one. In the American party system, most items are logically unrelated. This being so, "valid" choice would become possible only if the number of parties was increased to allow each party to stand for items that *were* logically related, if one issue became important to the exclusion of all the others, or if, by the elaboration of myth and ideology, pseudological relations were established among items.

The hope that the parties might commit themselves to carry out their programs is also illusory. A party could do this only if its leaders were able to tell the president and the party members in Congress what to do, and could discipline them if they failed to do it. Therefore, unless, like the Russians, we were to have two sets of national leaders, one in governmental office and another much more important one in party office, it would be necessary for our elected leaders—in effect, the president,

[18]One can imagine a set of symbols connected with a diffuse ideology dividing the society into two camps, and to a certain extent this exists. But it is hard to see in what sense this would present the electorate with "a valid choice." In other words, the existence of a body of nonsense which is treated as if it were a grand principle ought not to be regarded by reasonable critics of the party system as equivalent to the grand principle itself.

since only he and the vice-president are elected by the whole nation—to control the congressmen and senators of their party. This would be possible only if the president could deny reelection to members of Congress who did not support the party program. Thus, instead of merely bringing forward and electing candidates, as they do now, "responsible" parties would have to govern the country. We would have a parliamentary system with the president in a position somewhat like that of the British prime minister, except (a very important difference) that, not being a part of the legislature, he could not use it as a vehicle through which to exert his leadership.[19] The legislature would in fact have no function at all.

This great shift of power to the president would remedy another "defect" in the party system: its receptivity to the demands of interest groups.[20] With the president in full control of Congress, logrolling would cease or virtually cease. It would do so because no one could any longer make the president pay a price for assistance in getting legislation passed; the traders who now sell their bits and pieces of power to the highest bidders would have to lower their prices and would probably go out of business. With their opportunities for exercising influence vastly reduced, interest groups would be less enterprising both in their efforts to anticipate the effects of governmental action and in bringing their views to the attention of the policymakers.

The making of policy would thus pass largely into the hands of technical experts within the majority party, the White House, and the executive departments. These would be mindful of principles and impatient of interests. They would endeavor to make "coherent" policies, meaning, presumably, policies not based on compromise.[21] In all important matters, however, "the public interest" would prove an insufficient guide; the experts, when confronted with the necessity of choosing between

---

[19]The prime minister is the leader of his party outside as well as inside Parliament. Party leaders who are not also members of Parliament take no part in the running of the government, as the late Professor Harold Laski discovered when, as a leader of the Labour party, he presumed to give advice to Prime Minister Attlee. The party leaders discipline their followers by threatening to deprive them of renomination; accordingly most members of the House are "backbenchers" who participate in its affairs only as audience, and the function of the House as a whole is to criticize and advise the leaders of the majority party.

[20]Cf. Report of the Committee on Parties, *Two Party System*, 19–20.

[21]Ibid., 19.

alternatives that were equally in the public interest—that is, when no authoritative, ultimate criterion of choice existed for them to apply— would by the very necessities of the case have to balance the competing values as best they could, which means that they would have to fall back upon their personal tastes or professional biases.[22] Thus they would do badly (but in the name of "impartial administration") what is now done reasonably well by the political process.

The destruction of political traders and of local centers of power would mean also that the president's power would derive from somewhat different sources than at present. Instead of relying upon logrolling and patronage to get the votes he would need in Congress, he would have to rely upon direct appeals by the electorate. To some extent he might manipulate the electorate by charm and personality; TV and the arts of Madison Avenue would become more important in politics. But in order to get elected he would have to depend also, and to a greater extent, upon appeals to political principle or ideology. Whereas the political trader maintains his control by giving and withholding favors to individuals (a circumstance which makes his control both dependable in its operation and cheap), the president would have to maintain *his* by the uncertain and costly expedient of offering to whole classes of people—the farmer, the aged, the home owner, and so on—advantages that they would have only at each other's expense. If charm and the promise of "something for everybody" did not yield the amount of power he required to govern the country, the president might find it necessary to exploit whatever antag- onisms within the society might be made to yield more power. Class and ethnic differences might in this event serve somewhat the same function as logrolling and patronage do now. Mayor LaGuardia, for example, depended for power upon direct, personal appeal to the voters rather than upon organization. His charm and his support of "liberal" programs are well remembered. But it should not be forgotten that he depended also upon exploitation of ethnic loyalties and antipathies. According to Robert Moses,

> It must be admitted that in exploiting racial and religious prejudices LaGuardia could run circles around the bosses he despised and der- ided. When it came to raking ashes of Old World hates, warming

---

[22]This argument is developed in E. C. Banfield, *Political Influence* (Glencoe, Ill.: Free Press, 1961), Chapter 12.

ancient grudges, waving the bloody shirt, tuning the ear to ancestral voices, he could easily outdemagogue the demagogues. And for what purpose? To redress old wrongs abroad? To combat foreign levy or malice domestic? To produce peace on the Danube, the Nile, the Jordan? Not on your tintype. Fiorello LaGuardia knew better. He knew that the aim of the rabble rousers is simply to shoo into office for entirely extraneous, illogical and even silly reasons the municipal officials who clean city streets, teach in schools, protect, house and keep healthy, strong and happy millions of people crowded together here.[23]

That a president might rely more upon appeals to political principle does not at all mean that better judgments or results would follow. For the discussion of principles would probably not be *serious*; it would be for the purpose of securing popular interest and consent, not of finding a wise or right course of action. As long ago as 1886, Sir Henry Sumner Maine observed that democracy was tending toward government by salesmanship. Party and corruption had in the past always been relied upon to bring men under civil discipline, he said, but now a third expedient had been discovered:

> This is generalization, the trick of rapidly framing, and confidently uttering, general propositions of political subjects. . . . General formulas, which can be seen on examination to have been arrived at by attending only to particulars few, trivial or irrelevant, are turned out in as much profusion as if they dropped from an intellectual machine; and debates in the House of Commons may be constantly read, which consisted wholly in the exchange of weak generalities and strong personalities. On a pure Democracy this class of general formulas has a prodigious effect. Crowds of men can be got to assent to general statements, clothed in striking language, but unverified and perhaps incapable of verification; and thus there is formed a sort of sham and pretence of concurrent opinion. There has been a loose acquiescence in a vague proposition, and then the People, whose voice is the voice of God, is assumed to have spoken.[24]

Efforts to create "levity of assent," as Maine called it, will become more important in our politics to the extent that other means of bringing men under civil discipline are given up or lost.

[23] Robert Moses, *LaGuardia: A Salute and a Memoir* (New York: Simon & Schuster, 1957), 37–38.

[24] Sir Henry Sumner Maine, *Popular Government* (New York: Henry Holt, 1886), 106–8.

THE DANGER OF MEDDLING

A political system is an accident. It is an accumulation of habits, customs, prejudices, and principles that have survived a long process of trial and error and of ceaseless response to changing circumstance. If the system works well on the whole, it is a lucky accident—the luckiest, indeed, that can befall a society, for all of the institutions of the society, and thus its entire character and that of the human types formed within it, depend ultimately upon the government and the political order.

To meddle with the structure and operation of a successful political system is therefore the greatest foolishness that men are capable of. Because the system is intricate beyond comprehension, the chance of improving it in the ways intended is slight, whereas the danger of disturbing its working and of setting off a succession of unwanted effects that will extend throughout the whole society is great.

Democracy must always meddle, however. An immanent logic impels it to self-reform, and if other forces do not prevent, it must sooner or later reform itself out of existence.[25]

The logic of this is as follows. The ideal of democracy legitimates only such power as arises out of reasonable discussion about the common good in which all participate. Power that comes into being in any other way (e.g., by corruption, logrolling, appeals to sentiment or prejudice, the exercise of charm or charisma, "hasty generalization," terror, etc.) is radically undemocratic, and people inspired by the democratic ideal will therefore endeavor to eliminate it by destroying, or reforming whatever practices or institutions give rise to it.

No society, however, can be governed *solely* by reasonable discussion about the common good; even in a society of angels there might be disagreement about what the common good requires in the concrete case.[26] In most societies, far more power is needed to maintain civil discipline and protect the society from its enemies than can be got simply by reasonable discussion about the common good. Therefore the logical culmination of democratic reform, viz., the elimination of all undemocratic sources of power, would render government—and therefore the preservation of the society—impossible. Democratic reform can never

---

[25]For data and analysis pertinent to the discussion that follows, see James Q. Wilson, *The Amateur Democrat* (Chicago: University of Chicago Press, 1962).

[26]See Yves R. Simon, *The Philosophy of Democratic Government* (Chicago: University of Chicago Press, 1951), Chapter 1.

reach this point, of course, because, before reaching it, democracy itself would be destroyed and the impetus to further reform removed.

So far as it does succeed, however, the tendency of democratic reform is to reduce the power available for government. Such loss of power as occurs from the elimination of undemocratic sources of it will seldom be offset by increases in power of the kind that arises from reasonable discussion about the common good. Since there is a point beyond which no increase in democratic power is possible (the capacity of a society to engage in reasonable discussion about the common good being limited), reform, if carried far enough, must finally reduce the quantity of power.

There is, then, a danger that reform will chip away the foundations of power upon which the society rests. But this is not the only danger. A greater one, probably, is that in making some forms of undemocratic power less plentiful, reform may make others more plentiful, and by so doing set off changes that will ramify throughout the political system, changing its character completely. If, for example, politicians cannot get power by the methods of the machine (corruption, favor giving, and patronage), they may get it by other methods, such as charm, salesmanship, and "hasty generalization." The new methods may be better than the old by most standards (they cannot, of course, be better by the standard of democracy, according to which *all* power not arising from reasonable discussion about the common good is absolutely illegitimate); but even if they are better, the new methods may not serve as well as the old, or may not serve at all, in maintaining an effective political system and a good society.

Reform is, of course, far from being the only force at work. Compared to the other forces, some of which tend to produce competing changes and others of which tend to check all change, reform may be of slight effect. This is certainly true in general of such reform as is sought through formal organizations by people called *reformers*. It is much less true of reform in the broader sense of the general view and disposition of "the great body of right-thinking people." This kind of reform is likely to be of pervasive importance in the long run, although its effects are seldom what anyone intended.

Jefferson may have been right in saying that democracy cannot exist without a wide diffusion of knowledge throughout the society. But it may be right also to say that it cannot exist *with* it. For as we become a better and more democratic society, our very goodness and democracy may lead us to destroy goodness and democracy in the effort to increase and perfect them.

CHAPTER 3

Party "Reform" in Retrospect

*It can hardly be believed how many facts naturally flow from
the philosophical theory of the indefinite perfectibility of man
or how strong an influence it exercises even on those who, living
entirely for the purposes of action and not of thought, seem to
conform their actions to it without knowing anything about it.*

ALEXIS DE TOCQUEVILLE

In a paper written almost twenty years ago, I maintained that a political
system is an accident, and that to meddle with one that works well is the
greatest foolishness of which men are capable.[1] Nevertheless, I said, a

---

From *Political Parties in the Eighties*, ed. Robert A. Godwin (Washington, D.C.:
American Enterprise Institute, 1980), 20–33. Copyright 1980 by The American Enterprise
Institute for Public Policy Research. Reprinted by permission.

I put the word "reform" in quotation marks to call attention to its misuse by those who
think any change that aims at improvement is a reform even if it makes matters worse.
Properly speaking, a change is a reform only if it makes matters better in the manner
intended. As Burke said in his *Letter to a Noble Lord*, "to innovate is not to reform."

[1] "In Defense of the American Party System," *Political Parties, U.S.A.*, ed. Robert
A. Goldwin (Chicago: Rand-McNally, 1961), 21–39. In this section I have relied
heavily upon data from (in alphabetical order): Herbert E. Alexander, *Financing
Politics* (Washington, D.C.: Congressional Quarterly Press, 1976); Herbert B.
Asher, "The Media and the Presidential Selection Process," in *The Impact of
the Electoral Process*, vol. 3, ed. Louis Maisel and Joseph Cooper (Beverly Hills/
London: Sage Electoral Studies Yearbook, 1977); Jeane Jordan Kirkpatrick, *Dis-
mantling the Parties* (Washington, D.C.: American Enterprise Institute, 1978);
Everett Carll Ladd, Jr., *Where Have All the Voters Gone?* (New York: W. W.
Norton, 1978); Gary R. Orren, "Candidate Style and Voter Alignment in 1976,"
in *Emerging Coalitions in American Politics*, ed. Seymour Martin Lipset (San
Francisco: Institute for Contemporary Studies, 1978); Gerald M. Pomper, "The
Decline of Partisan Politics," in *The Impact of the Electoral Process*, ed. Maisel
and Cooper; Austin Ranney, *Curing the Mischiefs of Faction* (Berkeley: Uni-

democracy will always meddle, because its logic legitimates only such power as arises from reasonable discussion about the common good in which all participate. A democracy will therefore try to reform away all power from other sources and—since power arising from reasonable discussion is never enough to govern—democracy must eventually reform itself out of existence. My argument referred especially to the American party system, which had produced good results precisely because of its alleged defects, that is, its lack of correspondence to the democratic ideal. Eliminating these "defects," I concluded, might "set off changes that will ramify throughout the political system, changing its character completely."

In this essay, I begin by describing the changes (most of them unintended and unwanted consequences of efforts at reform) that have occurred in the national party system. I turn then to a consideration of how these changes in the party system seem to have affected the political system as a whole, conjecturing that the changes have indeed ramified and may prove to have changed the character of the system completely. Finally, I show that it was mainly the "logic of the democratic ideal" that brought about the changes.

THE "OLD" PARTY SYSTEM AND THE "NEW"

In the days of the "old" party system—from Andrew Jackson's time to about the mid-1950s—both national parties were loose confederations of state parties that came alive every four years to nominate presidential and vice-presidential candidates and then to wage compaigns for them. Some state parties existed in name only, but most were loose alliances of city machines, state and local officeholders, labor unions and other interest groups, and some wealthy individuals.

To be taken seriously as a contender for the presidential nomination, one had to be a leading figure in a major state party organization or have the backing of someone who was. The state leaders, many of whom were governors or senators, were political professionals who typically had worked themselves into positions of power by faithful service to the party. A few

---

versity of California Press, 1975); and Austin Ranney, "The Political Parties: Reform and Decline," in *The New American Political System,* ed. Anthony King (Washington, D.C.: American Enterprise Institute, 1978).

party leaders in each state, usually in some sort of convention—a "smoke-filled room"—chose the state's delegates to the national party convention.[2] To win the presidential nomination, therefore, an aspirant had to put together a winning coalition of state leaders. If he happened to be the party leader, perhaps the governor, of a large state, he would have as many as 15 percent of the number needed to win "in his pocket"—obviously a considerable advantage in the coalition-building process. Early in the present century, some states, in response to complaints of "boss-ism," substituted the direct primary for the convention as a way of choosing delegates. (In 1916 at least twenty-two states had presidential primaries, but the number had declined to about sixteen by 1936.) However, state conventions, which is to say state party leaders, chose well over half the delegates as recently as 1952, and therefore safely controlled the nominating process.

The nominee of a major party could be sure of a substantial vote even if he did nothing but sit on his front porch until election day. Most voters "belonged" to one party or the other and would vote almost automatically for the candidate of their party. In the 1950s, about 90 percent of those in national samples identified themselves with a major party.

Having come to life to nominate its candidate, a national party remained active long enough to raise and spend money on his campaign. In the days before radio, it relied mainly on meetings with oratory, full-page advertisements in the big-city newspapers, leaflets, brochures, buttons, and direct mail (in 1912 postcards went to each of the 1.6 million voters in the state of Pennsylvania). The national parties sometimes gave subventions to state and local party organizations to pay for house-to-house canvassers, whose main job was to see that the party regulars got to the polls.

---

[2]This does not necessarily mean that the state party leaders failed to choose the most popular candidates. William H. Lucy has shown that since 1936 (when polls began), the active candidate who led the final preconvention poll was nominated nineteen of twenty times (Lucy was writing in 1973) and all but two of these nominations were on the first ballot. "Whether under the old procedures or the new, the relationships since 1936 between polls, primaries, and nominations are consistent with the view that presidential politics have tended toward national publicity and moved away from reliance on factional alliances and tangible incentives as means of winning the nomination." "Polls, Primaries, and Presidential Nominations," *Journal of Politics* 35 (1973): 837, 847.

Being the choice of a winning coalition of state party leaders, the candidate who made it to the White House would bring with him the support of at least some key figures in the House and Senate. He could never be a real party "boss"; the separation of powers and the traditional congressional distrust of the president prevented that. He could, however, exert a good deal of influence (for example, on the selection of House leaders). As V. O. Key remarked in the first (1942) edition of his famous textbook, "the party system, when it operates properly, overcomes the handicaps to governance imposed by the separation of powers and furnishes a common leadership and a bond of loyalty by which the President and Congress may work together."[3]

A president could take it for granted that he would be renominated for a second term. If he had lost his popularity with the country or had antagonized a powerful faction of his party, he might have some opposition, but, because of the favors his administration could give or withhold in the states and because the state party leaders would consider it disastrous for the party to repudiate its leader, he could have his way if he insisted. (From 1860 on, the only elected presidents not renominated for a second term were two who declined—Hayes in 1880 and Roosevelt in 1908.)

That was the "old" party system. The "new" one is strikingly different. Today an aspirant for the presidential nomination need not be a state party leader or have the support of one. What is essential today is that he show promise of being able to win an enthusiastic, but not necessarily overwhelmingly large, popular following.

This change results from a sharp increase in the number of states having presidential primaries: from seventeen in 1968 to thirty in 1976. Now nearly three-fourths of the Democratic and more than two-thirds of the Republican convention delegates are chosen in primaries. Even in the states that retain the convention system, party leaders are not as free as before to choose—and control—delegates. Obviously, a would-be nominee will court voters, not party leaders. If he does well enough with the voters, he may have the nomination won weeks before the convention is called to order.

His problem being to show that he can win primaries, the aspiring candidate will need to build an organization suitable for that purpose— one or two advisers with a talent for this sort of entrepreneurship, and

[3]*Politics, Parties and Pressure Groups* (New York: Thomas Y. Crowell, 1942), 520.

several technical specialists: fund raisers, pollsters, direct mail advertising experts, television and radio producers, organizers of volunteers, legal advisers, and so on.

Since this will cost a lot, the candidate will, a year or so before he announces his candidacy, file with the Federal Election Commission the names of the members of an "exploratory committee" which will make the preliminary arrangements necessary to establish his eligibility for government preconvention campaign funding. (To be eligible, he must have raised $5,000 in contributions of $250 or less in each of twenty states.) Under present regulations, a candidate for the nomination may raise and spend $2.2 million for fund raising and another $5.5 million for campaign expenses. He may not accept individual contributions of more than $1,000. The government will match contributions of $250 or less per person up to a total of $5.5 million. Thus his preconvention expenses may be as much as $13.2 million. (If he elects to spend his own money, the law imposes no limit.)

As he moves from one primary to the next, the candidate is followed by a troop of newspaper and television reporters and cameramen. Much depends upon them. They are, as David Broder has remarked, "talent scouts" and "handicappers."[4] If a columnist says that a candidate's prospects are improving, that may help them to improve, for the report may bring more contributions, more volunteers, and more attention from the media, the effect of which may be to improve his standing in the polls, which in turn may reinforce the opinion that his prospects are improving, which in turn. . . . In the same way, of course, a reporter's impression may set the candidate's campaign on a downward spiral and bring it to an untimely end.

In making his appeal to primary voters, the candidate will consider very carefully just how partisan to be. He knows that about one-third of all voters—one-half among the young—disclaim any party affiliation. If he campaigns as a strong party man, he may drive many of these "independents" into the arms of his opposition in the general election. Even with the voters who belong to his party he must watch his step: if he lets himself be embraced too warmly by a state party leader, he may be charged with being a tool of the "bosses." In some primaries, crossover voting is allowed: that is, a voter registered in one party may have the right to vote in the primary of another. The turnout in primary elections

[4]Cited by Asher, "The Media," 213.

is small (usually from 25 to 35 percent of those eligible). Since what the candidate needs is a small but dependable army of enthusiasts, his rational strategy may be to make a conspicuous display of his independence by attacking the regulars of his own party. "Insurgency," Gerald Pomper has written with reference to the campaigns of Reagan, Carter, and Brown, "is no longer the crusade of political Don Quixotes; it is the most likely path to the political kingdom."[5]

Once nominated, the candidate—not the party—is eligible to receive a flat grant of government funds for his general election campaign expenses. (The party will have received a government grant, but only for convention expenses.) If he accepts the federal funds, as he almost certainly will, the candidate may not spend, or coordinate the spending of, any other funds. His party, however, may spend up to two cents per voting-age citizen (about $3.2 million) on his behalf. Individuals and groups—corporations and labor unions, for example—may spend as much as they please provided their activities are independent of his.

The newly elected president is not likely to be on close, cordial terms with the leaders of his party in the House and Senate. They and the other state party leaders had no part in choosing him as the nominee, and he had very likely been at pains to keep a certain distance from them during the campaign. From their standpoint, he may be an "outsider," even an "amateur." If the president and his party's congressional leadership overcome the handicaps to governance imposed by the separation of powers, it will obviously not be because the party system has furnished "a common leadership and a bond of loyalty."

Under the "new" system, the president cannot count on renomination. On the contrary, he must expect to fight for it and, perhaps more often than not, to lose. He will have a considerable advantage in the new round of popularity contests because of the constant attention he has been given by the media. On the other hand, he will be held accountable for whatever has gone wrong. That rather small and unrepresentative part of the public that votes in primaries is likely to have lost enthusiasm for him, since it is impossible for him to have done all that the enthusiasts expected from him in office. But even if the polls show there is little chance of beating him, some will try. With government financing of campaigns, there is little to lose, and of course there is something, if only notoriety, to gain.

[5]"Decline of Partisan Politics," 28; see also Orren, "Candidate Style," 146–51.

SOME CONSEQUENCES FOR THE POLITICAL SYSTEM

The party system is, of course, a component of the larger set of arrangements that is the political system. It is characteristic of any system that its elements are interrelated in such a manner that a change in the state of any one produces changes in all the others. "Reform" of the party system may therefore be expected to produce changes in features of the government that are in a sense remote from the parties themselves.

How has the political system been affected by the movement from the "old" to the "new" party? The question does not admit of a really satisfactory answer. For one thing, it is impossible to specify the particular causes, very likely many and diverse, that together have produced a particular effect; party "reform" was certainly a contributing cause of some changes, but how can one say whether its contribution was large or small? For another thing, the effects in question are seldom readily apparent; they may appear at unexpected places within the political system and at unexpected times: the most important may not appear for many years, until some unusual strain is put upon the political system.

Obviously, then, the list that follows must be regarded with much caution. It is a set of conjectures, the plausibility of which the reader will have to judge.

1. The near exclusion of state party leaders from the process by which presidential candidates are selected has gone far toward transforming the American political system from one of representative to one of direct democracy. As contrived by the Founders, the system distributed power among a numerous and intricately graded set of authorities who spoke for, and helped to form the opinions of, diverse publics. The Founders meant government to be responsive to widely shared, strongly held, and informed opinion, and they looked to the leaders of the many publics to organize and express this opinion.[6] State and local party leaders came in time to be among the most effective of these leaders.

---

[6]"Washington did not believe that it was either possible or practical to supply the people in advance with the information that would enable them to advise wisely before a decision was made. And after the decision, they would first have to be persuaded that the issue was adequately grave to command their deepest attention. Then, so that they would not come up with vaporous ideas, they would have to be made to understand what were the possible alternatives. They would have to be given time to study the matter. If the people then expressed in some finite manner—through a spontaneous flood of mass meetings or through the

These and other intermediate authorities having become fewer and less influential now, "public opinion" has become more and more what the individual, who may neither know nor care about the matter at hand and who need feel no obligation to consider the public interest, says in response to questions framed by the Gallup Poll or put before him or her in the privacy of the voting booth. Obviously, there is more reason now than there was in 1789 to fear that direct democracy—the democracy whose opinion is prepared by the newscaster and recorded by the pollster— will fail to produce governments capable of protecting the society from enemies foreign and domestic.

2. Insofar as democracy is direct rather than representative, people must be held in civil discipline by one central authority rather than by numerous intermediate ones. As the power of these other authorities declines, that of the central one must increase. The new power must come from different sources than the old and will therefore be of a very different character. Whereas local authority derived largely from respect for persons and institutions whose qualities were known more or less at first hand, central authority must depend essentially upon the arts of mass merchandising—on "image" rather than character. As a presidential public relations adviser remarked not long ago, what matters is less the content of policy than the way it is packaged. In this change, two dangers are apparent: first, that in order to generate the power it must have to govern, the central authority will come to neglect serious deliberation in favor of hasty generalization or "the trick of rapidly framing and confidently uttering general propositions" creating "levity of assent"; second, that this trick will succeed too well and tyranny will result.[7]

---

ballot box—disapproval of the government's actions, it would be 'disgraceful' for the government not to follow their lead." James Thomas Flexner, *George Washington*, vol. 4 (Boston: Little, Brown, 1972), 480.

[7]The quoted words are from Sir Henry Sumner Maine, *Popular Government* (New York: Henry Holt, 1886), 106–8. In my article cited before, I recalled Maine's observation that party and corruption had in the past always been relied upon to bring men under civil discipline but that now (1886) a third expedient had been discovered:

> This is generalization, the trick of rapidly framing, and confidently uttering, general propositions on political subjects. . . . General formulas, which can be seen on examination to have been arrived at by attending only to particulars few, trivial or irrelevant, are turned out in as much profusion as if they dropped from an intellectual machine;

3. The "new" system makes probable the election of a president who is radically unacceptable to a substantial majority of the electorate. Under the "old" system, there was always the likelihood that mediocrities would be nominated: one could be confident, however, that they would be *moderate* mediocrities—the party professionals would see to that. Now it is likely that unrepresentative enthusiasts, voting in primaries, will produce extremist candidates in both major parties. Even if, under the pressures of office, the elected extremist adapts to the realities of the situation by moving toward the center, the consensual basis on which freedom and order so critically depend may in the meanwhile have been badly damaged.

4. The changes in the party system have decreased somewhat the power of the presidency in relation to that of the Congress. A president under the "new" system normally will not enter the White House as the leader of a coalition that includes principal figures in the House and Senate and, as often as not, he may be an "outsider." Under the best of circumstances, it takes some time for a new president to establish the basis of understanding and trust with the leaders of his party in congress on which the success of his efforts at leadership depends. His position vis-à-vis Congress—indeed, vis-à-vis all those with whom he must deal—is much weakened by the fact that he may fail to win renomination: a president who has, say, a 25 percent chance of serving for eight years will presumably be taken only half as seriously as would one who has a 50 percent chance. The possibility of challenging the president for the nomination will be ever-present in the minds of some key figures in Congress. This will make it all the more difficult for the president to exert leadership there. Those who see themselves as possible challengers will refrain from supporting presidential positions they may later want to attack. The president, knowing that his strongest party associates are potential opponents,

---

and debates in the House of Commons may be constantly read, which consisted wholly in the exchange of weak generalities and strong personalities. On a pure Democracy this class of general formulas has a prodigious effect. Crowds of men can be got to assent to general statements, clothed in striking language, but unverified and perhaps incapable of verification: and thus there is formed a sort of sham and pretence of concurrent opinion. There has been a loose acquiescence in a vague proposition, and then the People, whose voice is the voice of God, is assumed to have spoken.

will avoid using them in ways that might improve their standing in the polls.

5. The diminution of the president's influence with the leaders of Congress has opened new opportunities for the exercise of influence by special interest groups. This, in some measure, explains the phenomenal increase in the number of Washington lobbies in the past thirty years— from about 2,000 to about 15,000. (The main cause, however, has surely been the extension of federal intervention into every nook and cranny of national life.) Insofar as the competitive bidding of interest groups replaces presidential leadership, the policy outcome will be delay, stalemate, and contradiction. Government will be more "conservative," that is, less able to act, but the flood of special interest legislation will continue to rise. Presidential influence would be needed to stop it—more influence now than in the "old" days because congressmen who once depended upon the party label and the party precinct captains to keep them in office must now look elsewhere for the large sums needed to court the issue-oriented voter. (The average House chairman received $45,000 from political action committees in the 1978 elections.)

6. Public confidence in and respect for government will decline. The "new" system gives members of the president's party the incentive to undermine confidence in him and in his administration, in the hope of taking the nomination from him. There is a limitless supply of grounds (albeit mostly technical) for charging violation of the campaign financing laws and therefore "corruption." An "outsider" president will be unable to establish his ascendancy over other party leaders. There will be an appalling increase in the amount of special interest legislation and a relentless thrust of the bureaucracies toward self-aggrandizement. These and other causes will make the citizen ever more angry and frustrated with government and ever less disposed to think of it as the defender of liberty and justice.

THE ACCIDENTS OF "REFORM"

The transformation of the "old" party system into the "new" happened largely by accident. No one intended the old system to be weakened, let alone dismantled. On the contrary, the intention of the reformers was to strengthen it by making it more democratic.

One very important set of "reforms" was made possible by court decisions which transferred power to make rules for delegate selection from the state parties and legislatures to the national conventions. Half a century before, reformers had given up as hopeless ideas which, because of the changed attitudes of the courts, now became practical. In the exercise of its newly won power, the Democratic National Convention did not mean to substitute primaries for conventions nor to exclude state party leaders from the process by which presidential nominees were chosen. The eighteen "guidelines" proposed by its Commission on Party Structure and Delegate Selection and revised by two subsequent commissions, were intended only to encourage participation, especially by women, young people, and members of minority groups, in all stages of the process. It was clear that some contenders for power, especially those associated with Senator George McGovern, would gain by this. The intention of the reformers, however, was to make the process of selection more democratic.

That the transfer of power from the states to the national conventions would cause many state party leaders to withdraw entirely from the delegate selection process was not foreseen. It is safe to say that none of the reformers of 1968 guessed that in less than ten years, a candidate would win the nomination well before the convention. It is safe to say, too, that none of them realized that by greatly increasing participation in the selection process (from 11 million in 1968 to 30 million in 1976) they might bring about the choice of a nominee less representative of the party rank and file than one chosen by a few party professionals.[8]

Another important set of "reforms"—those relating to campaign finance—were intended to make the system more democratic by eliminating from elections, so far as was feasible, the special influence of wealth. Public funding of the candidates, so it was thought, would guarantee nearly equal opportunity to all. That the Supreme Court, by striking down key sections of the law on First Amendment grounds, would actually increase the relative importance of "professional" contributors (interest

[8]Austin Ranney has said that the purpose of those who framed the new rules for the Democratic party in 1969–1972 was "not to make the party more combatready" or to make it more representative "of all elements of the New Deal coalition" and "certainly not of party notables or regulars or contributors." The reforms, he writes, "were intended to maximize the representation of 'purists,' not of 'professionals.' " "Comment of 'Changing the Rules Changes the Game,' " *American Political Science Review* 68 (March 1974): 44.

groups of all sorts) and of "family giving" by the rich, could not have been foreseen. It seems highly implausible, on the other hand, that the professional politicians did not realize they were grievously damaging the parties by voting to give campaign funds to candidates rather than to parties. Yet, the record seems to support this view.

The "reforms" of the late 1960s and early 1970s were the culmination of more than a century of efforts to bring party politics into correspondence with the democratic ideal. The ruling elite always favored—in principle if not in practice—a politics that would settle issues on their merits rather than "give everyone something." The ethos of this elite, toward the end of the nineteenth century, inspired the municipal reform movement, whose program included nomination by petition as well as the initiative, recall, and referendum. From city government, the reformers moved to state government: open primaries were adopted (as was noted before) by many states in the early years of the twentieth century. Apparently it was the impossibility of establishing uniform rules for selection of delegates under the legal doctrines then prevailing that caused many reformers to lose enthusiasm for presidential primaries.[9]

The nonpartisan style of the elite steadily diffused into the middle class. As incomes increased and more and more boys and girls went to high school and even to college, the size of the middle class grew both absolutely and in relation to the working class. All those who had read a civics book knew that "bosses" were always "corrupt" and that the enlightened citizen would vote for "the best man regardless of party." In the 1930s and 1940s there was (judging from the polls) no ideological difference between the college educated and others. There seem to have been, however, marked differences in political style. The relatively well educated were made uncomfortable by Harry Truman's affinity for professional politicians of the Kansas City variety and by his unabashed partisanship; at the same time they were respectful of Dwight Eisenhower's "man above party and above faction" stance and admired Adlai Stevenson's elegant superiority to the party "hacks." In 1952 and again in 1956, these two candidates, each in his own way, strengthened the already widely held view that parties were obsolete.

The Republican party was gravely damaged by Eisenhower's neglect. The Democratic party was hurt by Stevenson in an altogether different

[9]Robert C. Brooks, *Political Parties and Electoral Problems* (New York: Harper and Brothers, 1923), 279.

way. Enormously attractive to young liberals, he inspired many of them
to try to enter party politics in the large cities in the hope of democratizing
it. The "amateur democrat," as James Q. Wilson described the type in
the early 1960s, "sees the political world more in terms of ideas and
principles than in terms of persons. Politics is the determination of public
policy, and public policy ought to be set deliberately rather than as the
accidental by-product of a struggle for personal and party advantage.
Issues ought to be settled on their merits."[10] It was the "amateur democrat," moved now to the national scene, who, following the 1968 convention, devised and won acceptance for the "guidelines" that brought
the Democratic party close to the democratic ideal and close, as well, to
destruction.

In 1960 amateurs of a different sort demonstrated how little the
party professionals had come to matter. The nine men who met at his
father's Palm Beach home to plan a strategy for John F. Kennedy's nomination were not party leaders: they were relatives, friends, and employees
(only one was a professional politician)—persons attached to him as a
person, not as a party figure. At that time, only eleven states chose all
their convention delegates in primaries (four others chose some). Accordingly, Lawrence F. O'Brien went on the road to tell the key state leaders
of the candidate's qualities. The main thrust of the Kennedy strategy,
however, was not in this direction; rather, it was to make a showing in
the polls and the primaries. Television, which had begun to reach mass
audiences early in the Eisenhower years, had now come of age, changing
the situation fundamentally by making it possible for a candidate—albeit
only for one who was well financed—to address the voters over the heads
of the party leaders.

Appealing over the heads of the party leaders would have done a
candidate no good in the "old" days of strong party loyalty and strong
precinct organizations. It was because partisanship had declined sharply
that Kennedy could use the media with great effect. (In 1960 only 21
percent of the Survey Research Center's sample of voters identified themselves as "strong" Democrats; 33 percent were either "weak" or "independent Democrat," and 8 percent were outright "Independent.")
Kennedy's success may therefore be viewed, in large part, as a consequence of the decline of partisanship that had long been underway.

[10]*The Amateur Democrat* (Chicago: University of Chicago Press, 1962), 3.

This decline is sometimes accounted for in part by the greater salience to voters in the 1950s and 1960s of issues—especially race, the
Vietnam war, and the environment—that bear little relationship to the
historical bases of Republican-Democratic division: namely, taxes,
employment, farm prices, and the like. This is plausible. It is likewise
plausible that causality may have run in the opposite direction as well:
that the weakened attachment to party was a necessary and perhaps
sufficient condition of the emergence of the new issues; parties, both as
organizations and as bodies of received wisdom, may have functioned to
"decide" what was, and what was not, to be the subject matter of politics.
There is still another possibility. Both phenomena—the decline of partisanship and the salience of the new issues—may be effects of a single
cause: the general moralization of public life that has followed from mass
acceptance of views that once were the exclusive possession of an educated
elite.

THE PROSPECT

A good many politicians, journalists, and even some political scientists have begun to think twice about the wisdom of the "reforms" that
have been made in the party system; there is widespread agreement that
the parties have been seriously weakened without being made more democratic. No one, however, proposed trying to undo the damage to the
parties by (for example) restoring control over the selection of convention
delegates to the states or by ending government funding of campaigns.
The reason for this is not that the damage cannot be repaired (although
clearly the "old" system cannot be put back together again); rather, it is
that such proposals would not be taken seriously. Enthusiasm for pressing
further and faster toward direct democracy remains unabated. Further
changes are being made in the Democratic party's delegate selection rules
to "wipe out the last vestiges of the 'winner-take-all' system."[11] government funding of both House and Senate campaigns is under active consideration in Congress, support is growing for a constitutional amendment
to establish a national presidential primary (a Gallup Poll has found that
the public favors popular election of the president by five-to-one), and

[11]*The New York Times,* February 25, 1979, 17.

Senate hearings have been held on bills to amend the Constitution to provide both the referendum and the initiative.

There is a deep irony in the fact that, during the Bicentennial period in which we celebrate the achievement of the Founders, we also complete the undoing of it. The system that they established was based explicitly upon the understanding that man is a creature more of passions than of reason, that the problem of the lawgiver is to find ways of holding him in check, and that this can be done only by making ambition counteract ambition. It was based also upon the consequent understanding that government must have limited objects: to promote the life, liberty, and the pursuit of happiness of its citizens, *not* to right all wrongs. The fundamental fact of today is that man is seen, not as he is, but as he ought to be. Liberalism, which in one version or another has become our civil religion, imagines that he is above all reasonable and therefore capable, through the acquisition of knowledge (actually information and technique), of solving all problems and of progressing toward perfection. So long as this delusion persists, we shall be impelled to reform out of existence—at whatever cost to civilization—not only the "defects" of our political system but whatever else may, or may seem to, impede the establishment of heaven upon earth by the exercise of reason and good will.

PART II

Democratic Policymakers

The trends described in the previous section, namely erosion of the principle of enumerated powers and the weakening of party leadership by reform, contributed to a proliferation of new federal grant-in-aid programs. In this section the fragmented political system is seen trying to cope with further fragmentation.

In the mid-1950s, when there were about 130 federal grant-in-aid programs, President Eisenhower made a serious effort, which was almost wholly unsuccessful, to reduce the number by turning back to the states activities that were properly theirs. Ten years later, when the Model Cities Program got underway, the number of grant-in-aid programs had increased to about 400. After another five years, when the Nixon administration put forward its revenue sharing plan, it had reached about 500. Whether the system would allow itself to be coordinated was obviously a question.[1]

As the reader will see, coordination of its many urban-aid programs, although always one of the principal purposes of the Model Cities Program, was by no means its only purpose. In its rather brief life, Model Cities played many parts, two or three before its birth and others, some simultaneously and others seriatim, later. More than anything else, confusion of purpose destroyed the agency. In the account given here, the

---

[1]"The number of federal categorical grant-in-aid programs available to state and local governments stood at 392 as of January 1, 1984, according to a tally made by the ACIR staff. This represents a decrease of 142, or about 27%, from the 534 counted three years earlier." *A Catalog of Federal Grant-in-Aid Programs to State and Local Governments: Grants Funded FY 1984* (Washington, D.C.: Advisory Commission on Intergovernmental Relations, December 1984), 1.

confusion is seen to result not from some aberration of government but rather from the normal working of a political system in which power is very widely distributed.

Of the several political impossibilities that were demonstrated by the Model Cities Program one of the most evident is that of inducing, or compelling, federal agencies to work together toward some well-defined common purpose. There are of course strict statutory limits to the changes that agencies may make in their programs. But political imperatives are at least as important: every agency is under the necessity of fighting to preserve, and when possible to increase, its "turf," for if it fails to do so its support among interest groups and in Congress will evaporate. If a president had nothing else to do he could reduce the chaos somewhat by "knocking heads together." But a president, alas, has many other and more important things to do.

The second item in this section is a transcript of a rather informal lecture given in November 1972 to the Institute for Urban Studies of the University of Maryland. Here the author, who was chairman of the President's Task Force on Model Cities, tells how that body went about making a quick evaluation of the program. (The Task Force's report was published by the Government Printing Office in August 1970 under the title Model Cities: A Step Towards the New Federalism). *The reader who has taken an elementary course in policy analysis may well be amused at the crudity of the procedures described. He should ask himself, however, whether a political leader is not likely to want and demand recommendations "at once" and whether the questions that he wants answered are not likely to be unanswerable except on a largely subjective basis. How often will a policymaker have use for a proper study, the results of which, probably inconclusive, will not be available for two or three years, very likely after the policymaker's problem—and perhaps the policymaker himself—has passed from the scene?*

The last several paragraphs of the lecture to the Institute for Urban Studies represent a digression. They recall the response that the writer made when the White House asked for comments on the task force as an advice-giving device.

There is reason to believe that when the Nixon administration took office no one in the White House knew of, let alone had read, the memoranda left by the Johnson administration's administrators of Model Cities for their successors. (See the section of Chapter 4, "Making a New Federal

Program," entitled, "Two Voices of Experience.") As it turned out, however, the new administration acted very much as if it were following the advice left by its Democratic predecessors. It made no effort to establish a "superagency" to coordinate the others. It moved instead toward the consolidation of "categorical" (special purpose) grant-in-aid programs into "block" (general purpose) ones. And it pressed successfully for "revenue sharing": the distribution of federal funds to states and cities with no strings attached.

In theory, revenue sharing, as proposed by President Nixon, was to be "a New American Revolution . . . a peaceful revolution . . . in which government at all levels will be refreshed, renewed, and made truly responsive." In practice it turned out to be a minor disturbance at most. The same powerful political forces that had brought so many grant-in-aid programs into existence not only prevented their going out of existence but even added somewhat to their number. Some categorical programs were consolidated into block grants, but an unmanageable number remained. Revenue sharing became a reality, but on not much more than a token scale and with much uncertainty about its future. In the future, the essay concludes, the federal government will have a larger role in the spending as well as in the raising of state and local revenues. The import of the essay, and indeed of the whole section, is that it is now impossible to restrict the scope of the national government or even to bring its complexities under presidential management. If there is a New American Revolution, it will not be of the sort anticipated by President Nixon when he proposed revenue sharing.[2]

[2]For an up-to-date account of policy choices made with respect to the design and use of federal grants-in-aid to state and local governments by presidential administrations, the reader is referred to Lawrence P. Brown, James W. Fossett, and Kenneth T. Palmer, *The Changing Politics of Federal Grants* (Washington, D.C.: Brookings Institution, 1984).

CHAPTER 4

Making a New Federal Program
Model Cities, 1964–68

During the evening of the first full day of Lyndon B. Johnson's presidency—at 7:40 P.M. on November 23, 1963, to be precise—Walter Heller, chairman of the Council of Economic Advisers, came to tell him that three days before his assassination President Kennedy had approved a suggestion that the council and the Bureau of the Budget develop a program to alleviate poverty. They were thinking, Heller said, in terms of pilot projects to be tried in a few cities. The president was enthusiastic, but he wanted something "big and bold." A program for just a few cities could never be propelled through Congress and, in any case, it would be regarded as another example of "tokenism" by black leaders whose growing anger was a matter of concern to him. A few weeks later, in his first State of the Union Address, he declared "unconditional war on poverty in America." The Office of Economic Opportunity (OEO) was created almost at once.

This initial attack was to be followed by offensives along a broad front. The Vietnam War was expected to fizzle out shortly and the federal treasury to overflow with a "peace dividend" as the economy expanded. No doubt, too, the president was confident that after the November election he would have comfortable majorities in both houses of Congress (as it turned out, he had the largest ones since 1937). Under these circumstances, he needed big ideas to put before the new Congress—a legislative program that would be distinctively his.

From *Policy and Politics in America: Six Case Studies,* ed. Alan P. Sindler (Boston: Little, Brown & Co., 1973), 124–158. Copyright 1973 by Edward C. Banfield.

The following pages tell about one of the biggest and boldest of the programs that resulted—how the Model Cities effort was planned; how, through the intervention of the president himself it was pushed through a reluctant Congress; how its administrators grappled with dilemmas and constraints that were partly inherent in the conception of the program and partly in the harsh realities of the governmental system; and how, by the time the Johnson administration left office, the program had become very different from what its originators had intended it to be.

NEEDED: THE BEST THOUGHT

In May 1964, the president told a University of Michigan audience that it was in the cities, the countryside, and the classrooms that the Great Society was to be built. While the government had many programs directed at these "central issues," he did not pretend to have the "full answers." To get them he would establish "working groups" to make studies and hold White House conferences. "We are going," he said, "to assemble the best thought and the broadest knowledge from all over the world to find those answers for America." That there *were* "answers" to large social problems and that thought and knowledge could "find" them he seemed to take for granted.

The president knew that John F. Kennedy had used task forces in his 1960 campaign. Because these groups had been too heavily weighted with scholars, the president thought, "very few suggestions emerged that were practical enough to exploit." He intended his to have a broad balance of "thinkers" and "doers." The presidential assistant who played the leading part in making up the Metropolitan and Urban Affairs Task Force was Richard N. Goodwin, a young lawyer who had been in the Kennedy entourage and was now a Johnson speechwriter. Although there were "doers" in the group that he assembled, its most active members turned out to be mostly "thinkers."

One of these was the chairman, Robert C. Wood, an amiable man just entering middle age who was a professor of political science at M.I.T. Before beginning his scholarly career he had served for three years in the Bureau of the Budget. His later work dealt with practical matters largely (his most recent book—*1400 Governments*—was an account of the intricacies of governmental organization in the New York region), and he had been active in the Kennedy advisory group during the 1960 election campaign. He could, therefore, be considered both a "thinker" and a "doer." The other members of the task force were Paul Ylvisaker, director of the public affairs program of the Ford Foundation; Raymond Vernon, a Harvard economist who had directed the study of the New York metropolitan region; Jerome Cavanagh, mayor of Detroit; Saul B. Klaman, director of research for the National Association of Mutual Savings Banks; Ralph McGill, editor of the *Atlanta Constitution*; Dr. Karl A. Menninger, chief of staff of the Menninger Clinic; and four members of the faculty of the University of California at Berkeley: Nathan Glazer, a sociologist; Norman Kennedy, associate director of the Institute of Traffic and Transportation Engineering; Martin Meyerson, dean of the School of Environmental Design; and Catherine Bauer Wurster, professor of city planning. William B. Ross, a career civil servant from the Bureau of the Budget, was executive secretary. Goodwin was White House liaison.

The task force had less than three months in which to work. By the middle of November 1964, Goodwin, along with the assistants assigned to the other task forces, would have to begin "feeding" the president paragraphs for use in the various messages he would send to Congress in January 1965. Despite the lack of time, the group produced a report running to seventy double-spaced pages with an additional ten pages summarizing its recommendations. Often such reports consist largely of hot air; this one did not.

The report was politely but severely critical of the existing government programs for the cities. "Too frequently they operated within narrowly defined agency boundaries that fragment logically related services." This was true even of the newly created OEO ("admirable" but "designed to help primarily the temporarily disadvantaged"); the task of this and the juvenile delinquency programs "must be broadened to include health, education, recreation—the entire gamut of social development."

It went on to make many far-reaching proposals for change: the federal government should give block grants[1] to enable the cities to provide a wide range of services to all of their citizens as well as to care effectively for the poor; it should also make special grants for the support

---

[1] A "block" grant is one that may be used for any purpose by the recipient or, at least, for a very wide range of purposes. By contrast, a "categorical" grant may be used only for a specified—often very narrowly specified—purpose.

of community facilities (health stations, small parks, and so on); both the block and special grants should be contingent upon a *local* (this word was underlined) Social Renewal Plan (these words were capitalized); the government should also expand its technical assistance in many fields of public administration; and it should support planning by the cities and endeavor in other ways as well to strengthen the office of mayor (or manager).

In polite—and therefore not very vigorous or vivid—language, the report said that most urban renewal projects had served the interests of downtown department stores and real estate operators and that public housing projects had not made a dent in the slums. If middle-income people were to be lured back into the deteriorating areas of the larger, older cities, a new strategy would have to be used. Timid efforts would be worse than none. "What is needed is an intervention so large and so profound as to alter the image of a neighborhood."

There were many other recommendations, including building large new cities from the ground up, reorganization of housing finance, aid to rail rapid transit, and creation of an Urban Affairs Council headed by the vice-president. The emphasis, however, was on making federal efforts more comprehensive, improving their coordination, and giving local governments and the citizens most directly affected a greater share of control over the federal government's undertakings. The task force expected, it said (and these words were underlined),

> that the block service grants, the special local facility grants, the social renewal plan and the revised urban renewal plan would enable the mayor or other public bodies where appropriate, especially in larger cities, to bring together heretofore separate activities into a comprehensive strategy for local action.

Another report, this one prepared in the Bureau of the Budget and taking the form of a 100-page book, went to the president in May 1965. It reviewed various efforts to coordinate the many categorical programs through the voluntary efforts of the agencies themselves. (These included the Neighborhood Development Programs and One-Stop Welfare Centers.) These efforts had been almost fruitless and extremely time consuming. The president, the Bureau of the Budget suggested, might deal with the coordination problem in one or the other of two ways: either by giving the cities block grants or by instituting a program under which they would be helped to make comprehensive plans that, when approved by the White House, would be implemented with federal funds.

THE TWO-PAGE APPENDIX

Whether because the president thought that they smacked more of "thinkers" than of "doers" or for other reasons, neither the task force nor the Bureau of the Budget report seems to have influenced his 1965 legislative program. Ironically, a proposal that was *not* among the task force's many recommendations turned out to have great importance. In the course of its discussions, at least three members of the task force (Wurster, Meyerson, and McGill) had more or less independently suggested that the federal government "adopt" two or three large cities and in addition build a brand-new one in order to show what could be accomplished by well-conceived, large-scale, concerted effort.

A concrete proposal along these lines had come to the task force in the form of a two-page memorandum titled "Demonstration Cities" written by Antonia Chayes, a former student of Wood's, in collaboration with her boss, Leonard Duhl, a psychiatrist who was head of the Office of Planning of the National Institute of Mental Health. Mrs. Chayes and Dr. Duhl proposed that all relevant federal agencies join in sending a team of experts to three large cities to make comprehensive plans for dealing with their particular problems in a fundamental way. When the plans had been made, agencies would concentrate their resources in a coordinated effort to carry them into effect. This might involve a good deal of slum clearance and rebuilding as well as the repair of much dilapidated and blighted housing. The emphasis, however, would be shifted from the "bricks and mortar" approach of the urban renewal program to a "social and psychological" one involving the improvement of school and health facilities, provision of neighborhood recreation centers, better police–community relations, participation by neighborhood residents in decisions affecting them, and whatever else offered some hope of raising morale and creating self-confidence among those who lived in the poorest districts.

This proposal may have been influenced by what was then happening in New Haven. Beginning in 1959, that city had sharply redirected its efforts from slum clearance to a wide range of efforts to improve the opportunities of the chronically poor, especially the young among them. What was happening in New Haven probably influenced the conception of "youth development projects" that were begun in New York City in 1961 with the encouragement and support of Robert Kennedy, then the attorney general. The central ideas behind these developments—the prevention (as opposed to the "cure") of poverty, especially among the young, through concentration of efforts by a wide range of public and private agencies in order to change the character of an entire district in accordance with a carefully made plan—were also the basic principles of the proposal that Heller carried to President Kennedy and then, a few days later, to President Johnson. (A Bureau of the Budget memorandum of November 21, 1963, had recommended "human investment expenditures through education, training and health" directed at "particular problem groups" or "where the yield will be high," with a focus on "human development" rather than physical facilities.)

The task force did not recommend the "adoption" of any cities; it attached the Chayes–Duhl memorandum to its report as an appendix, however. When Mayor Cavanagh showed the report to fellow Detroiter, Walter Reuther, the president of the United Automobile Workers, it was the appendix that caught Reuther's eye. He and the mayor arranged to have a brochure prepared—"Detroit, A Demonstration City"—which they put before Robert C. Weaver, the head of the Federal Housing and Home Finance Agency, and later before the president. Presumably this is what Mr. Johnson refers to in his memoirs, *The Vantage Point*, where he says that in May 1965, Reuther gave him a memorandum warning of the "erosion of life in urban centers."

ANOTHER TASK FORCE

When Reuther's memorandum reached him, the president was beginning to think about his legislative program for 1966. He had directed his special assistant, Joseph A. Califano, a 33-year-old lawyer who had recently moved to the White House from the office of the secretary of defense, to "perform the spadework on a full-scale domestic program." "Rebuilding the slums," the president thought, represented "the first challenge." To help meet it, a new task force was created late in the summer. It was to work closely with Califano and have its recommendations ready by the end of November 1965.

Wood was to be chairman of the new task force, but "doers" were to dominate it. As the president wrote in his memoirs, "Business, labor, the construction industry, and the Congress were represented on that committee. The academic world was also represented. . . ." Reuther was the labor representative. Other members were Whitney Young, executive director of the National Urban League and a national Negro leader, Ben W. Heineman, president of the Chicago and Northwestern Railroad, Edgar Kaiser, president of Kaiser Industries, Senator Abraham Ribicoff (D., Conn.), Kermit Gordon, president of the Brookings Institution, William L. Rafsky, an economist who had served as program coordinator of the Philadelphia city government, and Charles M. Haar, a Harvard Law School professor who had written on city planning. Apart from the last three and Wood himself, none had any special knowledge of urban affairs.

This time there was to be no liaison with the Bureau of the Budget. The task force would get technical advice from its own professional staff, drawn from persons outside the government. Perhaps the differences between this task force and its predecessor reflected the president's belief that "big" and "fresh" ideas practical enough to be "exploited" would not come from scholars or career civil servants. As he explained later in *The Vantage Point*, he considered task forces better than the standard method of generating legislative proposals, which consisted of taking them from the departments and agencies and filtering them through the Bureau of the Budget and the White House staffs. He believed the bureaucracy was preoccupied with day-to-day operations, dedicated to the status quo, and not equipped to solve complex problems that cut across departmental jurisdictions. Scholars, in his view, had a different but equally serious limitation—that of impracticality.

In seeking to avoid these disadvantages, the president ran the risk of incurring others. For example, leaving agency interests out of account in the design of a new program might mean that the program would be sabotaged by them later on. Apparently he was prepared to take such risks. From his standpoint, the crucial thing seems to have been to present Congress and the electorate with a program that they would regard as a bold and promising response to "the urban crisis." If this was the president's intention, it must have been enormously strengthened in August 1965, when—after the decision to set up another task force had been made—the black enclave of Watts in Los Angeles erupted in a long and destructive riot.

The terms of reference that the president gave the task force were very general. They did, however, include one particular: "Pick up on the demonstration idea."

PICKING UP ON THE IDEA

The task force was not fully constituted until early October 1965. Like its predecessor, it would have to do the best it could in weekend meetings over a period of about two months. Its proceedings (again like those of its predecessor) would have to be carefully hidden from the press. (The president's insistence that his advisers work without publicity resulted, his critics said, from his desire to monopolize the limelight. His own explanation, given later in his book, was at least as plausible: advisers tend to be more candid and critical as well as less cautious when they work on a confidential basis; moreover, when plans leak, opposition is given time to form.)

To preserve confidentiality and to avoid the bureaucratic interests which the president so much distrusted, the task force avoided contact with the agencies concerned. Even Robert D. Weaver, who seemed likely to head the soon-to-be-created Department of Housing and Urban Development (HUD), was not officially informed of what was going on. (As will appear, he managed nevertheless to find out.)

The professional staff of the task force wrote some two dozen background papers, all of high quality (they cost somewhat more than $70,000), but little or no use was made of them. The policy alternatives discussed in the papers were not considered sufficiently practical. It was all very well to document the need for metropolitan area governments, for example, but creating such governments was beyond the power of Congress and the president. Consolidating the 200 or more categorical grant-in-aid programs into a relatively few would doubtless (as the previous task force had stressed) improve coordination, but there was no point in studying this either, for Congress could not be persuaded to allow consolidations and, anyway, the press and public would not be much impressed by that sort of achievement. Califano stressed that the president wanted ideas that would yield spectacular results and yield them *while he was still in office*.

The members of the task force soon found out that they differed on important matters. Reuther wanted to bulldoze slum districts out of existence and replace them with "model communities." Young was opposed to this. Slum clearance, he pointed out, generally turned out to be Negro clearance; he favored tearing down or rebuilding housing where absolutely necessary, but he insisted that the emphasis go on measures like improvement of schools, provision of health services, job training, and protection of civil rights. Kaiser shared Young's preference for the "social and psychological" rather than the "bricks and mortar" approach.

Wood thought of the new program as having the purpose (among others) of coordinating existing programs; he saw it as a way of exerting leverage on the federal agencies that administered the categorical grants as well as on state and local agencies. In this way vastly more resources would be brought to bear on the causes of urban poverty and distress. (This, it will be recalled, had been the main thrust of the previous year's report.) Gordon, who recently had been director of the Bureau of the Budget and who had served with Wood on the other task force, was also much concerned with achieving coordination.

The differences were mostly of emphasis (no one envisioned a program *solely* to clear slums or *solely* to improve coordination), but they were important because they entailed further differences. For example, the "social and psychological" approach implied giving the residents of slum and blighted neighborhoods a large measure of control over programs, for only in this way, it was supposed, could the morale and self-confidence of the poor be improved. "Coordination," however, implied putting control firmly into the hands of elected officials. Wood was as irked as anyone by the arrogance urban renewal officials had often shown in their dealings with local people, and he had no doubt that residents of affected neighborhoods should be consulted. But to give them *control*— that was out of the question.

The task force received a memorandum from the director of the Bureau of the Budget, Charles L. Schultze, which impressed it deeply. (As one of the president's staff, he was not subject to the ban against consultation with operating agencies.) He suggested that there might be a threshold below which public investment in poverty areas yielded little or no return. He had a hunch, he said, that by spending at a "saturation" level the net return might be increased by several orders of magnitude. But since this was only a hunch he favored trying "saturation" in no more than five or ten cities while making careful studies of the results. What he had in mind was an experiment, he told an interviewer later, not a national program, still less a national crusade.

Schultze's "hunch" was one that a first-rate economist might be expected to have. It is worth noting, however, that in 1963, as assistant director of the Bureau of the Budget, he had welcomed essentially the same ideas when he presided over the group which worked up the proposal for demonstration programs in several cities that Heller carried to

President Johnson. He himself had then suggested appointing a "federal czar" in each area to open doors to Washington decision makers and to "knock heads together when necessary to get cooperation" from local agencies. This idea also impressed the task force favorably.

The task force considered briefly giving one city "the complete treatment," but then decided not to, partly from fear that the administration would be charged with "tokenism" and partly because it would be impossible to choose between Detroit, Cavanagh's and Reuther's city, and Chicago, the city of the even more powerful Mayor Daley. Perhaps five cities should be chosen, someone said. Senator Ribicoff answered that five would not be nearly enough. Members of Congress would not vote large sums for a program which they knew in advance would not benefit their constituents. It was therefore a political necessity that enough cities be chosen to assure majorities in both Houses.

The task force began its report with a ringing declaration: 1966 could be the year of the urban turnabout in American history. For this to occur a dramatic new approach embodying three principles was required: *concentration* of available and special resources in sufficient magnitude to demonstrate swiftly what qualified urban communities could do and could become, *coordination* of all available talent and aid in a way impossible where assistance was provided across the board and men and money had to be spread thin, and *mobilization* of local leadership and initiative to assure that the key decisions were made by the citizens living in the cities affected. These principles were to be applied through a national program of city building in which specially qualified communities executed plans of such size and scale as to transform existing urban complexes into new cities, by tying together physical and human resource programs, by developing and testing methods and programs, and by bringing to bear all the techniques of which technology was capable.

The task force proposed that all cities be invited to propose action programs designed to have a major impact on the living conditions of their citizens. After the most promising of these proposals were identified by a special presidential commission, the qualified communities were to prepare more detailed proposals. Each was to establish an appropriate administrative mechanism, not necessarily a formal unit of government, to assemble leadership and resources both public and private. A federal coordinator, the task force said, should be assigned to each city to bring all relevant federal aids together. Once selected as qualified, a city would receive two types of federal assistance: (1) the complete array of existing grants to the maximum extent authorized by law and on a priority basis, and (2) supplemental grants (at a ratio of 80 percent federal, 20 percent local) to make up any difference between what was available under existing programs and what the demonstration would cost.

The report recommended that there be sixty-six demonstration cities: six of over 500,000 population, ten of between 250,000 and 500,000, and fifty of less than 250,000. In the largest cities the typical program should build or rebuild about 24,000 units of housing, in the medium-sized ones about 7,500, and in the smaller ones about 3,000. In all cases the objective should be to eliminate blight completely in the designated area and to replace it with attractive, economic shelter in a neighborhood with amenities essential to a full life.

If there was anything new in these proposals it was the supplemental grants. Wood referred to these as "bait" and "glue" because they would attract federal grants to the cities and also bind them together in workable programs. (These images, incidentally, had been used in 1963 by the Bureau of the Budget officials who worked up the proposal that, much altered, became OEO.) In private, out of earshot of department and agency heads, the supplementals were also spoken of as "clubs," and this indeed was the principal function they were intended to serve and what made this proposal different from earlier ones. A demonstration city, knowing that it would get from HUD any funds that might be lacking to carry out its approved plan, would be in a position to bargain with the other federal agencies and to compel them to accept coordination on the basis of its locally made, HUD-approved plan on pain of being left out altogether from the biggest and most glamorous undertakings. Thus city governments, in partnership with HUD, would bring about the coordination that had so far proved unachievable. The result of all this, Wood hoped, would be to "suck the system together."

HURRY TO THE HOPPER

The task force's proposals did not get the close scrutiny from affected agencies, the Bureau of the Budget, and the White House staff that is usually given to policy proposals before they are publicly endorsed by the president. LBJ met with the task force once to say that they had his wholehearted backing, but it was well understood that he did not want to be bothered with specifics. That was the job of Califano mainly, and he, of course, had countless other matters to attend to. Without much deliberation, then, the president committed himself in the State of the Union Address (January 26, 1966) to "a program to rebuild completely, on a scale never before attempted, entire central and slum areas of several of our cities. . . ." The next day he swore in Weaver and Wood as secretary and undersecretary of HUD, and a few days later he sent to Congress a special 4,000-word message on city development which began with words strikingly like those of the task force report: "Nineteen-sixty-six can be the year of rebirth for American cities." He went on to promise "an effort larger in scope, more comprehensive, and more concentrated than any before." The following day, the Cities Demonstration bill was introduced in both the House and the Senate.[2]

The bill followed the task force report very closely. It declared improving the quality of urban life to be the nation's most critical domestic problem; its purposes were to enable cities "both large and small" to plan and carry out programs to rebuild or revitalize large slum and blighted areas and to improve public services in those areas, to assist cities' demonstration agencies to develop and carry out programs of sufficient magnitude "in both physical and social dimensions," "to remove or arrest blight and decay in entire sections or neighborhoods," substantially to increase the supply of low and moderate cost housing, "make marked progress" in "reducing educational disadvantages, disease and enforced idleness," and "make a substantial impact on the sound development of the entire city."

Demonstration programs were to be "locally prepared" and approved by the local governing body and by any local agencies whose cooperation was necessary to their success; they were to provide for "widespread citizen participation," "counteract the segregation of housing by race or income," and meet "such additional requirements as the Secretary may establish."

The secretary was authorized to make grants covering 90 percent of the cost of planning demonstration programs and 80 percent of that of administering them and, when a city's program was approved, to make

[2]A second bill, implementing other recommendations of the task force, was introduced at the same time; it provided "incentives for planned metropolitan development." Eventually the two bills were combined, becoming different titles of an omnibus bill. For present purposes this development, as well as the substance of the metropolitan planning measure, can be ignored.

grants ("supplementals"), amounting to 80 percent of the nonfederal contributions to the program, for the support of any or all of the projects in
a city's program. In each demonstration city, an office of the federal
coordinator was to be established with a director designated by the secretary. This gave HUD no authority over other departments, however,
and the bill required it to consult each affected federal agency before
making a demonstration grant. The bill did not say how many cities were
to have demonstration programs, what the mix of city sizes was to be, or
how funds were to be apportioned. Appropriations were to be authorized
in "such sums as may be necessary," and the program was to terminate
in six years.

Despite the speed with which the task force's recommendations
moved into the legislative hoppers, they did not escape *all* high-level
review. Weaver had arranged, doubtless with Wood's assistance, to have
one of the task force's professional staff "leak" its doings to him, so that,
when finally he was appointed secretary, he knew enough about the new
program to be sure that (as he put it later) he "could live with it." HUD
lawyers drafted the bill and this gave him a chance to go over it in detail
and even to add a declaration that the new program would not supersede
urban renewal. He was sure that without some such language, the bill
would have no chance of passage.

Another cabinet member who was given an opportunity to express
his views was Willard Wirtz, the secretary of labor. He told Califano that
he did not think that the program would work. Concentrating federal
spending in certain cities for purposes decided upon by the local officials
might be desirable, but the hard fact was that practically all appropriations
had to be allocated as Congress specified in statutes. Moreover, as a
practical matter, department and agency heads had to exercise what little
discretion the laws left them in ways acceptable to the congressmen and
the interest groups upon whose goodwill they depended.

HEARINGS, AND A FORECAST

Hearings began at the end of February 1966 before the Subcommittee on Housing of the House Committee on Banking and Currency.
Secretary Weaver was the first of some seventy-five witnesses whose
testimony (not all of it concerning the bill) took four weeks and filled
1,100 pages.

When he entered the hearing room, Weaver believed that the president had in mind five to ten demonstration cities. (This, incidentally, was also the understanding of Schultze.) Before the session was called to order, however, it became apparent to him that congressmen expected there would be about seventy. After hasty consultation with Wood, he let that figure go unchallenged. He could see, he said later, that some of the congressmen would not discuss, let alone support, a program involving only five or ten cities.

After describing the provisions of the bill ("the most important proposal in the president's program for rebuilding America's cities"), Weaver responded to questions. He assured the chairman that many cities had indicated interest in the program, that its creation would not lead to any decrease of urban renewal funds for any city, that the federal coordinator's function was merely "to sort of lubricate the process" of reaching into the existing federal grant programs. When the ranking minority member expressed fear that the coordinator would leave local officials no discretion, Weaver explained that the demonstration program would be developed by the locality *before* the coordinator entered the picture. Spending decisions, he emphasized, would be made by the local government in accordance with a plan made by it.

A subcommittee member pointed out that a city's program had to meet "such additional requirements as the Secretary may establish."

"Do we have to have that type of criteria?" he asked.

That provision, Weaver replied, would be used "very, very lightly," and "no major criteria will be added which is not in the substitute statute."

The witnesses who followed Weaver were mayors. Addonizio of Newark was first, then Cavanagh of Detroit, Lindsay of New York, Daley of Chicago and (eventually) a dozen others. The mayors, of course, were enthusiastic at the prospect of getting more federal money for their cities, but they thought the sum proposed much too small. Lindsay said that as a rough guess the program to be "really meaningful" in New York would cost $1 billion for the "bare bones" on the physical side and another $1 billion on the human side. He proposed that the entire $2.3 billion for the six-year program be appropriated at once so that the cities could enter into long-term contracts right away. Chicago, Daley estimated, needed between $1.5 and $2 billion. Cavanagh's hopes ("dreams," he called them) were by far the most extravagant ("grand," he said). He envisioned building "a town within a town." "If we think," he told the subcommittee, "of a nation with a population of India by the end of the century and if we think of most of us living in cities, then we must not only dream grand dreams but we must also make them come true."

Opposition to the bill came chiefly from the National Association of Real Estate Boards, the U.S. Chamber of Commerce, the Mortgage Bankers Association of America, and the National Association of Manufacturers. The chamber of commerce man remarked that with huge subsidies a few favored cities would show that they could get certain things done. What, he asked, would this demonstrate to cities in general? Other opponents doubted that the federal agencies could be induced to coordinate their programs, and one said that there were not enough skilled workers to rebuild the cities within a few years on the scale contemplated.

Although the bill seemed to be faring well in the hearings (in his opening remarks, the chairman termed the president's special message "inspiring" and declared that "our job is to give the Administration the legislative authority it needs to get the job done"), afterward the chairman and the ranking minority member agreed to report the bill out amended so as to authorize grants only for planning. Such a move would have amounted to killing the bill. Their decision cannot be accounted for by the weight of the testimony they had heard. Perhaps it reflected distress at the rising costs of the Vietnam War or a feeling that the administration should be satisfied with the many programs for the cities that had already been enacted. Wood, however, suspected a more particular cause: a committee staff man, miffed at not having been given a high post in HUD, had persuaded the chairman to scuttle the bill. Whatever the reasons, the outlook was now bleak. "All signs on Capitol Hill," the *New York Times* reported in mid-May 1966, "suggest that the program is dead."

THE PROPHECY PROVES SELF-DEFEATING

When he read that in the *Times*, the president was furious. Calling in Vice-President Hubert Humphrey and his assistant for congressional relations, Lawrence O'Brien, he told them to find a way to save the bill. If the most important item in his urban program could be dumped by a House subcommittee, his prestige would suffer, and the rest of his legislative program would be in great danger.

O'Brien telephoned the chairman of the subcommittee and told him that the president's bill would have to be reported out intact—"or else." This was unusual language to use, but O'Brien considered it necessary and justified. In due course (on June 23, 1966), the bill was reported with only two amendments of importance. One, which Weaver had proposed, authorized an additional $600 million to be earmarked for urban renewal within demonstration city areas (the renewal "add-on," this was called); the other eliminated the office of federal coordinator.

Reproached in private by the ranking minority member of the sub-committee for having failed to abide by their agreement, the chairman replied as one politician to another. "There are times," he said, "when a man changes his mind and there are times when he has it changed for him. . . . I had mine changed for me."

Five days later, the full committee reported the bill favorably with other minor changes. "Metropolitan expediters" were provided in place of the coordinators. The committee report said that they were to have a "clearinghouse" function and were not to pass upon applications. "This," it declared, "is to be a local program." It also declared that the new program would not "in any way change the flow of funds, as among cities, under existing grant-in-aid programs." This could be taken as a mere statement of fact—the fact Wirtz had pointed out to Califano—or as a warning to federal administrators not to use what little discretion they had in such matters to bring about the concentration of effort that, in the opinion of some—Budget Director Schultze, for example—was the pro-gram's main justification.

Despite this progress, the bill was now stymied. It could not reach the floor of the House without clearance from the Rules Committee, seven members of which were strongly for it and seven strongly against. The swing vote was that of a southerner who objected to the provision requiring that demonstration programs promote racial integration.

Meanwhile, the vice-president and O'Brien had been rounding up support for the bill. They got the AFL–CIO to lobby energetically for it. They organized a committee of twenty-two business leaders, including David Rockefeller and Henry Ford II, to issue a statement calling for its passage. With other high figures in the administration, they had scores of private meetings with members of Congress to make explanations, listen to objections, and, when absolutely necessary, offer assurances that a place would not be overlooked when the time came to select the "winning" cities. By the end of the summer, this promise had been made to more than 100 legislators, which was less than a third of those who asked for it.

By late summer, the Senate hearings had long since been completed (they began April 19 and lasted only a few days); the bill would by now have been brought to the Senate floor but for the fact that the senators who normally managed housing legislation were not willing to manage it.

Senator John Sparkman (D., Ala.), the chairman of the Subcommittee on Housing on the Banking and Currency Committee, although surprised that after passage in the previous session (as he described it) "one of the most comprehensive housing bills of all times" the president was proposing important new legislation, had said when he opened the hearings that his first reaction to the bill was favorable. The objective— coordination of federal programs—was one he approved. He was up for reelection, however, and therefore too busy to lead the fight. Senator Paul Douglas (D., Ill.), who stood next to him in seniority, would not lead it either; he also was busy campaigning and, besides, his staff was critical of the bill.

O'Brien then turned to Senator Edmund Muskie (D., Me.), who was much interested in intergovernmental relations, and found him reluctant. He, too, had received a critical report on the bill from his staff (among other things, it was badly drafted and did not provide adequately for "institutional change"). Moreover, Maine, having no large cities, could not expect much benefit from it. After an all-night session with Muskie's legislative assistant, in the course of which representatives of the White House, HUD, and the Bureau of the Budget accepted certain modifications in the bill, the senator agreed to be its floor manager. On August 9 the full committee reported the bill favorably.

Because of concessions that had been made to Muskie and others, the Senate bill differed considerably from the one bottled up in the House Rules Committee. The Senate bill reduced the term of the program from six years to three, authorized smaller amounts for supplemental grants and the renewal "add-on," did not require cities to plan for racial and economic integration, emphasized the importance of local initiative, and instructed HUD to give "equal regard to the problems of small as well as large cities." During the all-night session in Muskie's office, it had been agreed that supplemental grants would be distributed according to a formula based on population, amount of substandard housing, percentage of families with income below $3,000, and percentage of adults with less than eight grades of schooling. This enabled three Maine cities— Bangor, Augusta, and Portland—to qualify. The formula was not written into the bill, however.

The committee bill reached the floor of the Senate on August 19. In the ensuing debate (which took parts of two days) it was made unmistakably clear that the small cities were to get their share of benefits.

George McGovern (D., S.D.): Do I understand the Senator correctly
that in his judgment some reasonable consideration will be given
to the allocation of funds to our smaller cities, and that not all the
funds will go to a few great metropolitan centers?

Edmund Muskie (D., Me.): I would say to the Senator that I would
expect that we would get demonstration city programs in these
small areas, and I would be disappointed if we did not.

After the defeat of several threatening amendments, the Senate
passed the bill by a roll-call vote, 53–22. Under the threat of a procedure
(passage of a resolution for which twenty-one days' notice had been given)
that would bypass it, the House Rules Committee finally (twenty-one
days after the filing of such a resolution) granted a rule. After acrimonious
debate in the course of which seventeen amendments were voted down
and twenty others (all agreed to by the administration) were adopted, the
bill was passed by a vote of 178–141 in a late-night session on October 14,
1966. (The lightness of the vote is accounted for by the fact that the 14th
was a Friday not many days before an election; many members had left
for their constituencies before the vote was called.) The conference committee report favored the Senate version, although conceding some details
to the House. The Senate accepted it quickly on a voice vote. In the
House, there were moments of uncertainty, but in the end it was approved
by a margin of fourteen votes.

The president must have been well satisfied when on November 3,
1966, he called the chief actors to his office to witness the signing of the
bill and to receive the souvenir pens that are distributed on such occasions.
In all essentials and in most details, he had got what he had asked for.
What was more, he had got it at the right time—election day was only a
week away.

BY WORDS WE ARE GOVERNED

If anyone had hoped that the hearings and debates would resolve,
or even call attention to, the contradictions in the program proposed by
the task force, he would have been disappointed. Some penetrating questions had been raised: Could the agencies administering the categorical
programs concentrate on the demonstration cities? Would they if they
could? Was it possible to have both "citizen participation" and control by

elected officials? Could there be a "partnership" between HUD and local governments?

The matters that had been of most interest to the few congressmen who had familiarized themselves with the specifics of the bill were mainly of symbolic importance, however. What did it matter whether an official was called a "coordinator" or an "expediter"? (That provision had been put in the bill in the first place as a "loss leader" to attract support from the "good government" movement, Wood later told an interviewer.) Actually, the secretary needed no special authorization to employ assistants to try to bring about coordination. Dropping the requirement that city programs "counteract the segregation of housing by race or income" was also of purely symbolic significance. Cities were still free to counteract racial segregation if they wished, and no one had ever supposed that a city making only a perfunctory gesture in this direction would be denied funds on that account.

In its final form the bill authorized only $900 million to be used over two years rather than $2.3 billion to be used over six years. This also made no practical difference; HUD had assured the subcommittees that it could not use more than that amount in the first two years anyway.

A few minutes before the president reached for the first of the pens with which he would sign the bill, Weaver told him of a congressman's objecting to the name *demonstration cities* on the grounds "we got enough demonstrations already . . . we don't need any more"—a reference, of course, to the riots of the summer before. The president took the point seriously. As he signed the Demonstration Cities and Metropolitan Development Act of 1966 into Public Law 89–754, he referred to the new program as Model Cities. It would, he said, "spell the difference between despair and the good life."

AN ADMINISTRATOR'S IMPRESS

H. Ralph Taylor, the man chosen to administer the new program, had been the first executive director of the New Haven Redevelopment Agency (New Haven, it will be recalled, was widely regarded as having led the way in urban renewal) and had later worked for AID in Latin America. As assistant secretary of HUD for Demonstrations and Intergovernmental Relations, the Model Cities program would be his main but by no means his only responsibility. As his principal subordinate, he chose Walter Farr, a young lawyer who had also been with AID in Latin America. Farr was to be the operating head of the Model Cities Administration (MCA).

The staff was small by the standards of the government—about forty professionals in Washington and seventy in regional offices—but it had an unusually high proportion of its personnel in the upper salary ranges. Morale was high. Taylor and Farr got along well and conferred often, and relations between them and their subordinates were easy and cordial. As a new agency, MCA was regarded with apprehension by the long-established housing and renewal agencies of HUD. Farr and the young men around him were dubbed "Taylor's Green Berets" by the standpatters in these agencies.

Although he had to spend most of his time in the first hectic months haggling with HUD administrative officers and the Civil Service Commission and meeting with mayors and others who were impatient for decisions, Taylor nevertheless managed to put the impress of his views on the program. The task force report could not, in his opinion, be taken seriously: its authors "did not understand housing." No one who *did* understand it would talk of rebuilding whole districts of the large cities: there were not enough carpenters, plumbers, and electricians to do the job, and, anyway, it was obvious that Congress would not appropriate the scores of billions of dollars that would be required.

What Model Cities *could* do, Taylor thought, was, first, to document the vast extent of the cities' needs so that, when the Vietnam War ended, the nation might be persuaded to make a commitment on the scale required, and, second, to prepare local governments for effective action when the nation finally made that commitment.

His experience in New Haven had convinced him that excessive fragmentation of governmental structure was the major obstacle to getting things done. This was being made steadily worse by the almost daily increase in federal efforts—most conspicuously OEO's Community Action Program—which bypassed local governments. He was determined that Model Cities should not add to the governmental confusion. It would help the cities plan but not impose plans on them. As one of his Green Berets wrote later,

> We had a great deal of faith at first in the latent power of the people "out there" in the cities, in their ability to come up with innovative solutions that could be replicated throughout the country if they were left alone with a nice piece of federal change. We saw them as

co-workers in an urban problems laboratory. Our job was to furnish equipment, assistance, support and occasionally a tube of catalyst. Theirs, was to innovate.

For Taylor, at least, "the people out there" were not primarily the residents of the slum and blighted neighborhoods. Model Cities, he told his staff, was to be "a mayors' and planners' program."

When the Model Cities bill was still before Congress, HUD had set up a committee under the chairmanship of William Rafsky (a member, it will be recalled, of the task force) to advise on policies to be followed when it passed. In the committee's recommenations, Taylor found much to support his views. It urged HUD to insist upon high-quality programs. Before receiving a planning grant, a city should be required to describe and analyze the physical and social problems of its poor areas, lay out five-year goals for dealing with them, and tell how it proposed to attain the goals. In addition, it should answer a long list of questions about its government and related matters. If its application did not meet HUD's stringent standards—if, for example, its projects were not innovative—HUD should demand something better. The cities making the best applications would receive grants for a year's further planning. An "action program" involving supplemental grants could begin only after satisfactory completion of "the planning phase."

The decision to require a one-year "planning phase" had been made at a very early stage: Weaver had mentioned it in his testimony to the House subcommittee. (For that matter, the planners of the 1963 anti-poverty proposals had intended to have the "pilot" cities spend a year planning and, as they put it, "tooling up.") Taylor, by following the recommendations of the Rafsky Committee and establishing an application process that might require several weeks or months, was in effect delaying the "action phase" still further. When the mayors realized how much delay there would be, they could be expected to protest vigorously. This would be awkward from a public relations standpoint, but Taylor was willing to pay the price in order to get information to document the extent of the cities' needs. With this in view, he established what he called a "high sights" policy: applications were to be made on the assumption that *all* of a city's problems were to be dealt with *within a few years.*

Fearful lest support for his agency in Congress melt away if more than a year passed without *any* "action," Taylor considered giving the cities small supplemental grants for "immediate impact projects," such as improved street cleaning, as soon as their applications were approved.

Against the public relations advantages of this there were, however, two offsetting disadvantages: first, such "cosmetic" grants would do nothing to cure the cities' fundamental ills, and, second, they would distract the cities from the "in depth" planning that was crucial. Reluctantly, Taylor decided to withhold the supplementals until planning had been done properly. Somewhat to his surprise, Weaver and Wood, who would have to bear the brunt of any criticism, agreed enthusiastically to these decisions.

In December 1966, MCA mailed out 3,000 copies of its program guide, a sixty-page brochure entitled "Improving the Quality of Urban Life." This explained that although Model Cities programs were to reflect local initiative, they had to conform to certain statutory requirements. The programs had to be of sufficient magnitude to (among other things) "make marked progress in serving the poor and disadvantaged people living in slum and blighted areas . . ." and "a substantial impact on the sound development of the entire city." Therefore the programs would have to deal with "all of the deep-rooted social and environmental problems of a neighborhood" and "make concentrated and coordinated use of all available federal aids."

The guide emphasized another standard which, although not derived from the statute, was in Taylor's opinion fully consistent with it—namely innovation. "Cities," it said, "should look upon this program as an opportunity to experiment, to become laboratories for testing and refining new and better methods for improving the quality of urban living."

The guide was soon followed by *CDA Letter No. 1*, "Model Cities Planning Requirements," which was the first of many formal instructions from MCA to city demonstration agencies. This letter made the "high sights" policy explicit.

> Cities must set their sights and their goals high. Model cities programs should aim at the solution of all critical neighborhood problems which it is within the power of the city to solve, and should be designed to make as much progress as possible toward solutions within five years. In most cases, it should be possible within five years to make all necessary institutional and legal changes at the state and local level and to initiate all necessary projects and activities which will, when carried to completion, achieve the long-range goals set by the city.

Applications, the instructions said, should describe and analyze each major problem, explain how and why each developed, and give a preliminary judgment as to what should be done about them. Cities were to decide for themselves what to do, but they must show that they followed proper procedures in making their decisions. "Program analysis," MCA stated, "is the foundation on which the entire Model Cities undertaking rests."

"Program analysis" was fashionable in Washington at this time. In fact, the president had in 1965 instructed all agencies to institute PPBS—the Planning-Programming-Budgeting System—a procedure for analyzing and evaluating program alternatives that had been used in the Defense Department and much acclaimed by the press. The popularity of PPBS had little or nothing to do with MCA's enthusiasm for planning, however. Some of the principal Green Berets were graduates of city planning schools and, as such, products of a different planning tradition, one that derived from the city beautification movement rather than from economics. Taylor seems to have accepted on faith their claims to having skills that, if given scope, would cure the ills of the cities. Moreover, he saw in comprehensive planning a means of collecting information that would impress the country with the extent and seriousness of its ills.

Knowing that there would not be funds to support the full program proposed by seventy-five cities (it had been decided to choose this number from a "first round" of applications; a "second round" would follow shortly), Taylor decided that the cities would have to concentrate their efforts on a single model neighborhood of not more than 15,000 persons or 10 percent of the city's population, whichever was larger; New York and a few other of the largest cities were to have more than one model neighborhood.

This decision left city officials wondering why they were required to describe and analyze the problems of *all* poor neighborhoods. It also raised the question why huge sums should be poured into one neighborhood while others nearby and equally poor got nothing. And it reduced whatever possibilities there were for collaboration between Model Cities and those federal and other agencies whose programs operated citywide. But Taylor really had no alternative. Given that there had to be seventy-five cities to begin with and more very soon, to make a "substantial impact" on a unit even as large as a neighborhood was hardly possible.

The essential decisions having been made, Taylor and some of his assistants set out in January 1967 as a "flying circus" to explain the program to local officials, answer their questions, offer help in filling out applications, and urge that applications be made early. "Analyze yourselves and your cities," Taylor told his audiences. "Walk the alleys and the streets and talk to people and get to the roots of the problem."

NEEDED: NEW STANDARDS

Applications soon poured in, but Taylor and his assistants were disappointed when they read them. One Green Beret thought that less than a dozen showed any understanding of the experimental nature of the program. The projects proposed were not innovative, they did not constitute comprehensive or coordinated programs, and it was evident that they had been made with little or no consultation with the residents of the affected neighborhoods or even, in some instances, with the mayor and city council. "Our lab partners," he wrote, "didn't seem to get the hang of it."

With a view to improving the quality of future applications (the "second round" would begin within a few months) as well as of the first year's planning by the "winning cities," Farr brought his staff into a conference with consultant specialists. (Dr. Duhl, the psychiatrist who with Mrs. Chayes had written the 1964 memorandum proposing Demonstration Cities, was one.) The result of these and other deliberations was revision of the instructions on planning and citizen participation.

The new planning instructions reflected the staff's conviction that most cities were incapable of "real" ("comprehensive") planning. Accordingly, the Five-Year Plan was deemphasized, becoming a "five-year forecast." City programs were henceforth to be based on analysis of *not quite* everything:

CDA Letter #1 (the original version, October 1967)

Each city's program should be based on a systematic analysis of all relevant social, economic, and physical problems which describes and measures the nature and extent of the problems, identifies their basic underlying causes, examines the inter-relationship between problems, and indicates the critical changes which must be made in order to overcome these problems.

CDA Letter #4 (the revised version, July 1968)

The problem analysis should cover all significant problems but the depth of the analysis can vary according to the significance of the problem and data available. High priority problems should receive the most attention the first year. Future planning should direct attention to these significant problems not adequately covered during the first year of planning. Although this section and its contents will vary according to local conditions and according to local understanding of problems, it should not avoid significant and historical causes of deprivation and inequality.

In a foreword by Taylor and in several footnotes, the new instruction described itself as "guidelines" which were to be interpreted "flexibly." The main purpose of the letter ("40 pages of agonized noncommunication," one of its authors—the Green Beret already quoted—called it) was to *explain* the nature of planning, not to tell the cities what they must do.

A revised statement of policy on citizen participation was issued in order to keep peace within the staff. In the spring of 1967 Taylor had agreed to substitute "meaningful citizen participation" for "widespread" in a revision of the program guide. Later, when the first planning grant application began arriving it was evident that in most places citizens were not being consulted. The citizen participation specialists of MCA, OEO, and HEW then jointly wrote Taylor and Farr, the heads of MCA, proposing that the residents of model neighborhoods be given the right to choose their own representatives, have a "genuine" policy role, be provided a budget from which they could hire technical assistance of their own choice, and be compensated for their time and travel.

Farr knew that there would be trouble if these recommendations were not accepted. Citizen participation, however, was one of the few matters Weaver insisted on having brought to him for decision. Weaver believed that the recommended changes went far beyond what Congress had intended by the phrase "widespread citizen participation," but he recognized that concessions were necessary to mollify a part of the staff. Accordingly *CDA Letter No. 3* (it was dated November 30, 1967, but distributed to the cities about a month before) informed applicants that in order to build "self esteem, competence and a desire to participate effectively" on the part of neighborhood residents, there would have to be "some form of organizational structure . . . embodying" . . . the residents "in the process of policy and program planning and program implementation and operation." The leadership of such a structure "must consist of persons whom neighborhood residents accept as representing their interests," and these must have "clear and direct access" to the CDA's decision making. Cities were to provide the residents' representatives with technical assistance "in some form" and were to compensate them when (but only when) doing so would remove barriers to their participation.

The compromise did not please the community participation specialist, who resigned before it was issued, but the rest of the staff was apparently satisfied.

THE CASE OF COMPTON, CALIFORNIA

One of the cities to file an application was Compton (population 56,000), a predominantly Negro suburb of Los Angeles. Following the common practice of small cities, the city manager engaged a firm of planning consultants to prepare the application. This cost $5,000 and assured its failure. A city could not learn about its problems by employing outsiders, MCA wrote. The rejection came to a new manager, James Johnson, who put everything else aside in order to draw up a new application himself. The task took three weeks.

When he had accumulated 300 pages ("I put everything I could think of in it"), he sent the application to Washington. After several months' wait, he was surprised and elated to learn that Compton had been granted $110,000 for further planning. His elation faded somewhat, however, when he found that he might not use the money to work out the details of the projects that he had outlined in the application. His 300-page document, MCA told him, was not a plan, but a "plan to plan." He would now have to set in motion a planning process that would be carried on for a year by the City Demonstration Agency (CDA) and in which the residents of the model neighborhood would be "meaningfully" involved. Only when this was done and Compton's problems had been analyzed in depth, its goals defined, a five-year forecast made, and a first-year action program derived from it could he hope for an "action grant."

The action grant—assuming the city eventually qualified for one— would not necessarily be in an amount sufficient to support more than a fraction of the first-year action program. At best, Compton could get only its pro rata share of whatever supplemental grant funds Congress appropriated.

To Johnson it appeared that MCA was requiring that the planning grant be used to prepare a rewrite of the 300 pages that he had already submitted: he doubted that the CDA and the citizen's advisory board could find much of anything to say about Compton's problems that he had not said in his application. It displeased him also that the further planning would have to be done by the CDA. This would tend to cut him, the city manager, off from a process which, if it were to lead to action, required his intimate participation. Perhaps in a large city such participation by a chief executive was impossible or undesirable, but why not let a city decide such matters for itself?

Johnson was also dubious about the citizen participation requirements. As in other black communities, politics in Compton was in a state of ferment; people who had had little or no experience in public affairs, many of them young and militant, were searching for leadership. In these circumstances, "citizen participation" was likely to mean endless talk at best. It was exasperating, too, Johnson thought, that the program could operate in only one neighborhood. Anyone could see that Compton's problems were citywide—indeed, *more* than citywide.

CHOOSING THE "WINNERS"

The "first round" applications were ready by the MCA staff in Washington, which prepared brief appraisals of each. These included summary information about each city's political leadership, administrative capability, and understanding of its problems, along with a judgment as to the innovativeness of its proposals. These appraisals were read by an interagency committee of which Taylor was chairman. It found some applicants clearly unsuitable. The others were given an "advanced" review. After considering a proposal for assigning numerical weights to a long list of sharply defined criteria, the interagency committee fell back upon an unsystematic "common-sense" procedure.

A list of sixty-three cities recommended by the committee was forwarded to Weaver in June in the expectation that it would be quickly approved. It was not, however. A month or two later Weaver had spent a bad morning answering questions before an appropriations subcommittee which was considering MCA along with other HUD items. It was evident that the congressmen did not like the idea of appropriating $650 million without knowing where the money would be spent or (except in very general terms) for what purposes. When they were told that only eight cities had so far submitted completed applications, some of them were surprised and annoyed. Recalling their attitudes, Weaver decided not to announce the "winning" cities until the bill had been passed. This, he supposed, would be fairly soon: appropriations bills were normally passed before the start of a fiscal year, July 1. This time, however, there was disagreement over many items in the bill and, although passage seemed likely from week to week, it did not actually occur until November.

The delay in announcing the "winning" cities, the reasons for which could not be made public, caused the backers of some promising projects to lose interest in them. The MCA staff, however, thought that on the whole the effect of the delay was beneficial because most cities had not yet developed accounting procedures that met its standards.

All but nine of the sixty-three "winners" were in Democratic congressional districts. This was not surprising, since most sizable cities were Democratic. Asked by the press how much politics there had been in the selections, Weaver said, "As little as possible."

This was undoubtedly true. Taylor had resisted political interference because he knew both that his staff would lose respect for him if he "went too far" and that the prestige of the program—perhaps even its existence—would be threatened by including cities whose projects were ill conceived or certain to fail for other reasons, such as administrative incapacity. In some instances, however, political influences could not be withstood. The White House was sure that Eagle Pass and Laredo, Texas, met all possible criteria. Smithfield, Tennessee, was the hometown of the chairman of the House subcommittee that passed on HUD appropriations, and Pikesville, Kentucky, was the abode of another chairman. Montana, the Senate majority leader's state, had two especially deserving cities. Maine, Muskie's state, had three.

Some voices—Republican ones, of course—had been saying all along that Model Cities was just another Democratic pork barrel. To prove them wrong, at least one large city represented in Congress by a Republican had to be on the list. Happily, one—Columbus, Ohio—was found. A few cities that could not have qualified by nonpolitical standards were on the list (Laredo had not even submitted an application). In general, however, the "political musts" met MCA's standards as well or better than other applicants. Of the 150 cities eventually chosen, less than a dozen, Taylor said later, were accepted solely because of political pressure. If he had succumbed to it very often "the pressure would never have stopped . . . and the program would have been ruined from the start."

When he was asked how much politics there had been in the selections Weaver might have said as truthfully, "Less than necessary." For, despite his precautions, Congress in late 1967 gave Model Cities less than half what had been asked for—$200 million for supplemental grants and $100 million for the renewal add-on. The only item *not* cut was for additional planning grants—$12 million.

This meant that Congress was ready for a "second round" of applications, bringing the number of Model Cities to 150. By halving the
appropriations and doubling the number of cities, it was making ridiculous
Wood's continued declaration that the purpose of the program was "to
determine the 'critical mass' for real change in the problems of human
and physical deterioration in our cities."

MORE DISAPPOINTMENTS

The first-year programs, which began to reach MCA in the fall of
1968, nearly two years after the passage of the act, were found to be even
worse than the applications had been. This at least was the judgment of
the Green Beret who had helped write the "40 pages of agonized noncommunication" that *CDA Letter No. 4*. The cities, this man wrote later,
were "willing to play our silly little game for money," but they did not
understand that MCA was trying to get them to experiment and innovate;
they found it easier to accept regulation than to make use of freedom.

Very likely he was right. The CDA planners, however, worked
under certain constraints that should be noted. For one thing, the amount
of the planning grant ($150,000 for most cities) was not nearly enough for
"in depth" fact gathering and analysis of a great many problems. It was
inevitable, too, that careful analysis would sometimes knock the props
out from under proposals of the sort that the planners were expected to
bring forward. In Binghamton, New York, for example, one survey of
opinion in a model neighborhood revealed that more than 75 percent of
blacks considered their housing either "good" or "excellent," and another
survey showed that 78 percent of those polled (black and white) were
satisfied with the public schools.

However hard they might have tried, the cities would have found
it remarkably difficult to devise programs that were "innovative," "comprehensive," or "coordinative," and perhaps impossible to devise ones
that were all three simultaneously. The words themselves were ambiguous
in the extreme. (In Atlanta, researchers asked seventy local officials what
they understood by "coordination" and got seven distinctly different
replies.) Even if the objectives had been made perfectly clear, the cities
might not have been able to devise programs to achieve them because
of the inherent difficulty of the task. Innovation, for example—devising

measures that are new, not unreasonably expensive, acceptable to the public, and that produce the effects intended—was not as easy as Washington seemed to think. (In Boston, where the Ford Foundation created a committee of "experts" to find "new ideas" for helping the poor, a long effort yielded next to nothing.)

It was probably impossible to achieve simultaneously and in a high degree all the standards MCA set because, however they might be interpreted, there were tensions if not incompatibilities among them. The more "comprehensive" a program, the less "innovative" it was likely to be; to the extent that it was "innovative" it would probably make "coordination" more difficult. (Taylor's determination not to worsen the fragmentation of local government led him to insist that projects be administered through *existing* agencies, but innovation, sometimes if not generally, required creation of new ones.) And the more "innovative" a project the less its chance of making a "substantial impact," for new efforts usually have to start small. Similarly with citizen participation: the more seriously MCA took it, the more it had to sacrifice in terms of its other goals.

CDA directors seized different horns of these dilemmas. Some bore down hard on "comprehensiveness," appointing dozens, or even scores, of task forces that produced hundreds of projects which were offered as a "program," although they were really a sort of civic laundry list. Others emphasized "coordination" and drew up schemes for exchanging information among agencies by means of committees, conferences, and the distribution of memoranda. Those who placed their bets on "innovation" usually hired consultants to think up ideas for them.

Whichever way they turned, the CDAs had to take account of something that *CDA Letter No. 4* overlooked: local politics. As perceived from Washington, a city government was an entity capable, if sufficiently prodded and when provided with a grant, of making decisions in a rational manner, that is, of comprehensive planning. City officials, including of course CDA members, knew, however, that only in a rather limited sense did such a thing as a city government exist; for them the reality was bits and pieces of power and authority, the focuses of which were constantly changing. Bringing the bits and pieces together long enough to carry out an undertaking was a delicate and precarious operation requiring skills and statuses that few persons possessed. To those who saw governing from this perspective, MCA's rigamarole about program analysis, goals statement, and strategy statement did indeed appear a silly game. "If the Feds tell us to jump through hoops for their money," a CDA director said, "we'll jump through hoops."

INTERAGENCY RELATIONS

After two years of trying, MCA could not claim much success in "sucking the system together." When the act was passed, there had been in the larger cities a rush by local bodies to offer cooperation in the spending of the large sums that Model Cities was expected to control, but this enthusiasm faded fast. Some mayors (and city managers) used the CDA to strengthen their management. Others, however, more or less deliberately let their Model Cities program become the preserve, or playground, of persons who claimed to speak for the poor. The CDA letter strengthening citizen participation requirements, which appeared at the end of the second summer of urban riots (1966), made such mayors all the more inclined to (as an MCA official put it later) "turn the whole thing over to the noisiest citizens' group in order to keep them quiet." It is probably safe to say that where mayors *tried* to use the program to improve coordination they found it useful, and where they did *not* try the result was confusion worse confounded.

Supplemental grants, it will be recalled, were intended by the task force to serve as both "bait" and "glue"—"bait" in that they would enable a city to attract categorical grants from federal agencies anxious to try things that they could not try elsewhere; "glue" in that they could be used to bring diverse projects (federal, state, and local) into internally consistent wholes. In this way, MCA and the city governments, working in partnership, were expected to bring about coordination.

To establish the relationships with federal agencies required for this, Taylor had sent them drafts of the program guide for review and comment. Except for OEO, which urged MCA to adopt its "confrontation" tactics in dealing with city agencies, their responses had been perfunctory. He had persisted, however. When the time came to select the first-round cities, he created regional interagency committees to make recommendations. These played little or no part in the selection process, probably for lack of time to give them training, but (as has been explained) a Washington interagency committee *did* pass upon the applications.

After the selections had been made, the interagency committees, both regional and Washington, were given the task of reviewing the cities' plans and giving them technical assistance. Since the funds for model cities' projects would presumably come mainly from departments and agencies other than HUD, it was necessary to bring these units into close contact with the CDAs, the local bodies making the plans.

That these arrangements sometimes did not work as they were supposed to is evident from the experience (perhaps not typical) of the Texarkana, Arkansas, CDA when it asked its regional interagency committee for advice. "We expected a dialogue on the purpose and potential impact of individual projects and groups of projects," a CDA official wrote. Instead, the federal agency representatives met first in a session from which the locals were excluded; later the locals found themselves answering questions about budgetary and other details of particular projects. It was apparent that most of the "Feds" had not read the city's five-year plan; they were interested only in its first-year action program and only in those parts of it that related to their agencies' programs. "Federal employees," the Texarkana official concluded, "are generally limited to specific project categories." This, of course, was one of the very conditions that the Model Cities program had been created to correct.

It is not surprising that the interagency committee members were not interested in Texarkana's five-year plan: each was there primarily to protect and promote the interest of *his* agency. It was all very well for MCA, which had cast itself in the role of coordinator, to preach the virtues of coordination. The other agencies did not have the same interest in the matter: on the contrary, they had reason to view coordination with some suspicion; its success would tend to subordinate them to MCA.

Relations between MCA and OEO were always difficult. In theory, CDAs did planning, whereas CAAs (Community Action Agencies of OEO) operated programs. Naturally, CDAs contended that *all* programs in model neighborhoods should fit harmoniously into their plans; naturally, too, CAAs insisted that they had a right—indeed a responsibility—to extend any of their programs to the *whole* of any city.

What most concerned Weaver, Wood, and Taylor was the failure of supplemental grants to serve as "bait," "glue," or "clubs." "Supplemental" soon proved to be a misnomer. Most CDAs made little or no effort to get categorical grants; instead they asked for supplementals to finance projects that could have been financed with categoricals.

There were several reasons for this. CDAs found it less trouble to use supplementals because they did not have to find out about, and adapt to, the detailed restrictions with which Congress had loaded the categoricals and also because they were more used to dealing with MCA, their parent agency, than with others. The administrators of the categorical programs thought—or claimed to think—that model cities (the "winners" in the national competition) would be generously funded by MCA and that therefore their own grants should be given to less favored places. Thus when the Seattle CDA allocated $35,000 in supplemental funds to expand an OEO legal services program, OEO at once reduced its contribution to the project by that amount. From OEO's standpoint, this was the sensible thing to do: since the legal services program was now going to be supported at what it (OEO) deemed an adequate level, it would put its $35,000 into something else.

As Secretary of Labor Wirtz had pointed out to Califano (special assistant to LBJ), administrators of categorical grants had very little discretion in the allocation of them. For example, HEW, the department most disposed to cooperate with model cities, had appropriations of $6.4 billion in 1968, but only $181 million of this—roughly 3 percent—was not committed by statute or otherwise.

By the spring of 1968, it was evident that vigorous intervention from the White House would be required to establish MCA in a coordinating role. The president's prestige was now very low (he had announced that he would not be a candidate); even so, a few well-chosen words from him would greatly strengthen MCA's influence with the other agencies. Getting him to say the words was a problem, however. Only Weaver, the cabinet member, was in a position to approach him on the subject, and he sensed that to do so might irritate him and do MCA's cause more harm than good.

Accordingly, Weaver, with Wood, Taylor, and Farr in tow, went to see Califano and Schultze, the White House assistants most responsible for such matters. Both of them readily agreed to request in the president's name that the relevant departments and agencies earmark part of their discretionary funds for use in model cities neighborhoods.

Taylor meanwhile was doing what he could to get state governments to give the model cities neighborhoods more of what they got in federal grants as well as more from their other revenues. At his suggestion, Weaver invited nine states (later others as well) to participate in the review of "second round" applications. These efforts, however, yielded little. State agencies had not much more freedom of action than did federal ones, and, besides, they, too, had their own axes to grind.

THE PRE–INAUGURATION DAY RUSH

Two years after the passage of the Model Cities Act, Richard Nixon was elected as president, and Weaver, Wood, and Farr made ready to leave the program to successors who, it seemed likely, would liquidate it. As matters stood on election day, this would have been easy. As yet only planning grants had been made, and not as many of these as intended. The "first round" cities were now approaching the end of their "planning year," but none had submitted its first-year program. Consequently, not a dollar of supplemental grants had been obligated.

To make liquidation of the program awkward for the new administration, MCA put the "first round" cities under pressure to make program submissions by December 1 so that funds could be committed to them before Inauguration Day. Sixteen cities met the deadline, and the programs of nine were approved. Selecting the remaining "second round" cities was easier, and the quota was filled by the end of December.

The Nixon administration would be in office for six months before the beginning of the new fiscal year, but, unless great inconvenience and confusion were to ensue, the Johnson administration had to prepare a tentative budget proposal for fiscal 1970. In the previous two years, Congress had appropriated nearly $1 billion for Model Cities grants. Only a few million of this could be obligated, for lack of cities whose submissions were ready, or almost ready, to be approved. Nevertheless, the Johnson administration's 1970 budget proposed another $750 million for Model Cities. It was, of course, good politics to put the Nixon administration in the position of appearing unsympathetic to the poor and black by trimming from the Johnson budget funds that could not possibly be spent.

Weaver, Wood, Taylor, and Farr, although loyal members of the Johnson "team," were, it is safe to say, less interested than the White House in creating embarrassments for the next administration. They did, however, want to do what they could to assure the continuance, growth, and success of the Model Cities program. Taylor approved only nine of the sixteen program submissions because he depended upon a normal bureaucratic evaluation process, which took a certain amount of time and turned up only nine submissions that met the established standards. (One partial exception may be noted: Seattle's submission was taken out of turn because it was thought expedient to show West Coast Republicans that they had a stake in the continuance of the program.)

This was in fact a time when politics receded farther into the background than usual. "We were leaving office," one of those most concerned said later, "therefore we had nothing to lose by being *not* political."

TWO VOICES OF EXPERIENCE

Before they emptied their desks, Taylor and Farr wrote memoranda to be read by their successors.

Farr said that substantial progress had been made in strengthening local governments and in creating a federal organization that could respond quickly to the cities' needs. But so much time had been required to develop workable relations between city governments and neighborhood groups that little had been left for planning. Innovation had been hard to "pull off," both for lack of money and because of the commitment not to increase the fragmentation of local government. Coordination, when it occurred, had more often been information sharing than program integration, and the federal agencies (except those in HUD itself) had made only nominal efforts to help the cities develop high-quality programs.

The promise of the program, Farr said, could not be realized unless there was a "quantum leap" in the amount of technical assistance given the cities and in the attention given to the evaluation of their results. The likelihood of MCAs ever having enough personnel to do these things adequately was, he judged, zero. But without vast expansion and improvement along these lines the Model Cities program would amount to little more than block grants. This might be valid and appropriate, but it was not what Congress intended.

Taylor, in his memorandum, revealed that two years of trying had left him convinced that local governments could not be "tooled up" from Washington: the federal government simply did not have the capacity. Moreover, the categorical system was an impediment to local planning: except as they had latitude to make choices, local governments were without incentives to plan. The number of categorical programs, he thought, should be reduced by at least one-half through consolidation; HUD should be made a "Department of Urban Affairs" with authority to coordinate federal efforts in urban areas and revenue sharing (an arrangement by which the federal government would regularly pass a fixed percentage of its revenue on to the states and cities to use as they saw fit) should be instituted.

The role Taylor envisioned for Model Cities was a limited one, and it was based upon assumptions that the rest of his memorandum seemed to contradict. MCA, he thought, should be the Department of Urban Affairs' instrument for developing and testing new techniques for improving local public services; in particular it should try to discover whether, by the use of block grants and the concentration of resources, local officials could be induced to take responsibility for the solution of local problems.

CONCLUDING OBSERVATIONS

There was not in 1965, when the second task force did its work, the slightest possibility of a federal program being brought into existence which could accomplish any of the various large purposes that Model Cities was supposed to serve. The president and his advisers, including the members of the task force, knew perfectly well that the boundaries of political possibilities precluded measures that would change the situation fundamentally. Block grants, not to mention revenue sharing, were out of the question. (As Wood remarked later, "Congress was not about to collect taxes and then let the states and cities spend them with no strings attached.") Bringing the more than 200 categorical programs under central control was equally unfeasible.

It was also politically impossible to concentrate federal funds and efforts in a few large cities, let alone to "saturate" a few. Even if there had been no Vietnam War, the scores of billions that would have been required to make a "substantial impact" on the physical slum in the dozen or so largest cities would not have been appropriated and, even if it were, men could not have been found to do the work. Even to *stop* doing things that manifestly did not work—urban renewal, for example—was beyond the bounds of possibility; one had to expand what did not work as the price of trying something that *might*.

These constraints were all apparent to the designers of the Model Cities program. There were other constraints which a charitable reader may suppose were *not* apparent to them and are visible now only by the light of hindsight. That local political arrangements in the United States preclude anything remotely resembling "comprehensive" or "rational" planning is a fact familiar to all practical men and documented repeatedly by journalists and even by scholars. That inventing social reforms that are likely to work and that voters are willing to have tried is very difficult to do anywhere (and almost impossible to do in an American city where "veto groups" abound) is a fact hardly less obvious. As for citizen participation, very little judgment or experience was needed to see that this would be costly in terms of other goals—planning, coordination, and innovation.

Even if these constraints had not existed, the new program would have been the product of (at best) the "educated guesses" of persons with experience and judgment but without any sort of technical expertise on the basis of which they could answer the really important questions. That the "social and psychological" approach would achieve worthwhile results, for example, was mere conjecture. No "facts" or "tested theories" existed from which a remedy for any of the ailments of the cities could have been derived.

There *did* exist, however, a fairly good example of the very program that the designers of Model Cities were trying to create, one which, had they examined it closely (time did not permit), would have revealed the futility of what they were about to attempt. New Haven (as was remarked previously) was a model city before Model Cities was invented. Its circumstances were peculiarly favorable: small (population 150,000) and without much poverty, it had (what was most unusual) a mayor who not only won elections but surrounded himself with able assistants, some of whom were unusually talented at devising innovative measures to improve the lot of those for whom poverty had become a way of life. While the Wood task force was at work, New Haven had more than twenty-five agencies, public and private, participating in about sixty programs that were coordinated by a special body. Federal and other spending there was close to a "saturation" level, amounting in 1965 to nearly $7,000 for each family with an income under $4,000.

If Wood and his associates had studied this ready-made "pilot program" they would not have found evidence that it was making a "substantial impact" on the "quality of life" in the city. But even if they *had* found convincing evidence of this, the significance of the finding for other cities would have been questionable. Weaver was optimistic, to say the least, when he told the House subcommittee that successful demonstrations in a few cities would be copied by thousands of others. This might happen if the demonstrations were of new methods that could be easily applied, although political or other circumstances might stand in the way even then. But if the nature of a demonstration were such as to preclude its general application—if, for example, it required extraordinary or administrative skills—then Weaver's hope could not be realized. It was more than likely that any measures that could make a "substantial impact" would be so costly in money and other resources (e.g., political leadership) as to preclude their being generalized. To invest as much per capita in *all* cities as had been invested in New Haven was fiscally impossible.

Supposing, however, that from the example of New Haven or otherwise the task force had devised a simple and highly effective method for bringing about complete coordination of federal, state, and local efforts, public and private, while realizing simultaneously and to the full the ideals of comprehensive planning, citizen participation, and innovation. Is it likely that this would have "solved," or much alleviated, the problems of the cities? There is very little reason to think so. Conceivably, the result might have been the more effective pursuit of courses of action that in the long run would have made matters worse. (Suppose, for example, that such successes had, by making life in the poorest parts of the city less dismal, reduced the pressure to move to the suburban areas where job opportunities would certainly be better in the long run.)

However this may be, President Johnson's prediction that the Model Cities Act would "spell the difference between despair and the good life" was far fetched in the extreme. Possibly, however, his rhetoric and the act itself were not intended to be taken at face value, but rather were gestures intended to help create confidence in government and (what may have been the same in the president's eyes) in himself as a leader. If this is taken as the "real" purpose of the program, one may well wonder what was the nature and extent of the gains that accrued from holding out such promises and whether they were later offset—perhaps more than offset—by a loss of confidence when the promises proved unrealizable.

These questions will never be fully answerable, and it is probably fruitless even to speculate about them now. The forces that converged in the making of the Model Cities program are still at work, and what they will lead to eventually is anyone's guess.

SOURCES AND READINGS

The reader is asked to bear in mind that this account does not carry the Model Cities program beyond January 1968. It should not be assumed that the program and the conditions affecting it have remained unchanged.

The following were kind enough to read an earlier version of the case in its entirety: Jan Janis, Marshall Kaplan, William L. Rafsky, H. Ralph Taylor, Robert C. Weaver, and Robert C. Wood. Others who were helpful on particular matters were Antonia Chayes, William Cannon, James Johnson, Martin Meyerson, and Charles L. Schultze. It goes without saying that some of these may disagree with my interpretations.

The work is based largely upon interviews by the author with participants. He has also drawn upon interviews by Christopher DeMuth, for which privilege he is grateful. He is grateful also for much stimulating comment and criticism from Mr. DeMuth and from Lawrence D. Brown.

Only such source material is cited as is easily available in published form. This includes especially the hearings held in 1966 by the housing subcommittees of the Committees on Banking and Currency of both House and Senate. The House hearings are entitled *Hearings on Demonstration Cities, Housing and Urban Development and Urban Transit.* The Senate ones are entitled *Hearings, Housing Legislation of 1966.* President Johnson's special message on demonstration cities may be found in *Public Papers of the Presidents,* 1966, vol. I. For the reaction of academic specialists to the president's message the student should read the symposium of comment in the November 1966 issue of *The Journal of the American Institute of Planners.* In the writer's judgment, Professor James Q. Wilson was particularly prescient in his contribution to that symposium.

The debates on the bill are to be found in the *Congressional Record.* The Senate debates were on August 18 and 19 (vol. 112, pt. 5) and the House debates on October 13 and 14 (vol. 112, pt. 20).

For the ideas and efforts that preceded and presumably influenced the planning of the Model Cities program the student is referred to Richard Blumenthal's chapter "The Bureaucracy: Antipoverty and the Community Action Program," in Allan P. Sindler (ed.), *American Political Institutions and Public Policy* (Boston: Little, Brown, 1969), Peter Marris and Martin Rein, *Dilemmas of Social Reform: Poverty and Community Action in the United States* (New York: Atherton Press, 1967), and D. P. Moynihan, *Maximum Feasible Misunderstanding* (New York: Free Press, 1966). For a full and careful account of what was done in New Haven, consult Russell D. Murphy, *Political Entrepreneurs and Urban Poverty* (Lexington, Mass.: Heath Lexington Books, 1971). Former President Johnson's (retrospective) account of his administration (which is quoted or paraphrased several times in the case study) is of course of unique interest and value; his book is *The Vantage Point* (New York: Holt, Rinehart & Winston, 1971).

Some who played active roles in the Model Cities program or who observed its beginnings at close range have written about it. One short but valuable article is by Fred Jordan (the "Green Beret" much quoted in the text in this chapter): "The Confessions of a Former Grantsman," in *City*, Summer 1971. Planning in three model cities is described at length by the members of a firm of consultants employed by MCA to evaluate its efforts; see Marshall Kaplan, Sheldon P. Gans, and Howard M. Kahn, *The Model Cities Program: The Planning Process in Atlanta, Seattle and Dayton* (New York: Praeger, 1970).

The Model Cities program has been described as a "coordinating structure" in one chapter of a study of coordination in the federal system by James L. Sundquist with the collaboration of David W. David; their book is *Making Federalism Work* (Washington, D.C.: The Brookings Institution, 1969).

CHAPTER **5**

Evaluating a Federal Program

Evaluation of federal government programs is in a way a nonsubject because there hasn't been a serious evaluation of a federal program for reasons I'll go into a little later. My thought is to tell you how a particular federal program was "evaluated." The Model Cities Program is one that I have had some hand in evaluating, so that in a way, this is a case history, and, unfortunately, partly an autobiographical one.

To put the case study in perspective, I must begin by telling you how I came to be the chairman of the president's task force to evaluate the Model Cities Program. The reason for going into this, I think, is to give you some sense of what kinds of political or other credentials, or lack of credentials, an "evaluator" such as myself has.

In the early spring of 1968 before President Nixon had been nominated for the presidency, Martin Anderson called me. Anderson had been at the Harvard School of Business and had written a book called *The Federal Bulldozer*, which was quite critical of urban renewal. I had reviewed his book favorably. Although I had never met him, Anderson asked me if I would be interested in assisting Candidate Nixon's effort to get the nomination. I said no, I would not, because I had other things to do, and, apart from that, any candidate who took my advice was sure to lose. I wasn't sure that I wanted Nixon beaten before he started. I said if he wins, come around and perhaps my advice will be of some use, but in order to win you have to support things that I'm against.

Later on in the summer, a friend and neighbor of mine in Vermont by the name of Milton Friedman, whom you know as a conservative

Transcript of a lecture given to the Institute for Urban Studies, University of Maryland, College Park, November 2, 1972.

economist, asked me if I would be willing to serve as chairman of an
urban task force during the campaign. I said no, the reason being that I
was still too busy. I was hard at work on a book, and, anyway, I didn't
feel like being involved as a campaign politician. I regard myself as a
scholar, not a politician.

Nixon's academic backers were trying to bring academic people into
connection with him because at that time there were few professors who
had any first-hand acquaintance with Mr. Nixon or he with them, and
his supporters felt that there should be some ties between the candidate
and the academic community. While I declined to be chairman, I did
agree to serve on a committee. But then, as it turned out, John Mitchell,
who was running the campaign, killed the whole enterprise because he
decided, and I think quite rightly, that the academic people were likely
to come up with something dangerous from the standpoint of the candidate, and that therefore it would be better not to have any task forces
until after the election.

When the idea was resurrected following the election, I agreed to
serve as chairman of the one on urban affairs with the understanding—
which applied, incidentally, to all of the other task forces as well—that
the enterprise would be confidential. I could invite anybody I wished to
serve on the task force, and there would be no publicity of any kind;
recommendations would go to the president and his principal associates
and not to the press. So it was a far different proposition from the pre-
election task force.

I found that although most of the people I asked—they were all
from academic life—were not supporters of President Nixon or the Republican party, nobody refused. Every professor felt that if the president or
his administration wanted advice on this basis, for nonpolitical use, they
were under some obligation to do the best they could to provide it. In a
couple of months we prepared a set of recommendations to the president
on urban policy. And that was the beginning of my connection with the
administration.

The following October (1969), the White House decided to have
another round of task forces. This time the background was somewhat
different. Dr. Arthur Burns, who's now the head of the Federal Reserve
System, was at that time in the White House as counsel to the president
while waiting to be appointed to his present post. (It was well understood
that when Mr. Martin's term expired, Burns would replace him.) Burns

was more or less at loggerheads with Dr. Moynihan. Burns is an economist, a conservative, and was very much concerned about fiscal policy and fearful of inflation. He wanted to cut government spending. Moynihan, on the other hand, who as you remember had been in the Kennedy and Johnson administrations, was there to see that spending wasn't cut, but rather, expanded. It turned out unexpectedly that Burns and Moynihan developed a healthy respect for, indeed, affection for, each other, but they were nevertheless very much at odds in the approach to economic policy.

Burns, whom I'd known through Friedman, regarded me, I think, as on his side and therefore a reliable person to be in charge of the Model Cities Task Force. He probably surmised that I would be as keen as he on abolishing it. The administration did, however, solicit names from the heads of HUD for the prospective chairmen of task forces that had to do with HUD affairs, and it turned out (so I was later told) that HUD, for whatever reasons, also suggested me. That seemed to qualify me.

Model Cities was one of about twenty subjects on which task forces were organized. The earlier task force that I had chaired hadn't known anything about the Model Cities Program at the time, and neither had anybody in the administration, including Moynihan and Burns, when they first arrived. As a candidate, Mr. Nixon had taken no stand on the subject. But it was taken for granted from his general position that he wouldn't be at all keen on it and, on the contrary, would eye it with great skepticism. In fact, his first week or so of office, the president turned to Moynihan, I was told later, and said, "Let's get rid of that Model Cities Program." Moynihan did not say, "Yes, Mr. President," and go out and get rid of it. He may have said, "Yes, Mr. President," but, as is usual in such cases, he did nothing. He wanted to give the president a chance to change his mind. He stalled because he thought that promises made by the former administration, however ill advised, ought not to be lightly repudiated by a new administration. The American people should have some feeling of continuity from administration to administration. While it might be well to get rid of the Model Cities Program, it shouldn't be done lightly or until something had been found to take its place.

Meanwhile, the secretary of HUD, George Romney, had appointed an assistant secretary to administer the Model Cities Program, one Floyd Hyde, former mayor of Fresno, California. Mr. Hyde had been one of the few Republican mayors in the country, and to make him even more distinguished, had been one of the few who had taken an active interest in the Model Cities Program while he was a mayor. Well, when Hyde got to Washington and in the administrator's chair of the Model Cities Program, he was soon engulfed by the career bureaucracy of the Model Cities administration. He was primed to be their spokesman and to speak for the maintenance and continuity of the agency. So he became a very effective proponent of Model Cities and did much to convince Romney that it should play a part in the urban program.

The president by this time had been boxed in. He couldn't very well tell the new secretary of HUD, even if he wanted to, "I'll overrule you and get rid of Model Cities, even though you as the secretary think that it's a desirable thing."

But by October, it was evident that nobody in the administration knew what to make of the program. They didn't know what it was about. They didn't know where it could be fitted into a long-term urban program; they didn't know whether it was to be a major or a minor aspect of their program, if indeed, it was to be kept at all. Therefore, a task force was needed.

The second task force was created on a quite different basis from the first. This one was not to be a behind-the-scenes affair. The membership was to be announced and the report made public. Therefore the composition and procedure of this task force took on a different aspect in the eyes of the administration. They had to have a majority of Republicans, for one thing, or at least they had to be sure they didn't have a majority of Democrats. You may ask why. One reason seemed to be that they were afraid of being ridiculed in the press for having the Democrats formulate their programs. Also, there were regulars in the Republican party who seemed to think that being on a task force, a job which doesn't pay anything and takes up your nights and weekends, was kind of a plum that ought not be given to people who were not of the right party. And then there was the further necessity that since the composition was to be made public, it had to be representative. That is to say, they had to make sure that there was the right proportion of blacks, and the right proportion of people who represented state government, inner city business interests, labor, city managers, and so on.

From my standpoint, it was not too hard to arrive at a modus vivendi with the administration. It was very simple. I said, so long as I can have a few people whose judgment and knowledge I trust, I don't care who else is on the committee. So we went ahead on that basis, with the understanding that the administration would exercise a veto right over my choices and I over its choices. As it turned out, neither of us exercised any vetoes, and I got people that I rightly expected to be first rate.

One was Mayor Lugar of Indianapolis, a former Rhodes scholar, very young and dynamic, and happily for me, a Republican. Another was James Q. Wilson, my colleague and collaborator at Harvard. Another was Bernie Friedan, a city planner at M.I.T. who had been a member of the professional staff who had advised the Johnson administration task force that had originated the Model Cities Program. I wanted him not only because I had confidence in his knowledge of the subject but also because he represented continuity with the earlier effort. I had Bill Robinson, a black man who had been the administrator of welfare in Cook County, Illinois. And there was James Buchanan, a very eminent economist in the field of public finance. There were others who were less well known to me but who made their contributions in one way or another.

Curiously enough, everyone had to have F.B.I. clearance. This seemed absurd; it took a lot of time, and time was short because the task force hadn't been conceived until October. The report had to be drafted by January, at the latest, in the event that the president wished to make use of it in his State of the Union Message. Well, everything was stalled while the F.B.I. checked to find out if the proposed committee members were reliable enough to talk about the Model Cities Program.

We had about a month to work in, and we were starting from scratch. There were about twelve or fifteen of us, some completely unknown to each other, having very little in common, and some knowing absolutely nothing about the Model Cities Program. When we first assembled, we decided to ask the White House for the loan of the young man who, under Moynihan's supervision, had been liaison between the White House and the Model Cities administration. Christopher DeMuth had been a student of mine at Harvard. In a few months in the White House he had learned almost everything there was to know about the Model Cities Program. He knew more about it, I think, than the administrators, and we were fortunate in having him on loan.

DeMuth assembled very quickly all the essential background documents, which had been written by puzzled Nixon administration cabinet officers and their subordinates who were trying to figure out what Model Cities was and what to do with it. The questions that concerned them had been discussed in the Johnson administration, too. Was the Model Cities Program an effort to achieve coordination of federal, state, and local services? Was it an effort to teach the cities how to do city planning? Was it an effort to bring about institutional change? (That is to say, for example, to get the police departments in the cities to behave more decently toward the poor or the blacks.) Was it to stimulate the innovation of new programs for the poor in the city? What was it? We were told it was all these things, and it didn't seem possible that it could be all of them at once. There had to be some selection; some goals had to be defined.

Floyd Hyde in the Model Cities administration promised to cooperate and help us in any way. They had made a big thing from the very start about self-evaluation. Lots of money had been ear-marked for the evaluation of the program, but unfortunately they had not gotten around to evaluating anything as yet. They had been collecting data; they had a firm of consultants that had been monitoring their activities from the very beginning, and the consultants knew a great deal but, unfortunately, what they knew was mostly in their files or in the head of one of the partners, a young man named Marshall Kaplan, who turned out to be very useful, thoughtful, and a good source of information. We invited him to sit with the task force as a program advisor. He was sort of a holdover from the Johnson administration, and in fact he is still a consultant to the Model Cities Program.

Well, what did we do? How did we go about it? We began by asking the Model Cities administration for an account of what the cities had proposed for financing. What was it that they wanted to do? We asked for laundry lists, as we called them, from the cities—fifty- or sixty-word descriptions, project by project, of what each city thought it would do with Model Cities money if it got it. At this time only twenty-five cities had reached this stage, so that the task involved looking over several hundred projects proposed by these twenty-five cities. We had the list organized according to subject matter—crime, schools, education, labor, manpower training, and community organization. With a laundry list for each of these subject matters, we then enlisted the volunteer services of academic specialists who were asked to appraise the projects that were being proposed. We had an eminent man in each field make a hurried analysis and write us a memorandum. (We were now working with a deadline of a couple of weeks.) What did he think of the manpower programs that were being proposed? Was there anything innovative about them? Were they going to work? Did they represent improvements over what had already been done?

As it turned out, the evaluators as a whole were very dubious about these programs. But the Model Cities people swept their criticism aside and they said to me, in effect,

> You're making a mistake by trying to evaluate these projects because most of the projects the cities propose don't really matter. We're not mainly interested in getting good projects at this stage, though it would be nice to have them. Our purpose is to change the nature of decision making in the cities. We are trying to create in the cities an institutional process by which the cities will make decisions more rationally. In short, we are trying to strengthen the office of the mayor or the city manager and create under his aegis a planning body with a set of procedures that will result in the creation of new and innovative programs. We also want to create citizen participation bodies that will advise, and in a sense validate, what the planners propose. So you really have to judge us not so much by these projects as by what we do to change the nature of the decision-making process of the cities.

That was a hard one to deal with because, after all, how can you, within a month or so, evaluate the changes in the process of decision making in the cities? If they occur at all, they occur over a period of years. But the president of the United States wanted to decide before the first of January—shall we continue this thing, and if we continue it, shall we make it a big thing or a little thing, or *what* shall we do with it? Shall we emphasize coordination, or shall we emphasize something else? Obviously, the really important question from the standpoint of the Model Cities administration was one that could not possibly be evaluated on the basis of information because the relevant information did not exist and would not exist for several years to come.

We therefore fell back on evaluating projects, much to the dismay of the Model Cities professionals. We also worked up a questionnaire for mayors and city managers. We wanted to know how the Model Cities Program was perceived by them. Did they think it would help to strengthen their offices? Did it seem likely to help them plan? What were the trade-offs that they could see between Model Cities and other goals that they might have? If they had their choice between a certain amount of Model Cities money and a larger or a smaller amount of money to do something else, which would they choose?

Here again we found that we were up against some practical obstacles. How to reach the mayors? Well, you might say, send a questionnaire to the mayor in the mail and ask him to send it back. But you know perfectly well that when the letter arrives, the mayor's secretary will route it to the director of the City Demonstration Agency, the Model Cities agency's local man, and he will evaluate himself. We wanted *mayors* to answer the questionnaire, not their subordinates whom we were trying to evaluate. Then we thought, "Well, why don't we have somebody in the White House call the mayors direct on the phone and talk to them?" That had some advantages, but there were reasons against that too. The newspapers at this time were very wary about what the Republican administration was going to do to Model Cities, the poor, the blacks, and so forth. The worst was expected. We didn't want to give rise to the belief that the Nixon administration was going to kill the Model Cities program. It was very hard to ask these questions without expecting that the questionnaire would fall into the hands of the press and be used in ways that would be embarrassing to the administration. The press might wonder, why are they asking mayors whether they preferred Model Cities money or block grants—unless they intend to do away with Model Cities. So, we were stymied.

In the end, the individual task force members, myself included, interviewed mayors, hit or miss, the best we could; I interviewed Mayor Daley in Chicago. I had some prior acquaintance with him, and I found that he was his old, very shrewd self. He said that Model Cities was the best thing that ever happened. "We can't have too much Model Cities, etc." Well, what he meant, of course, was that the more money given by the federal government to Chicago, the better. And he didn't want to be put into a position of saying anything that could possibly be interpreted by anybody as an excuse for not giving money to Chicago.

We also interviewed the former administrator of the Model Cities administration and got some very helpful and thoughtful advice from a man who had been through the mill. He was not in a position to talk publicly about its failure or weaknesses, of course, but he could do so privately.

Finally, when we had a draft of the report, we turned it over to Floyd Hyde for criticism by the Model Cities administration. Up to this time we had been very skittish about our dealings with them despite Hyde's offers of cooperation. We thought that very likely, government agencies being what they are and human nature being what it is, the Model Cities administration would use the occasion to build fires in its regional and field offices to make it politically infeasible for the Nixon administration to do anything but maintain and enhance the Model Cities administration. We didn't want to give them the chance to get those fires going before we knew where we stood. Actually, it turned out we needn't have worried. Floyd Hyde was not a very bureaucratic bureaucrat. He didn't set any fires.

Eventually, the majority of the task force concluded that on the whole, the projects that the cities were proposing were probably no worse and maybe even better than similar programs for cities being developed under the so-called categorical programs. However, that isn't saying very much for the Model Cities Program because, in my judgment, these other programs weren't working; in fact, in many cases they were doing more harm than good. On the basis of our survey, we felt that the Model Cities administration had been entranced with the idea of teaching the cities how to plan, textbook fashion. That is, they took city-planning textbooks and people with degrees in city planning very seriously. They said, this is the way to do it; this is the procedure to be followed. And they said it at such length that the cities had spent, in every case, at least a full year making a comprehensive plan to solve all urban problems throughout the city. They had to identify all the problems, their causes, and solutions before they could be considered seriously for a loan. In our judgment, there had been altogether too much of this paperwork. The city of Chicago had sent in, I believe, about 1,700 legal-sized pages for its plan. Well, of course, Mayor Daley and his administration had known for years what they thought they were doing in Chicago and writing it down in this form was sheer hocus pocus and delay. It only made the whole program ridiculous in the eyes of the city managers and mayors. The Model Cities administration seemed oblivious of city politics. They didn't seem to realize that there were voters out there in the cities, that there were elective representatives with political power who had to be taken into account, and that a plan which made believe that racial and other conflicts didn't exist was a silly piece of nonsense and could not be taken seriously.

Some of us thought that on its merits, the Model Cities program probably should be abolished. It was not innovative. It was not effecting coordination, and it didn't seem likely to. It was not securing institutional change or any of these other things that it was supposed to be doing. But, it was a very popular program. Of course, most people didn't know what it was. If you asked the man on the street what the Model Cities Program was, he would be likely to say that it was a procedure to build brand-new technologically up-to-date cities in the hinterlands somewhere.

Well, as I say, the popularity of Model Cities, misconceived as it was, had to be taken into account. It was, and I think still is, a symbol for people who think that there is an urban crisis and who think that it can be solved by bureaucratic methods and by pouring money into the poor neighborhoods.

This was a consideration because a president ought not to shake the confidence of a large sector of the people by failing to do something they regard as essential to their welfare. He ought not to seem to repudiate a large part of the country. To offer a rather far-fetched analogy, if you were the president of a nation of believers in witchcraft, you would pretend, at least, to believe in witchcraft too. Because, otherwise, you couldn't expect to have the confidence of your people and be able to govern.

Another consideration of the majority of us on the task force who were conservatives was that the Model Cities administration, for all its faults, was a move in the direction of giving the cities much more freedom in spending money. Despite all this rigmarole about planning, the cities did have freedom as to the substances of program proposals. By and large, they were free to use the money as they wanted, and that seemed to us like a very good idea. Given that huge amounts were going to be poured into the cities anyway, it seemed much wiser to let the cities and the people of the districts affected decide for themselves how they wanted to use this money.

Model Cities was for this reason less wasteful than the categorical programs. Under those programs a city might find itself in the position of having to take a large sum for something it didn't really want, or—the worst thing that could possibly happen from the standpoint of the city and its politicians—not get the money at all. Suppose that the federal government is prepared to give $10 million to build a third bridge over the river. You don't need a third bridge, and you do need $100 thousand for training special teachers to work in certain schools. But, there isn't a program for the $100 thousand. It's $10 million or nothing, and if you don't take it somebody else will. You pay the same taxes, but some other city will get the $10 million. Well, naturally, you take the $10 million. If the money is given to the cities directly with no strings attached, then they can decide: Will we build a bridge or will we train teachers? That's where it seems to make sense to move in a direction of block grants or, as it's now called, *revenue sharing.*

When we finished our report we titled it *Model Cities: A Step toward the New Federalism.* New Federalism indicated that we thought that the rationale of the Model Cities program was in the direction of decentralizing the administration of federal programs and giving cities more say. The report was published by the White House under this title.

Well that's not quite the end of the story. Several months later after our report had been submitted, and I think before it had been made public, I was invited to meet with the members of the cabinet at Camp David. But I didn't get there because it turned out to be a bad day for helicopters. I did get to Blair House, across the street from the White House where the president's assistant, John Ehrlichman, and a few cabinet members concerned with these matters were trying to decide again what to do about Model Cities.

They had read the report, and they seemed to have thought about it. They took it seriously, but there were still some very very difficult questions unresolved in their minds. I remember George Schultz, who at that time was secretary of labor, felt that the program still lacked a rationale, that it didn't have a coherent, intelligible, and defensible purpose. He would ask, "What is the Model Cities Program supposed to be? Is it coordination? Is it institutional change? Is it planning? What is it?" And he never got a satisfactory answer. Nor could he see how it was possible in the long run to defend a program which operated in only 150 cities. If it's good for 150, why not for others?

More important still, as it turned out, Don Rumsfeld, who at that time was the head of OEO, gave a vigorous defense of OEO. He smelled a danger of the Community Action Program being absorbed into a new and reorganized Model Cities Program. That was something that we of the task force missed altogether—the problem of competition between these two agencies which had somewhat similar roles.

After a day's discussion, the people at Blair House decided, as such groups always do, "Let's have some more meetings. What we need is a continuing committee. We'll meet again in two weeks, same people, same place." I said, "Please count me out." It seemed to me that the remaining decisions were ones that people who were in responsible positions in government had to make. There was no room for a professor there.

About a year later, when I happened to be visiting some people in the White House, I asked how the Model Cities Program was going. One of the president's assistants said to me very seriously that the Model Cities program had taken more cabinet-level time than any other subject except Vietnam. I said, "That appalls me. If you fellows haven't got anything better to talk about than Model Cities, something's very wrong." He said, no, he thought not. "The Model Cities Program," he said, "is a very convenient way to approach the whole problem of the relations of the federal government to the states and the cities; in effect, we are discussing the problem of federalism."

THE TASK FORCE PROCESS

Afterward, one of the president's assistants wrote me asking for my comments on the task force process. It seemed that the president was interested in the opinions of the various task force chairmen concerning the task force idea. Did it make sense? Was it a useful way of approaching these matters? How could it be improved? What did we have to say?

I said that in appointing a task force, a president could have various purposes. If his purpose was to clarify a particular problem, then the task force ought to be secret because people are much more inclined to level with you when they know they're not going to have to have their words published or be put on the record. After all, they represent interest groups of one kind or another—labor and blacks, for example—and so they can't jeopardize their position with their constituents by telling the whole truth in public, but they can sometimes tell part of it in private. Regarding membership, I said, if the purpose is to get to the bottom of a matter, make it small; twelve or fifteen are too many; four or five would be much better. I also suggested that the committee should consist of so-called "experts"—that is to say, people who have some background knowledge of the subject matter—chosen by the chairman, not by someone on the White House staff.

Assuming still that the purpose is to clarify a particular problem, the chairman should be briefed by the political head, in this case, the president. We never knew from the time we started whether the administration was really willing to abolish Model Cities, or whether it was willing to go entirely the other way—to increase its appropriations by, say, a billion dollars. It would have been helpful to know where the boundries were. If you don't know where they are, the inevitable tendency is to set them too narrow. The assumptions that the nonpolitician makes about what the politician is willing to do are generally too confining. The politician is often bolder than his advisors suppose.

I said that if the purpose of the task force is not to clarify a particular problem but rather to generate public support, to help "sell" some program, or to resolve conflict about some idea, the task force should be made public—the whole thing should go on under the spotlight. The composition of the task force, accordingly, should be representative of the various interests which have something at stake. They should be there explicitly as ex officio representatives of whites or blacks, Catholics, or businessmen, or whatever. Experts should be out of sight, if not altogether out of the picture, because the question is not one of expertise. Accordingly, the task force membership should be rather large.

Now, unhappily, as a practical matter, usually both purposes are involved when a president starts a task force. What do we do in that case? Well, I would think that we then have the worst of both worlds. You have a group which is neither public nor private; that is, it's not really secret. It doesn't consist of true "experts" or true interest group representatives. It doesn't know why it exists or what it's doing.

I also suggested that in some cases it might be well to appoint two or more task forces and have them operate independently and competitively in an adversary process. For example, have one task force consisting of people you know to be pro-Model Cities, and another you know to be anti-Model Cities; let each make the best case it can, and then let them argue it out, perhaps in the presence of the decision makers. This kind of competition would bring out more aspects of the situation.

I said it was tempting to suggest that these task forces should be provided with more staff and, above all, more time to work on the problems. But the more I thought about it, the less sense that seemed to make because if the matter is one that can't be decided on the basis of general principles and in a brief time, then it ought not to be turned over to a bunch of amateurs at all. It ought to be decided in the government by the people who have been elected or have been appointed for that purpose and who have responsibility.

Frankly, I think that it is a very dubious proposition to make use of amateurs coming from outside, who are essentially dilettantes who have no real responsibility for the outcome. They're not accountable; they don't have to live with it; they don't have to explain the outcome to critical Congressmen or others afterwards.

CHAPTER   6

Revenue Sharing in Theory and Practice

How one evaluates revenue sharing will depend upon what one takes the central issues to be. Oddly enough, what must appear to many people to be *the issue*—namely, how to keep the cities and states from going bankrupt—is not properly speaking an issue at all.

Mayor Lindsay has long tried to give the impression that catastrophe lies just ahead unless the federal government provides "massive" additional financial support. Recently other political leaders have been saying the same thing. "Countless cities across the nation," Mayor Gibson of Newark told the press recently, are "rapidly approaching bankruptcy." Governor Rockefeller, after having been informed of the administration's latest plans, remarked that the federal government must do even more to prevent the states and cities from "virtually falling to pieces." Meanwhile, Governor Cahill of New Jersey was telling a joint session of his legislature that "the sovereign states of this nation can no longer supply the funds to meet urgent and necessary needs of our citizens, and institutions and our cities." A day or two later Senator Humphrey in a single sentence made two of the most outstanding rhetorical contributions. The cities, he said, are "mortally sick and getting sicker," and the states "are in a state of chronic fiscal crisis."

In fact, the revenue-sharing idea was, at its inception, the product of exactly such forebodings. Back in 1964, when Walter Heller and Joseph Pechman proposed it, many well-informed people expected that state and

From *The Public Interest*, no. 23 (Spring 1971), pp. 33–45. Copyright 1971 by National Affairs, Inc. Reprinted by permission.

local governments would soon be in serious financial difficulties, while at the same time the federal government would be enjoying a large and rapidly growing surplus. The war in Vietnam was expected to end soon, and, if federal income tax levels remained unchanged, the normal growth of the economy and the increase of population would yield large increases in revenue year after year. Thus, while the federal government fattened, the state and local governments would grow leaner and leaner. Because of rising birthrates and population movements, the demands made upon states and localities for all sorts of services, but especially schools, would increase much more rapidly than would their ability to raise revenue. Whereas the federal government depends largely upon the personal income tax, the yield of which increases automatically with incomes, state and local governments depend mainly upon sales and property taxes, which are inelastic. This being the outlook, it seemed sensible to make up the expected deficit of the state and local governments from the expected surplus of the federal government. Heller and Pechman proposed to do this by giving the states a claim on a fixed percentage of federal taxable income, subject to the requirement that a fair amount "pass through" the states directly to the cities. The idea quickly won wide acceptance. Both political parties adopted revenue-sharing planks, and in 1968 some ninety revenue-sharing bills were introduced in Congress.

What happened, however, was not what was expected. Federal expenditures rose unexpectedly (defense spending was cut, but increases in the numbers of persons eligible for social security together with higher payment levels and unexpectedly high costs for Medicaid took up the slack), and, because of the recession, tax collections fell off. Instead of a surplus the federal government faced a deficit. State and local governments meanwhile fared better than expected. Legislatures and electorates were surprisingly ready to approve new taxes and higher rates. In 1967, for example, the states increased their tax collections by 15 percent, and in 1968 they increased them by another 15 percent. Cities also found it possible to raise more revenue than they had expected. Between 1948 and 1969 state-local expenditures increased in real terms from 6.7 percent of gross national product to about 10 percent. For some time they have been the fastest growing sector of the economy. The credit rating of the cities has improved, not worsened. With respect to the fifty largest cities, only three (New York, Boston, and Baltimore) received lower ratings from Moody's Investment Service in 1971 than in 1940 and many (including Chicago, Los Angeles, and Cleveland) received higher ones.

Dangerous as even short-run predictions in these matters have proved to be, it therefore seems safe to say that no "fiscal crisis" looms for most states and cities. In 1975, according to an estimate citied in a recent article by Richard Musgrave and A. Mitchell Polinsky, state and local expenditures will reach $119 billion. Assuming that federal aid increases at no more than the normal rate of recent years, this will leave a shortfall of $17 billion. Eleven billion of this will be made up from normal borrowing. The remaining deficit of $5 billion, Musgrave and Polinsky say, "could be met by a 5 percent increase in tax rates at the state-local level, an increase which seems well within the reach of state-local governments. . . ." As more and more state governments adopt income tax laws, their revenues will be less dependent upon the vagaries of legislatures and electorates. Moreover, thanks to the recent decline in the birthrate, the principal item of state-local expense—schooling—will for at least a decade be considerably less than had been expected.

INABILITY OR UNWILLINGNESS?

The "fiscal crisis" issue is spurious if defined as the inability (economic, organizational, legal, or even political) of the states—and therefore in a sense of the cities, which are their legal creatures—to support public services at high and rising levels. It is real, however, if defined as their unwillingness to support many of these services at what most reformers deem minimum-adequate levels. Presumably what Governor Cahill meant to tell the New Jersey legislature was something like this: "Any proposal to raise state and local taxes to what everyone would consider satisfactory levels would surely be voted down." The issue, then, has to do with using federal revenue to raise the level of services above the level that, given the realities of state and local politics, would otherwise exist. In other words, it concerns the amount and kind of income redistribution that the federal government should undertake.

That the federal government, and not state-local ones, should be primarily responsible for any income redistribution has long been generally accepted. In recent decades this principle has been used to justify giving federal aid to states and localities in spectacularly increasing amounts. As De Grove has pointed out, the increase was tenfold in the last twenty years, fourfold in the last ten, and twofold in the last five. In 1970 federal aid to states and cities reached an all-time peak of over $24 billion. The question therefore is not *whether* they should be aided but (a) by how much and (b) on what principle of distribution.

In considering what is involved in this, it is necessary to distinguish those state-local needs that are in some sense national from those that are not. That millions of poor rural people have moved to the cities is not something that the taxpayers of the cities should bear the entire financial responsibility for; apart from fairness, there is another consideration—presumably the nation as a whole will be injured if these millions do not receive adequate school, health, police, and other services that only state and local governments can provide. This is one argument that may justify large additional federal support for the states even though they could—if they would—raise the necessary money themselves. Some state-local needs, however, are in no sense national. Most pollution control and much highway construction is in this category. Why, one may ask, should the people of New York be taxed to pay for cleaning up a river in Vermont? As Dick Netzer has remarked in his excellent *Economics and Urban Problems,* ideally such nonnational needs should be met by the development of regional governmental agencies, interstate in some cases and metropolitan in others, that can collect taxes and distribute benefits with a view to whatever public is affected—and to that public only. Unfortunately, such jurisdictions do not exist, and it is politically impossible to create them. There is, however, as Netzer points out, a substitute for them—namely the state governments. So when mayors and governors demand *federal* aid for nonnational purposes, they do not have a persuasive case. Not only is it unfair to shift the cost of essentially state-local benefits to the national public, it is also very wasteful, for when someone else is to pay the bill, the natural tendency is to be prodigal. (Since Uncle Sam is to pay, why not build the bridge or sewerage system twice as big and four times as costly as necessary?)

Still, there unquestionably do exist truly national needs which urgently require increases in federal aid. But aid to whom? Revenue sharing is not a self-evident proposition. In dealing with the redistribution problem, the Nixon administration itself has consistently put the emphasis on aiding individuals rather than governments. Shortly after taking office it exempted persons below the poverty line from paying federal income taxes. In its first two years its main effort was to bring into being the Family Assistance Plan, the effect of which would be to reduce those income inequalities that in large part constitute the "crisis of the cities."

Looking at revenue sharing from this standpoint, there is much to be said against it. Compared to the existing federal grant-in-aid programs, it would be much less redistributive.[1] The existing programs are redistributive because grants are generally awarded on the basis of some criteria of need. The shared revenue, on the other hand, would be distributed to states and cities on the basis of population and tax effort. This means that the wealthier states (in terms of per capita income) would benefit; the poorer states would not. Moreover, it is not likely that all of the money that went to the richer states would end up in the pockets of its neediest citizens. Such features of revenue sharing make for complications.

Under the present grant programs, for example, New Yorkers pay in taxes much more to the federal government than they get back in grants from it. In 1967, the per capita personal income tax paid from New York state to the federal government was $433, whereas the grants received in 1968 amounted to only $120 per capita. In North Dakota it was the other way around; there the per capita personal income tax payment was $177, and per capita grants were $357. On the basis of these figures, a New York politician (say Senator Javits) might decide that revenue sharing is a big improvement over the grant system. After all, under the administration's proposed plan, New York state, which accounts for 10.98 percent of the national income, would get 10.68 percent of the shared ("general") revenue, whereas under the present grant system it gets only (the figure is for 1968) 8.6 percent. On the other hand, an Arkansas politician (say Wilbur Mills, chairman of the House Ways and Means Committee) might conclude that revenue sharing is a very bad idea. Arkansas, which accounts for .67 percent of the national income and gets (1968) 1.46 percent of the federal grants would, under revenue sharing, get only .86 percent of the $5 billion.

In the last decade or so, the movement of poor people from the country to the city has made income redistribution an intracity, or rather intrametropolitan area, problem, as well as an interstate one. Of the $17 billion granted in the fiscal year 1968, $10 billion went to metropolitan areas. Some of the largest grant programs—especially OEO, Model Cities,

---

[1] What is under discussion at this point is the so-called "general" revenue-sharing proposal: that is, the $5 billion, to start with, that would go to states and cities with "no strings attached." This is to be distinguished from "special" revenue sharing, which is the grant system much reorganized and with $1 billion in "new" money added to it.

and Title I of the Elementary and Secondary Education Act of 1965—put money mainly or entirely in so-called poverty areas. From the standpoint of the people who live in these areas, revenue sharing is subject to exactly the same objection that it is subject to in Arkansas: that is, that it will give these areas less than they would get if the same amount were distributed under the existing grant programs. As Governor Sargent of Massachusetts has pointed out, the wealthy suburb of Newton would get $1,527,668 of the $5 billion that the administration proposes to share, whereas Fall River, a city that is really poor, would get only $827,760. One way to meet this objection, at least in part, would be to declare small cities, most of which are well-off suburbs, ineligible to share in the fund. Two years ago, the Intergovernmental Relations Advisory Commission suggested limiting eligibility to cities of 50,000 or more, but it has since been realized that this would leave more than 40 percent of all cities without an incentive to support the plan. The small suburbs are disproportionately Republican, of course, and this must also be taken into account by the Nixon administration.

That a state like Arkansas or a city like Fall River would rather have $5 billion distributed under the existing grant programs than under revenue sharing is quite irrelevant if, as some observers claim, Congress could not possibly be persuaded to increase the total of grants by any such sum. Those who think that the administration must "come up with something new" if it is to have any chance of getting "massive" new money for the cities will presumably conclude that $5 billion in revenue sharing is preferable to, at best, a few hundred million in grants.

Of course the choice need not be between revenue sharing and grants. There are indications that Congress might be willing to assume the costs of certain social programs, especially welfare. Insofar as the object is to redistribute income, it would certainly make more sense to allot the $5 billion to welfare than to revenue sharing. This has been the administration's position all along. Family assistance, not revenue sharing, was, and presumably still is, its first love.

A "NEW FEDERALISM"?

From the standpoint of the administration, the central issue is neither the alleged "fiscal crisis" nor the problem of income redistribution. Rather it is the direction in which the federal system is to develop. From his first statement on the subject (August 13, 1969), the president has emphasized the need to create what he calls a New Federalism. Revenue sharing, he said when he first proposed it, would "mark a turning point in federal-state relations, the beginning of the decentralization of governmental power, and the restoration of a rightful balance between state capitals and the national capital." By the end of the decade, he predicted, "the political landscape of America will be visibly altered, and state and cities will have a far greater share of power and responsibility for solving their own problems." In his recent State of the Union Message, he went even farther. He was proposing, he said, a New American Revolution:

> a peaceful revolution in which power will be turned back to the people—in which government at all levels will be refreshed, renewed, and made truly responsive. This can be a revolution as profound, as far reaching, as exciting, as that first revolution almost 200 years ago.

One of the things that caused the president to think along these lines was the rapid and continuing growth that has been—and still is—taking place in the number of federal grant-in-aid programs. As he pointed out in his 1969 message, this growth has been "near explosive"; between 1962 and 1966, he said, the number of categorical grant programs increased from 160 to 349. There was no reason to think that the rate of increase would slow down, much less stop, of its own accord; but unless it *did* slow down, categorical programs would soon number in the thousands. Revenue sharing was one of the means by which the administration hoped to slow it down. Instead of creating more categorical programs, Congress would be asked to give the money to the states and cities "with no strings attached." Along with revenue sharing, the president asked for authority to order consolidations of categorical programs, provided that Congress did not within sixty days disapprove his orders. He also proposed a Manpower Training Act which (among other things) would have permitted the consolidation of about 20 more or less competing manpower programs and would have given the governors of the states a good deal of control over the consolidated program.

All of these measures had a common rationale—to simplify the structure of federal aid to the states and cities in order to bring it under control and reduce waste. With hundreds of grant programs, each with its own laws and regulations, no central direction is possible. Cabinet officers cannot keep track of—let alone exercise policy direction over—the many and varied programs for which they are responsible. Governors cannot find out what federal money is coming into their states or what is being done with it. The largest cities employ practitioners of the new art of "grantsmanship," in some instances with great success; many small cities, however, finding that they must apply to scores, or even hundreds, of programs, each with its own special requirements and each administered by a different bureaucracy, have more or less given up any hope of getting much help. The system, if it can be called that, is as wasteful as it is frustrating. A state or local government cannot trade a project that is low on its priority list for one that is high. Perhaps it can get $20 million for an expressway that it does not want but not $200 thousand for a drug addiction project that it wants desperately. This involves a double waste: first in what is taken (local authorities can rarely refuse money that is "free") and second in the foregone benefits of desirable projects for which grants are unavailable.

The proposals in the State of the Union Message represent elaborations of those put forward in 1969. "General" revenue sharing differs from the revenue sharing then proposed only in amount and in the percentage (now 48) to be "passed through" to the cities. "Special" revenue sharing is (despite the confusing terminology) nothing but consolidation of categorical programs on an all-at-once, comprehensive basis rather than on a piecemeal one. As everybody presumably knows by now, a few categorical programs would be eliminated and most of the others grouped into six supercategories (urban community development, rural community development, education, manpower training, law enforcement, and transportation) each under a cabinet officer. Under "special" revenue sharing, grants-in-aid would be distributed among states and local governments on the basis of need as in the past, but the distribution would be according to an agreed-upon formula (or rather formulae because each supercategory would have its own) rather than as the result of (among other things) grantsmanship, endurance, "clout," and chicanery—criteria that cannot be excluded at present. Under the plans so far announced (in the nature of the case these are incomplete and somewhat tentative), some interests are bound to gain and others to lose. Mayor Hatcher of Gary, Indiana, for example, has complained that his city would lose about one-fourth of the $150 million a year that it now receives. To such complaints the administration has replied that it will hold in reserve a fund from which to make up any losses that local governments may suffer because of changes in distribution formulae.

ORGANIZED BENEFICIARIES

There is no doubt that state and local officials enthusiastically favor consolidating and simplifying the grant system and giving them (the officials) wide discretion in deciding the uses to which federal aid should be put. Congressmen, too, are well aware of the faults of the present system, and many have spoken out against it. Nevertheless, there is reason to think that proposals to change the system fundamentally will not prove acceptable now or later, no matter who proposes them or what their merits.

It must be remembered that every one of the categorical programs (the most recent estimate is 550) has its organized beneficiaries—not only those who receive grants but also those who are paid salaries for administering them. These beneficiaries have a much livelier interest in maintaining and enlarging their special benefits than the generality of taxpayers—unorganized, of course—has in curtailing them. If there happens to be a grant program for retraining teachers in secondary schools having high dropout rates, it is safe to say that there exists an organization that will exert itself vigorously to prevent the consolidation of that program with others. It is safe to say, too, that no organization exists to put in a good word for the consolidation of the teacher-retraining program with other manpower programs.

From the standpoint of organized interests, dealing with Congress and the Washington bureaucracies (a few key Congressmen and administrators are usually all that matter to any particular interest) is vastly easier and more likely to succeed than is dealing with the legislatures and governors of fifty states, not to mention the officials of countless cities, counties, and special districts. This consideration alone might well be decisive from the standpoint of organized labor, which knows just where to go and whom to see in Washington, even if the political complexions of the state and local governments were exactly the same as that of the national government. In fact, of course, they are not; some interests that are well received and can make themselves heard in Washington—organized labor, minority groups, the poor—would be ignored in certain state capitols and city halls.

Interest groups will not be the only, or probably the most important, defenders of the grant system, however. Congressmen—especially those on important committees—are fond of categorical programs for at least two reasons. One is that they constitute answers to the perennial question:

What have you done for me lately? A narrowly defined category is ideal from this standpoint. It is custom made to suit the requirements of some key group of constituents, and the congressman can plainly label it "from me to you." Revenue sharing, whether "general" or "special," altogether lacks this advantage. It gives benefits not to constituents directly but in wholesale lots to state and local politicians who will package them for retail distribution under their own labels, taking all of the credit.

Congressmen also like categorical programs because of the opportunities they afford to interfere in administration and thus to secure special treatment, or at least the appearance of it, for constituents among whom, as Jerome T. Murphy shows in his case study of the politics of educational reform (*Harvard Educational Review*, February 1971), state and local as well as federal agencies sometimes figure prominently. These opportunities are plentiful because the congressmen see to it that "ifs," "ands," and "buts" are written into the legislation in the right places, and because administrators are well aware that every year they must respond in public to whatever questions may be asked in appropriations and other hearings. Wanting to stay on the right side of those members of Congress with whom they must deal, administrators frequently ask them for "advice." Perhaps it is not too much to say that the categorical system constitutes a last line of defense against what many congressmen regard as the usurpation of their function by the executive branch.

As this implies, the present coldness of Congress to President Nixon's revenue-sharing proposals is not to be explained solely or perhaps even mainly on the ground that he is Republican and Congress is Democratic. The crucial fact is that his proposals would involve a large-scale shift of power from Congress to the White House. *No* Congress would like that, although sooner or later one may feel compelled to accept it.

Revenue sharing would also shift power to governors and mayors. To hear some of them talk, one might think that they would like to have the federal government dismantled and the pieces turned over to them. In fact, most of them are likely to find excuses for not accepting powers that may be politically awkward—and what ones may not?

In his valuable book, *The American System,* the late Morton Grodzins provides some evidence on this point. He tells at some length the story of the Joint Federal-State Action Committee, which President Eisenhower created in 1957 after a flight of oratory (". . . those who would be free must stand eternal watch against excessive concentration of power in government . . .") to designate federal functions that might be devolved to the states along with revenue to support them. The committee was a very high-level one; it included three members of the cabinet, the director of the Bureau of the Budget, and a dozen governors. After laboring for two years, it found only two programs that the federal government would give up and that the states would accept—vocational education and municipal waste treatment plants. As Grodzins explains, the difficulty was not so much that the federal agencies could not be persuaded to give up functions as that the governors would not accept them. They would not take the school lunch program, for example, because doing so would involve a fuss about parochial schools, and they would not take old age assistance because they knew that the old people's lobby would not like having it transferred to the states. Modest as the committee's two proposals were and strongly as President Eisenhower backed them, Congress turned them down.

Whether state and local governments would make good use of the federal funds if given them "with no strings attached" is much doubted by career civil servants in Washington and by what may be called the good government movement. State and city governments, it is frequently said, are in general grossly inefficient and in many instances corrupt as well. The charge is certainly plausible—one wonders whether New York City, for example, has the capacity to use wisely the large amounts that it would receive. The fact is, however, that no one really knows what the state and local governments are capable of. And even if their capacity should prove to be as little as the pessimists say it is, may it not even so be superior to that of the present system *as it will be in another decade or two?*

Administrators in Washington generally assume that the management capacity of state and local governments can be much improved by provision of special grants to strengthen the staffs of the chief executives and by teaching the techniques and advantages of comprehensive planning. The lessons of the last ten years give little support to this assumption, but the administration is nevertheless proposing a fund of $100 million for more such efforts. In my judgment the results are bound to be disappointing. It is the necessity of working out compromises among the numerous holders of bits and pieces of power on the state-local scene that is the main cause, not only of "inefficiency," but of corruption as well. Giving governors and mayors authority over the spending of federal funds would, by strengthening their political positions, reduce the amount of compromising that they must do and the amount of corruption that they must pretend not to see if they are to get anything done. In this way it would contribute more than anything else to increasing the *coherence* (to use the word that is favored among planners) of state-local programs. The $100 million in management assistance that the federal government proposes would probably work in the very opposite direction. In practice if not in theory, giving "technical assistance" usually means maintaining and extending the influence of the federal agencies.

THE PROSPECT BEFORE US

Insofar as there would be a real devolution of power to governors and mayors—and therefore, as the president said, to the people who elect them—the administration's proposals could bring the federal system closer to what the Founding Fathers intended it to be. In my opinion, this is a consummation devoutly to be desired. There is no denying, however, that the short-run effect of decentralization of power would be to take a great deal of pressure off those state and local regimes, of which there may be many, that have no disposition to provide essential public services on an equitable basis or at what reasonable people would regard as adequate levels. This is a powerful objection. It may, however, only be a temporary one. Within a very few years, the political arithmetic of every sizable city and every industrial state will be such as to give politicians strong incentives to take very full account of the needs and wishes of those elements of the electorate that have been, and in some places still are, neglected.

Still, given the political realities that I have mentioned—a public opinion that favors income redistribution but is divided as to how far it should go and how costs and benefits should be apportioned; tax boundary lines inappropriately drawn but not susceptible to being redrawn; hundreds of federal agencies having programs to protect; interest groups even more numerous and with more at stake in the status quo; interstate and intra-metropolitan area differences of interest; congressmen loath to see their powers diminished; governors and mayors equally loath to accept responsibilities that can be avoided—it is not to be expected that any quick or clear-cut settlement will be found for the issues that revenue sharing raises.

I expect that the federal government will continue to play a larger role in raising revenue for all sorts of purposes. As Julius Margolis has pointed out, the larger and more diverse the "package" of expenditure (or other) items that a government presents to its voters, the harder it is for the people to make their decisions on the basis of self-interest as opposed to ideology. This being the case, those who want to win accept-ance for proposals that would not be accepted if self-interest were the criterion always try to include them in packages that are sufficiently large. That is, they prefer to have decisions made on a citywide rather than a neighborhood basis, on a statewide rather than a citywide one, and on a national rather than a state one. It seems to me that the changing class character of the population reinforces this tendency. As we become more heavily upper middle-class, we are increasingly disposed to regard general principles (or ideology, as Margolis calls it), not self-interest, as the proper criterion.

If the federal government will have an ever-larger part in raising revenue, it is not likely to have an ever-smaller one in spending it. He who pays the piper calls the tune. To be sure, he may choose to permit, or even to require, others to do some calling when the number of tunes to be called is inconveniently large, and by so doing he may make everyone better off. The essential fact is, however, that the state governments can be what Governor Cahill called them—sovereign—only if they do what he says they cannot do—supply the funds to meet the urgent and nec-essary needs of their people. I myself am strongly in favor of the reforms that the president has proposed because I think they represent the largest improvement over the present situation that it is reasonable to hope for. I do not however, share his expectation that these reforms will bring about "a historic and massive reversal of the flow of power in America." Indeed, in the event—unlikely, I am afraid—that his proposals are accepted and carried into effect, I would be very surprised if first Mississippi and then New York did not discover that they are ruled as much as ever by national public opinion, acting through national institutions—the presi-dency, Congress, the Supreme Court—and that this opinion and these institutions, the White House most of all perhaps, will in the years ahead assert conceptions of the national interest more vigorously than ever.

PART III

Politics versus Administration

In his famous essay, "The Study of Administration"[1] Woodrow Wilson distinguished categorically between politics and administration. Politics, he said, has to do with the great plans of governmental action and is the province of statesmanship; administration is the detailed execution of these plans and is a science. The science of administration had developed on the Continent, especially in Prussia. In America, where public opinion was enthroned, there had been no place of executive expertise: "advance must be made through compromise, by a compounding of differences, by a trimming of plans and a suppression of too straightforward principles." The time had finally come, however, Wilson wrote, when the public would have to be made to see that self-government does not consist in having a hand in everything; administration should be sensitive to public opinion, but a body of throughly trained officials "is a plain business necessity."

Business necessity or not, American government still advances by compounding differences, trimming plans, and suppressing too straightforward principles. That is the import of this volume as a whole, and especially that of the essays in this section.

"Ends and Means in Planning" denies the validity of the categorical distinction between politics and administration ("ends" and "means"). It does this by showing that even under ideal circumstances no choice can be scientific ("rational"); there must always be an arbitrary or subjective element in choice, and usually accident and uncertainty play a large part in a choice process. By way of illustration, a brief account is given of an effort of the Chicago Housing Authority to select sites in a rational way.

[1]Woodrow Wilson, "The Study of Administration," *Political Science Quarterly* II:2 (June 1887).

The Chicago planners, like the Model Cities planners of the previous section, found that both the nature of a public organization (especially its having many vague and more or less conflicting ends) and the structure of government (especially the extreme fragmentation of formal authority) made "planning" indistinguishable from "politics."

The second essay of the section elaborates the difference between governmental and nongovernmental organization. The object of the essay is to identify the variables affecting corruption in government agencies. The approach is to ask how much and what kind of effort an efficient organization would put forth to prevent or limit corruption. From this perspective, striking differences appear between what a business organization and a public agency would do. The former, it is argued, would seek an optimal amount of corruption (i.e., it would try to maximize profit), whereas the latter would, perhaps not very effectively, seek to eliminate all corruption at any cost. Essentially the difference is that American public administration cannot be entirely separated from politics. The extreme fragmentation of authority that was mentioned before must be mitigated or overcome if anything is to be done, and this, by definition, involves corruption, for example, the intervention of a "machine" and its "boss." It will be seen that in this and other ways the dependence of a governmental organization on public opinion obliges it to violate the canons of administrative rationality.

The remaining two essays raise further doubts about the possibility of rational management under any circumstances. The first of the two points out that insofar as there is anything that may properly be called a science of administration it can be of use only to those relatively few persons who are concerned with the system of relationships that contitutes the organization, that is, with administrative management rather than the performance of substantive tasks. Even to such persons the value of the science of administration is very limited, especially with respect to their most essential functions: the making of judgments about matters of value, morality, and probability. It follows that what an executive needs is not science, or technical knowledge, but the kind of skill that is the art of prudent judgment. This, as the philosopher Michael Oakeshott has stressed, can only be acquired by experience.

Ironically, the final essay of the section sees in the postwar growth of "policy science" some danger that Woodrow Wilson's hopes may be realized. As the Founding Fathers so well understood, the great task is to create and maintain a public opinion devoted to freedom and capable of intelligent discussion. Given the nature of man—"ambitious, vindictive, and rapacious"—this requires statesmanship. Insofar as "policy analysis" supplants statesmanship, the successful working of the political system is jeopardized. For an informative discussion of the record of both policy science and statesmanship with respect to national defense, the reader is referred to an article by Stephen Rosen, a member of the National Security staff.[2]

[2]Stephen Rosen, "Systems Analysis and the Quest for Rational Defense," *The Public Interest* 76 (Summer 1984): 3–17.

CHAPTER 7

Policy Science as Metaphysical Madness

A statesman differs from a professor in a university; the latter has only the general view of society; the former, the statesman, has a number of circumstances to combine with those general ideas, and to take into his consideration. Circumstances are infinite, are infinitely combined, are variable and transient; he who does not take them into consideration is not erroneous, but stark mad—dat operam ut cum ratione insaniat—he is meta-physically mad.

EDMUND BURKE, *Speech on the Petition of the Unitarians*

In the past dozen years or so, policy-oriented social science research and analysis has become a growth industry in the United States. This has occurred in response to demand created by the spate of social welfare programs initiated by the Great Society and, for the most part, continued and expanded by the later administrations. Whereas in 1965 federal agencies spent about $235 million on applied social science research, in 1975 they spent almost $1 billion. Of the approximately $7.4 billion spent in these eleven years about two-thirds was under contract.[1] This brought into being several large independent research bodies, some quasi-public

From *Bureaucrats, Policy Analysts, Statesmen: Who Leads?*, ed. Robert A. Goldwin (Washington, D.C.: American Enterprise Institute, 1980), 1–19. Copyright 1980 by The American Enterprise Institute for Public Policy Research. Reprinted by permission.

[1]The figures are from Clark C. Abt, "Toward the Benefit/Cost Evaluation of U.S. Government Social Research" (Cambridge, Mass.: Abt Associates, Inc., 1976). Abt's paper includes a table giving expenditures by department and by year. According to Laurence E. Lynn, Jr., in 1976 the federal government invested more than $1.8 billion in "social research and development." In Lynn, ed., *Knowledge and Policy: The Uncertain Connection* (Washington, D.C.: The National Research Council, 1978), 1.

and others private, and it greatly increased the amount of university-based policy-oriented social research and the supply of social scientists. According to the 1970 census, the number of social scientists increased by 163 percent in the 1960s, an increase larger than that of any other major occupational group and nearly three times that of professional and technical workers as a whole.

The federal agencies' enthusiasm for policy-oriented research quickly communicated itself to the colleges and universities, which now take a lively interest in whatever may plausibly have the word policy attached to it.[2] Almost all the major universities have established schools to give graduate training in what is now called "policy science," and these have already turned out hundreds of Ph.D.'s. To be sure, not many of the graduates occupy high posts in government,[3] but it is reasonable to expect

[2] Lest this be thought an exaggeration, consider the following from a task-force report submitted to the president and the provost of the University of Pennsylvania by an associate dean of the Wharton School and published in the University's Almanac, January 15, 1974: "Concern with issues of public policy pervades the University of Pennsylvania. Indeed, it is so pervasive that it is impossible to provide anything approaching a full account of the various educational and research programs relating to public policy. Virtually the entire curriculum of the Law School involves public policy. So does much of the research at that school. The Annenberg School of Communications, the School of Social Work, the Graduate School of Education and the Schools of Medicine and Veterinary Medicine deal with public policy issues also. Research at the Schools of Engineering and Applied Science has a substantial policy content. The City and Urban Engineering program, the National Center for Energy Management and Power and the Transportation Studies Center illustrate interests of this sort. Course offerings at both the undergraduate and the graduate levels and extracurricular science and society programs in engineering are similarly focused. The new graduate program in telecommunications engineering and spectrum management exemplifies engineering interest in public policy.

"City and Regional Planning is a policy-oriented program in the School of Fine Arts. The undergraduate Urban Studies Program is operated from the Provost's Office and involves faculty from several schools. The Wharton School, in addition to the many policy-related educational and research activities of the Social Science departments currently therein, has within it the Fels Center, the Rodney L. White Center for Financial Research, the Leonard Davis Institute of Health Economics, the Master of Public Administration program, the Industrial Research Unit, the Labor Relations Council, the Multinational Enterprise Unit, the Busch Center, and the Management and Behavioral Science Center."

[3] As long ago as 1970 the Civil Service Commission listed 563 "senior executive civil servants associated with program analysis." Arnold J. Meltsner, *Policy Analysis in the Bureaucracy* (Berkeley: University of California Press, 1976), 15.

that within a decade or two they will dominate the upper echelons of the federal and state career services as well as those of some of the large cities.

The penetration of policy science into the executive branch has led to, or at any rate been paralleled by, a comparable penetration into the legislative branch. Congress now employs several thousand professionals, a significant and increasing proportion of whom are trained to do policy-related social science research or analysis. Some of these are employed by individual members and others by committee staffs; most, however, are in one or another of several recently established bodies: the Congressional Research Service (1970), the Office of Technology Assessment (1972), the General Accounting Office's division for program evaluation (1974), and the Congressional Budget Office (1974). There is now serious talk of creating an additional body—an "Institute for Congress"—to be privately funded at first and staffed by professionals "whose stature and ability would earn the deference of the members."[4]

The scale and pace of these developments suggest that the American governmental system may be undergoing profound change. As "policy scientists" come to dominate the bureaucracy, not only its decision-making procedures but its style and ethos will change. In addition, those policymakers—"politicians," who are good at taking circumstances into account (they are "statesmen" only if they also take a general view of society)—will find the bureaucracy more resistant than ever to control: policy science may make it a fourth branch, almost independent of the others. If the analytical techniques produced and propagated from the universities supersede the skills of the politician and (on the rare but all-important occasions when it is manifested) the wisdom of the statesman, the successful working of the political system will be gravely jeopardized.

THE METHODOLOGY OF REFORM

The sudden growth of the policy sciences can be seen as a by-product of the civil rights movement and the War on Poverty. In the 1960s, these brought hundreds of new governmental agencies into existence—all providing new job opportunities—and stirred the imaginations of those who

[4]Alton Frye, "Congressional Politics and Policy Analysis: Bridging the Gap," *Policy Analysis* 2 (Spring 1976): 276.

believed that government, if only it tried hard enough, could cure the ills of society.

Actually there has long been a symbiotic relationship between social science and social reform. In the 1880s, Frederick Winslow Taylor spread the gospel of "scientific management" to businessmen and, a little later, schools of business developed budgetary methods. Late in the century, chairs in social science were established, and by 1920 all self-respecting universities had social science departments. By then it was widely believed that government no less than business should—and therefore could—be expertly run. (The city manager movement got underway in 1914.) Naturally the social scientists in the universities were looked to as a principal source of expertise for the organization and management of government, and thus of society generally.

At the beginning of the century, according to historian Barry D. Karl, there developed a methodology of social reform consisting of variations upon three basic steps: first a core group of specialists and influentials, coming together perhaps at a meeting of a professional group, would define a needed social reform or "problem"; then a conference would be called to broaden the coalition by bringing in journalists, philanthropists, and political leaders; and, finally, a survey would be made and a document produced "containing all the information and interpretation on which reasonable men, presumably in government, would base programs for reform."[5]

This was the method used in 1929 when President Herbert Hoover appointed his Research Committee on Social Trends, whose 1,200-page report, Karl tells us, established the principles that "social" behavior came within the purview of the national government, that "science" could do better at framing programs of reform than could legislators or citizens, and that "social welfare" was as fit a subject for national debate as, say, currency reform or the tariff.[6]

In the 1960s this method was used again, and the principles were further extended in order to bring the social science establishment and the Great Society into mutually advantageous relations. This time the specialists and their allies acted through that most prestigious of

[5]Barry D. Karl, "Presidential Planning and Social Science Research: Mr. Hoover's Experts," *Perspectives in American History*, vol. 3 (Cambridge, Mass.: Charles Warren Center for Studies in American History, Harvard University, 1969), 350.
[6]Ibid., 348.

professional associations, the National Academy of Sciences. A report issued under its aegis in 1968 defined the view that reasonable men should take toward the claims of the social scientists to be brought into the policy-forming process:

> The federal government confronts increasingly complex problems in foreign affairs, defense strategy and management, urban reconstruction, civil rights, economic growth and stability, public health, social welfare, and education and training. The decisions and actions taken by the President, the Congress, and the executive departments and agencies must be based on valid social and economic information and involve a high degree of judgment about human behavior. The knowledge and methods of the behavioral sciences, devoted as they are to an understanding of human behavior and social institutions, should be applied as effectively as possible to the programs and policy processes of the federal government. Finally, the behavioral sciences, like the physical and biological sciences, require financial support from the federal government to continue to develop that knowledge and those methods that can lead to greater understanding of the basic processes of individual and group behavior.[7]

Although the report was remarkably adroit in the ambiguity, even confusion, of its wording, it succeeded in conveying the impression that social science had much to contribute to the making of sound policy. Its spirit, though not its letter, reflected the "social science utopianism" that, Karl says, was espoused by Hoover "to be a revolution against politics, committed to the rational, unemotional building of a new, scientific society."[8]

Policy science, in this perspective, appears as one in a long series of efforts by the Progressive Movement and its heirs to change the character of the American political system—to transfer power from the corrupt, the ignorant, and the self-serving to the virtuous, the educated, and the public spirited; and to enhance the capacity of the executive to make and carry out internally consistent, comprehensive plans for implementing the public interest. These were the motives that inspired the Pendleton Act of 1881, establishing a civil service system based on the merit principle; the Budget and Accounting Act of 1921; the President's Committee on Administrative Management in 1937 and the two Hoover Commissions, in 1949 and 1955; and the Council of Economic Advisers

[7]National Academy of Sciences, *Government's Need for Knowledge and Information* (Washington, D.C., 1968).
[8]Karl, "Presidential Planning," 408.

in 1946. They were the motives that inspired proposals to replace politicians with experts in the legislatures and to do away with political parties.[9] When these proved utopian, lesser reforms were advocated that were steps in the same general direction: for example, changes in the organization and practices of Congress to make it an assembly of statesmen deliberating upon the great issues, instead of one of politicians arranging deals and running errands; and changes to require the political parties to "bring forth programs to which they commit themselves."[10]

Today's proponents of policy science are not as naively antipolitical as were the reformers of a generation or two ago; they do not think of themselves as engaged in a "revolution against politics." The old bias is still there, however; witness the intention to provide Congress with a staff of professionals who will earn the "deference" of members (why not just their respect?). Now and then distaste for politicians and their ways is made explicit, as, for example, when an economist, after finding that the structure of Congress falls "enormously short" of what is required for an "ideal" legislative process, takes some comfort in developments to which the Congressional Budget Act of 1974 may give rise: "With a well-trained, nonpartisan professional staff in both the budget committees and the Budget Office, it will be possible to reduce congressional reliance on the hearings process with its domination by special interests and the executive branch."[11]

SOCIAL SCIENCE AND POLICY

The persistent efforts of reformers to do away with politics and to put social science and other expertise in its place are not to be accounted for by the existence of a body of knowledge about how to solve social

[9] For example, Herbert Croly, *Progressive Democracy* (New York: The Macmillan Company, 1914).

[10] Evron M. Kirkpatrick, "Toward A More Responsible Two-Party System: Political Science, Policy Science, or Pseudo-Science?" *American Political Science Review* 65(4) (December 1971): 965–90.

[11] Robert H. Haveman, "Policy Analysis and the Congress: An Economist's View," *Policy Analysis* 2(2) (Spring 1976): 242, 249. See also the lament of Howard F. Freeman (in his foreword to the work edited by Caro cited in footnote 16): "Political pressures continue to result in expeditious [expediential?] decisions, which are then modified by counter-pressures." This, he thinks, "makes it difficult to be optimistic about the future for the world of social action."

problems. There was a time when social scientists thought that eventually they would find laws governing behavior, and most of them seem to have persuaded themselves that the discovery of such laws somehow would make for more democratic, or at least more effective, government. Pending the discovery of such laws, what social research had to offer was not solutions but problems. *Recent Social Trends,* for example, the monumental report of the committee appointed by President Hoover, attempted to establish the facts of social life in a way that would display to the public and its leaders the hitherto unappreciated extent and nature of social problems, but if offered no "solutions."[12]

Now, tens of thousands of Ph.D. dissertations later, there are few social science theories or findings that could be of much help to a policymaker—so few, indeed, that when the would-be writer of a "Handbook of Behavioral Sciences for Policy Making" went through the 600-odd pages of the "inventory of scientific findings" put together some years ago at great expense to the Ford Foundation, the results were "insufficient for a short article, not to speak of a 'handbook.' "[13]

To be sure, some social science theories did have an important influence on the development of the new government programs in the 1960s: those of Lloyd Ohlin and Richard Cloward on "opportunity structures" and those of Gary Becker on "human capital," for example, entered significantly into the conception of the Great Society's poverty program. Policy science, however, consists of the application of methods and techniques, not of substantive theories.

For several decades social scientists had been developing ways of assessing the relative importance of causal factors where several operated

---

[12]U.S. President's Research Committee on Social Trends, *Recent Social Trends* (New York: McGraw-Hill, 1933).

[13]Y. Dror, in Irving Louis Horowitz, ed., *The Use and Abuse of Social Science* (New Brunswick, N.J.: Transaction Books, distributed by E. P. Dutton, 1971), 127. The inventory was Bernard Berelson and Gary A. Steiner, eds., *Human Behavior, An Inventory of Scientific Findings* (New York: Harcourt, Brace & World, 1964). For a more recent compilation, this one financed by the National Institute of Mental Health Research "to demonstrate practical use of generalizations from social science to enhance social practice and policy formation," see Jack Rothman, *Action Principles from Social Science Research* (New York: Columbia University Press, 1974). One can get an idea of the usefulness of the "propositions" in this book from the following: "Success in community intervention varies directly with the sheer amount of practitioner activity or energy applied to role performance" (p. 71).

simultaneously. Further statistical advances occurred during World War II, when engineers, mathematicians, and statisticians were called upon by the military services to find answers to a wide range of very practical questions: What, for example, was the optimal search pattern for locating a pilot down at sea? Wartime experience produced a set of techniques—operations research—the usefulness of which in dealing with a certain class of problems was dramatically demonstrated many times. The class of problems was, however, a sharply restricted one: objectives had to be well defined, operations had to be describable by a mathematical model the parameters of which could be readily estimated from available data, and the current practices had to be ones leaving ample room for improvement.[14]

During the war there were also important developments in statistical inference, probability theory, and what is now called game theory. These developments were readily assimilated into economic theory along with the methods of operations research. Although economists were relative latecomers to the scene (the Rand Corporation had been in business for some time before it hired its first economist, Charles Hitch), they soon became the main force in the development and application of theories of decision making. The rapid concurrent development of computer technology encouraged the elaboration of highly abstract theory by making practicable the working out of computations that had previously been prohibitively time consuming.

When in 1961 Robert McNamara became secretary of defense he brought Hitch and several of his Rand associates into the department, where they introduced the new techniques of formal policy analysis. President Johnson, impressed, it has been said, by McNamara's performance at cabinet meetings and also, one suspects, by the attention the Defense Department's "whiz kids" were receiving from the press, ordered all agencies of the executive branch to introduce "a very new and very revolutionary system" for program planning and analysis along the lines laid out by Defense. Most agencies found ways to avoid carrying out the order, which was soon rescinded by the Nixon administration. The idea of policy analysis, however, made an enduring impression on many bureau chiefs (perhaps because it offered them a means of establishing control over their subordinates) and also on those upper-echelon career civil

[14]See Robert Dorfman, "Operations Research," *American Economic Review* 50 (September 1960): 613.

servants—especially economists—whose exposure to the realities of the policymaking process had not yet made them complete cynics. Today most agencies have offices, headed in some instances by an assistant secretary, to clarify the agency's objectives, monitor its performance, and assess systematically the costs and benefits of alternative courses of action. It was partly in order to cope with the often highly technical reports produced by these analysts in the executive branch that Congress has added many analysts to its own staffs.

In the universities, economists, statisticians, political scientists, and others, excited by the new intellectual problems, challenged by opportunities to contribute to the solution of urgent social problems, and eager to share in the money and power of government, have hastened to develop policy science as an important field of graduate study. As one might expect, in most places the curricula developed for the prospective policy scientists consist largely of highly abstract methodological courses. Students without a considerable aptitude for mathematics cannot take these courses; that the student may have good practical judgment and a strong feeling for institutional realities will not overcome this fatal handicap. After all, the purpose of training in policy science is to improve upon practical judgment and to be able to substitute for it. It is not surprising, then, to find prospective students being told that they can hope to play an important part in public affairs if—but only if—they pass courses in formal analysis. (This presumably is what the Kennedy School of Government at Harvard means by a remarkable sentence in its *Official Register* for 1977–1978: "What the basic curriculum imparts to all individuals is essential to the effective functions of any individual who wishes to play an important role in the policy arena.") The curriculum of the Rand Graduate Institute is reasonably representative of that of most such schools: Microeconomics; Data Analysis and Statistics; Organizational Behavior and Analysis; Econometrics; Technology and Public Policy; The Scope of the Policy Sciences; The Adviser and Society.

THE ROLE OF POLICY SCIENCE

In the past fifteen years policy scientists have approached the policymaking process from several directions, none of which has brought them into intimate connection with it.

Perhaps the least successful role of the policy scientist has been that of proposer of new program ideas. Ideas that are really new are, of course, always hard to find, and, when one is found, it is very likely to prove either infeasible (perhaps because it requires skills or other resources that are not available) or politically unacceptable. At any rate, very few program innovations can be attributed to policy scientists. The Model Cities Program, for example, although preceded by the labors of two task forces, each abundantly supported by consultant specialists, turned out to be altogether different from what the planners had in mind.

Formal modeling—the development of sets of equations describing in quantitative terms the functional relationships in a system of behavior (for example, an economy)—is a mainstay of the policy scientist. There are models that purport to simulate the national economy, models that purport to simulate the impact of government policies on some part of the population (for example, of changes in welfare policies on welfare recipients), models that purport to simulate the effects of new transportation technology on regional growth, and so on. Unfortunately the models constructed by policy analysts are rarely operational. Unlike the operations researcher, whose problems characteristically involve technological relationships that are precisely measurable, the policy analyst typically models relationships that cannot be fully specified or exactly measured, and the results his equations yield—when they yield any at all—are therefore seldom of any help to the policymaker. "To the extent that it *could* answer questions," a model user complained, "they were questions that nobody was asking."[15]

Program evaluation—usually meaning the measurement of policy inputs and outputs with respect either to programs underway or to ones that are contemplated—has doubtless absorbed more time and money in the last decade than has all other policy research put together. The eruption in the 1960s of scores of new social programs, coinciding as it did with the vogue of policy research, led to serious, systematic efforts, often by "outside" research bodies, to measure the cost-effectiveness of the programs. Programs in health, manpower training, law enforcement,

---

[15]Gary D. Brewer, *Politicians, Bureaucrats and the Consultant* (New York: Basic Books, 1973), 165. See also W. Leontief's expressions of concern about the irrelevance, inadequacy, and "consistently indifferent performance in practical applications" of econometric models, in "Theoretical Assumptions and Non-observed Facts," *American Economic Review* 61 (March 1971): 1–7.

housing, and so on are now more or less routinely studied in the administering agencies or in independent bodies under contract to them and by the General Accounting Office (whose authority to make such studies was much extended by the Legislative Reorganization Act of 1970 and the Congressional Budget Act of 1974).

Generally speaking, these evaluations, especially those done by outside agencies, have shown the social programs to be ineffective, or at least far less effective than their proponents claimed. They have, however, had remarkably little effect on policy: one can think of no program that was terminated, or even very substantially revised, because of an evaluation by policy scientists. Findings that do not support "what everyone knows" or that run contrary to the interest of some politically important group (organized teachers, for example) are especially likely to be ignored. The testimony of Peter Rossi, the sociologist, is instructive:

> It is an article of faith among educators that the smaller the class per teacher, the greater the learning experience. Research on this question goes back to the very beginnings of empirical research in educational social science in the early 1920s. There has scarcely been a year since without several dissertations and theses on this topic, as well as larger researches by mature scholars—over 200 of them. . . . Results? *By and large, class size has no effect on learning by students, with the possible exception of the language arts.*
> What effect did all this have on policy? Virtually none. Almost every proposal for better education calls for reduced class size. Even researchers themselves have been apologetic, pointing out how they *might* have erred.[16]

The technical inadequacies of retrospective evaluation have caused policy scientists increasingly to call for experimentation. Economic reasoning, sophisticated analysis, sample surveys, and observational studies, a team of distinguished statisticians writes, will give some good "guesses . . . but we still will not know how things will work in practice until we try them *in practice.*"[17] Policy scientists want to try out policies under conditions that are carefully controlled in order to measure the effects of a change in a specified variable (the teacher–pupil ratio, say) on

[16]Peter Rossi, "Evaluating Social Action Programs," ed. Francis G. Caro, *Readings in Evaluation Research* (New York: Russell Sage, 1971), 278.
[17]John P. Gilbert, Richard J. Light, and Frederick Mosteller in *Evaluation and Experiment: Some Critical Issues in Assessing Social Programs*, eds., Carl A. Bennett and Arthur A. Lumsdaine (New York: Academic Press, 1975), 46.

the achievement of an objective (improved learning). Social experiments—"randomized controlled field trials"—are of course far more expensive than retrospective evaluations (six conducted thus far cost a total of $162 million, whereas the Westinghouse Corporation's evaluation of Headstart cost $585,000).[18] They are also difficult, sometimes impossible, to arrange, as the manipulations of the experimenters are often unacceptable to the subjects; and they are so time consuming—usually covering several years—that the situation is almost sure to have changed materially before the results are in. No experiment, moreover, can yield reliable information about long-term effects, although these may often be the most important. That welfare recipients' willingness to work is not affected much by the introduction of a negative income tax, for example, tells nothing of the effects that such a tax might have on the work motivation of adults who were children in families with guaranteed incomes. Finally, it seems likely that policy may prove as immune to the results of experimentation as to those of evaluation. "After making a head-piece," de Jouvenal reminds us, "Don Quixote tested it by striking it with his sword. The headpiece shattered. He reassembled it, but this time did not strike it, for fear of again losing a possibly worthless helmet."[19]

Recently policy analysts have been turning their attention to "implementation"—the systematic analysis of what is involved in carrying out a course of action. A leading practitioner, Alain Enthoven, formulates the key questions as follows: "Will the people or organizations affected really respond as assumed? What incentives motivate them? Is the proposed course of action compatible with the institutions that must carry it out?"[20] To illustrate what is involved, Enthoven recalls that in 1967 he advised Secretary of Defense McNamara to approve a "thin veil" ABM defense system designed to protect ICBM silos. The Army, which for years had been planning a national ABM system to protect cities, persisted with its plan despite the secretary's order in favor of the "thin veil" system. "A deeper insight into how the Army would actually respond to the decision,"

---

[18]N. P. Roos, "Contrasting Social Experimentation with Retrospective Evaluation: A Health Care Perspective," *Public Policy* 23 (Spring 1975): 254.

[19]Bertrand de Jouvenal, *The Art of Conjecture* (New York: Basic Books, 1967), 103.

[20]Alain Enthoven, eds. Richard Zeckhauser *et al.*, *Benefit-Cost and Policy Analysis: 1974* (Chicago: Aldine, 1975), 464.

Enthoven writes, "would probably have led to a different recommendation."[21] One wonders, however, how an analyst could have gained a deep
enough insight into how the Army would respond to justify a different
recommendation. Could a policy scientist have told the secretary that the
Army would have its way no matter what he (the secretary) might decide?
Dealing as it must with such extreme uncertainties, "implementation"
appears to be a most unsuitable subject for policy science.

LIMITATIONS OF TECHNIQUES

Enough has been said of these principal tasks of the policy scientist
to reveal sharp limitations on his techniques. Some of these are of such
a nature that they cannot be eliminated or even much reduced by better
theorizing or by further advances in computer technology. There can be
no "scientific method" for developing new program ideas, for example.
It will always be impossible to construct a formal model that will be of
use to policymakers when, as is invariably the case with the "important"
problems, one cannot identify all the crucial parameters or match them
with adequate data. No one will ever find a technique for discovering the
concrete implications of vague, contradictory, and fluctuating purposes.
There is no logic by which one can pass from axiological principles to
particular value judgments, and there can be no nonarbitrary way of
finding the optimal terms of trade at the margin among government
objectives when—as is always the case—they are not given to begin with.
Finally there is no "objective" way of making correct probability judgments: some ways of making such judgments are surely better than others,
but none can altogether exclude guesswork. Even if the policy scientist
could know precisely what constitutes "good housing," "good schooling,"
and so on, he could not know (except in cases so obvious as not to need
analysis) which policy alternative would yield the preferred set of consequences. In a world in which everything, including opinions as to what
is preferable, is subject to rapid change, this limitation must be of enormous importance. Despite his claims to method and technique, the policy

[21] Ibid., 464–65.

scientist must in all these matters make up his mind very much as the layman does and always has done.[22]

If to the inherent limitations on analytical techniques one adds the existential ones, policy science appears feebler still. Consider, for example, the practical difficulties in the way of getting reliable data on almost anything: for example, in 1960 and again in 1970 the Bureau of the Census failed to count one black male in ten; and in 1970 the bureau, having concluded that its 1960 and 1950 data on housing conditions were highly inaccurate, then collected none at all.

There are practical difficulties, too—sometimes insuperable ones— in getting policymakers to take the work of the analyst seriously, and these are likely to exist even if the analyst's work *deserves* to be taken seriously. Some arise from the analyst's failure to speak a language that the policymaker understands. To be sure, many of those who have been trained in the techniques of policy science can adapt to a policymaking setting by subordinating "science" to common sense. A policy scientist who lacks this flexibility, however, is likely to find that he can communicate only with other policy scientists. The political executive, whether elected or appointed, and the lawmaker and his staff, although intelligent and well informed, do not know now and are not likely in the future to know enough statistics to interpret the analyst's reports; indeed, the method and mode of thought of the anlyst are likely to strike the practical man as perverse, even ridiculous.

The widest gulf between the analyst and the policymaker is not one of communications, however. The more important fact is that what is of primary importance to the former is generally of little or no importance to the latter. Typically the agency head is chiefly concerned with maintaining and enhancing his organization, and therefore with things that may make a good impression on those (the White House, congressional committees, interest groups, media, and so on) who can help or hurt in

---

[22]After noting that "there is almost no scientific knowledge concerning the long term effects of punishment on the amount of crime" and these effects may be the most important, an analyst of the U.S. Department of Justice concludes: "For the foreseeable future, careful *a priori* reasoning, descriptive evidence on human nature and criminogenic processes, and common sense will rightfully remain the principal sources of evidence in the debate over criminal justice policy." Philip J. Cook, in U.S. Department of Justice, "Punishment and Crime," Working Paper, Economic Research Program, Office of Policy and Planning, August 1976, Opp-ERP 76-3, p. 54.

this; the analyst's words will carry weight with him only when and insofar as they are useful in his day-to-day task of fending off the agency's enemies and bringing its friends into a closer embrace. The elected official's case is not essentially different: typically his main concern is in being reelected, and to spend time and effort on matters that do not promise to improve his position with his constituents by the time of the next election—six years hence at the most—is a luxury he rarely can afford. The conclusions ("hypotheses") of a study of the responses of a Senate committee and of officials of the Food and Drug Administration to policy analysis are therefore not at all surprising: "Congress is almost totally impervious to systematic analysis, particularly in the short run."[23]

If the policymaker himself is impervious to policy analysis, its impact on *policy* may nevertheless be great. Indeed, the proliferation of policy science is making policy problems more numerous and complex. David Cohen and Janet Weiss show this in their review of the "torrent" of research done on schools and race since the *Brown* v. *Board of Education* decision. One study, they found, led to another that was more sophisticated, and then to still another, and so on. The quality of research improved as the process went on, but the outcome was usually not greater clarity about what to think or do, but, instead, a greater sense of complexity, a shifting in the terms of the problem, and more "mystification" in the interpretation of findings. "One thing is clear from this story," Cohen and Weiss conclude, "the more research on a social problem prospers, the harder it is for policymakers and courts to get the sort of guidance they often want: clear recommendations about what to do, or at least clear alternatives." At its best, they say, social research "provides a reasonable sense of the various ways a problem can be understood and a reasonable account of how solutions might be approached."[24] Perhaps one is justified

---

[23] David Seidman, "The Politics of Policy Analysis," *Regulation* 1 (July/August 1977): 35. "The consensus seems to be," writes Carol H. Weiss, "that most research studies bounce off the policy process without making much of a dent on the course of events." She adds, however, that several studies "suggest that the major effect of research on policy may be the gradual sedimentation of insights, theories, concepts, and ways of looking at the world." See "Research for Policy's Sake: The Enlightenment Function of Social Research," *Policy Analysis* 3 (Fall 1977): 532, 535.

[24] David K. Cohen and Janet A. Weiss, "Social Science and Social Policy: Schools and Race," ed. Carol H. Weiss, *Using Social Research in Public Policy Making* (Lexington, Mass.: D.C. Heath and Company, 1977), 67–83.

in concluding (as they do not) that it is easily possible to have too much of a good thing: that an analytical society may increase its problems while decreasing its ability to cope with them.

THE EFFECTS OF POLICY SCIENCE

What has been said so far should relieve any reader who might have feared that the policy scientists are exercising undue influence. In fact, they have very little influence, certainly very little of a direct kind. What someone said of the decisions resulting in Medicare and Medicaid—that they were the result of negotiations between "Wilbur and Wilbur" (Congressman Wilbur Mills and Health, Education, and Welfare Secretary Wilbur Cohen) and were not directly related to any research—may doubtless be said of almost all the important decisions made with regard to foreign affairs, energy, welfare, and the rest. When still a Rand analyst, James R. Schlesinger gave "two cheers and a half" for policy analysis, it would, he said, "shake up many a stale mill pond." But he went on to assert—as he himself has recently demonstrated—that democratic policies would remain unchanged: "a combination of pie-in-the-sky and a-bird-in-the-hand."[25]

The political institutions handed down by the Founding Fathers have proved remarkably resistant to all efforts to make political life more rational. Perfectly aware that the great task of government is to give political leadership—to create and maintain conditions that foster the growth of a public opinion capable of intelligent discussion and of eventual agreement—the Founders were also perfectly aware that that task could never be fully accomplished. The nature of man, as they understood it, precluded the replacement of politics by reason. "Men," Hamilton warned in *Federalist* 6, "are ambitious, vindictive, and rapacious." They were susceptible to some improvement but not to a great deal: certainly they

[25]James R. Schlesinger, "Systems Analysis and the Political Process," *Journal of Law and Economics* 11 (October 1968): 297. Formal analysis is itself sometimes a political weapon. In *Models in the Policy Process* (New York: Russell Sage, 1976), 337, Martin Greenberger, Matthew A. Crenson, and Brian L. Crissey write: "The use of models to dramatize or publicize particular points of view is overshadowing their use for the enlightenment of policymakers." See also Howard Pack and Janet Rothenberg Pack, "Urban Land Use Models: The Determinants of Adoption and Use," *Policy Sciences* (March 1977): 79–101.

could not, as the *philosophes* supposed, be brought to perfection. Struggle and conflict, however mutually disadvantageous, were ineradicable, and therefore the problem of the statesman was to find ways of containing them, not of eliminating them. In the system of checks and balances that they devised, the Founders responded to the political realities of their day (to the conflict between large states and small and the North and South particularly) and to what they knew would be the continuing fact of political struggle.

That the structure of the federal system has remained thus far sufficiently fragmented to ensure the supremacy of a more or less democratic politics may lead us to overlook or underestimate the importance of tendencies that have long been at work, that are now accelerating, and that have changed and will change further the essential character of our political system. Modern America, according to Robert E. Lane, has for some time been moving in the direction of becoming a "knowledgeable society"—that is, one in which, more than in other societies, men inquire into the basis of their beliefs, are guided by objective standards of truth, devote considerable resources to getting and interpreting knowledge, and employ this knowledge to illuminate and perhaps modify their values and goals as well as to advance them. In support of this view, he notes, for example, that from 1940 to 1963 federal government expenditures for research and development increased from $74 million to $10 billion (he was writing in 1966; by 1976 the figure has risen to $22 billion); that from 1953 to 1963 expenditures for research and development by colleges and universities increased from $420 million to $1,700 million (in the following ten years they increased to $3,395 million); and that in the seven years from 1957 to 1964 the number of Ph.D.'s conferred annually increased from 1,634 to 2,320 in the life sciences (to 3,611 in 1975) and from 1,824 to 2,860 in the social sciences (to 11,040 in 1975).[26] That between 1965 and 1975 the production of social science Ph.D.'s increased fourfold while that of physical science Ph.D.'s decreased is surely a measure of the effect of the Great Society's social reform on a crucial component of "the knowledge industry."

If one assumes (as Lane does not) that the experience of going to college tends both to make one disaffected with social and political institutions and to give one a naive confidence in the possibility of improving

[26]Robert E. Lane, "The Decline of Politics and Ideology in a Knowledgeable Society," *American Sociological Review* 31 (October 1966): 650, 653.

them by some sort of social engineering, data on the increase in the number of college graduates are especially relevant: in 1900 there were 19 college graduates per 1,000 persons twenty-three years of age or older; in 1940, 81; in 1960, 182; and in 1976 (estimated), 259.[27] By 1985, about 20 percent of the employed population of the United States will have graduated from college—enough, surely, to affect profoundly the character of the electorate.

A principal consequence of growth in the direction of the knowledgeable society, Lane thinks, has been a shrinkage of the "political domain" (where decisions are determined by calculations of influence, power, and electoral advantage) relative to the "knowledge domain" (where they are determined by calculations of how to implement agreed-upon values rationally and efficiently). Politics, Lane acknowledges, will not cease to exist even in the most knowledgeable of societies, but as our society becomes more knowledgeable political criteria decline in relative importance, and professional, problem-oriented scientists come to have a larger say. This, of course, entails differences in the nature of policy itself. One such difference is in the very *consciousness* (Lane's emphasis) of a problem.

> The man in the middle of a problem (sickness, poverty, waste and especially ignorance) often does not know that there is anything problematic about his state. He may accept his condition as embodying the costs of living. . . . Often it takes years of dedicated agitation to make people aware that they live in the midst of a problem.[28]

The curious thing about modern times, Lane remarks, is the degree to which the government undertakes to do what used to be done by the agitator; consciousness of a problem may in the knowledgeable society come *first* (his emphasis) to the scientific and governmental authorities. Knowledge thus "creates a pressure for policy change with a force all its own"; it "sets up a disequilibrium or pressure which requires compensating thought or action."[29]

---

[27] *Statistical Abstracts*, 1976, Table 231.

[28] Lane, "The Decline of Politics," 659. "The problem of liberal reform," Charles Frankel has written, "has become that of dramatizing for an increasingly comfortable people, the existence of problems that are not immediately visible and which it takes an exercise of imagination to recognize." *The Democratic Prospect* (New York: Harper & Row, 1962), 164.

[29] Lane, "The Decline of Politics," 662.

Although he tries hard to avoid making value judgments (he is, after all, a social scientist writing for a professional journal), one gets the impression that Lane thinks our society improves by becoming more knowledgeable: now that scientific and governmental authorities take the lead in discovering and defining social problems, surely they will be brought to solution faster. That, it would seem, is the implication.

Lane's confidence in the scientific and governmental authorities is misplaced, however. This is evident from the examples he gives of "important findings" in the social sciences that have been produced by the scientific apparatus of the knowledgeable society: (1) The United States ranked sixteenth among nations in the rate of infant mortality in 1961 (ignoring the fact that the United States defines infant mortality more inclusively than do certain other countries). (2) It would cost about $10 billion a year to raise all the individuals and families now below a subsistence income to that level (the phrase *subsistence income* is, of course, meaningless; but even apart from that, the statement is misleading because measures to raise incomes of the "poor" to some acceptable level would inevitably attract many newcomers—how many depending upon the level set—into the "poor" category). (3) The reinforcing experience for convicted criminals while in jail results in high rates of recidivism (many other causes of recidivism are in fact more important). (4) Pollution of soil with arsenic pesticides causes cancer in schoolchildren (this taken from Rachel Carson's *Silent Spring*). (5) The more an individual interacts with persons of another race or ethnic group the less likely he is to be prejudiced against them (could it be that persons who are not prejudiced are more likely to interact?).

Why are these "findings" important? Surely not because they constitute, or point to, "solutions" of policy problems. They are important as propaganda: by creating dissatisfaction they will lead to change. "Knowledge *and what is regarded as knowledge* [emphasis added]," Lane says, "is pressure without pressure groups. . . ." The influence of professionals and their associations, he acknowledges, is "not all good," but it is, he thinks, "generally responsive to the needs of society."[30]

One may well reach a contrary judgment: that professionals, because of their commitment to the ideal of rationality, are chronically given to finding fault with institutions ("bringing to public consciousness" new social problems) and, by virtue of their mastery of techniques of analysis,

[30]Ibid.

to discovering the almost infinite complexity and ambiguity of any problem. Like the social researchers of a generation or two ago, the policy
scientist contributes problems, not solutions. But whereas in the past the
problems appeared manageable to men of common sense and were understood to lie in the domain of the politician or statesman, now they have
been shown to be too complicated for ordinary people to deal with, and
they are, more and more, believed to lie in the domain of the policy
scientist.

It is a dangerous delusion to think that the policy scientist can
supplant successfully the politician or statesman. Social problems are at
bottom political; they arise from differences of opinion and interest and,
except in trivial instances, are difficulties to be coped with (ignored, got
around, put up with, exorcised by the arts of rhetoric, etc.) rather than
puzzles to be solved.[31] In coping with difficulties, formal analysis may
sometimes be helpful, but it is not always so. (Would anyone maintain
that in the Convention of 1787 the Founders would have reached a better
result with the assistance of a staff of model builders?) Except in those
rare instances where the problem is mainly a puzzle rather than a difficulty, the policy scientist is likely to exhibit a "trained incapacity" for
performing what are the essential tasks of political leadership. These are,
first, to find the terms on which ambitious, vindictive, and rapacious men
will restrain one another, and, beyond that, to foster a public opinion
that is reasonable about what can and cannot be done to make the society
better. One cannot perform these tasks merely on the basis of general
ideas or methods; one must have the ability, not taught in schools of policy
science, of taking circumstances—infinite, variable, and transient—into
consideration. What the political leader requires is not policy science but
good judgment—or, better, the union of virtue and wisdom that the
ancients called prudence.

---

[31]The distinction between difficulties and puzzles is elaborated by T. W. Weldon,
The Vocabulary of Politics (Gretna, La.: Pelican, 1953).

CHAPTER 8

Corruption as a Feature of Governmental Organization

This is an exploratory paper, the purposes of which are to identify the principal variables having to do with corruption in governmental organizations in the United States and to point out some significant relationships among them. The paper begins by setting forth a conceptual scheme for the description and analysis of corruption in all sorts of organizational settings. This is applied first to the "typical" business and then to the "typical" governmental organization. (The reason for introducing the business organization into the discussion is to create a contrast that will highlight the characteristic features of governmental organization.) In the concluding section some dynamic factors are noted.

THE CONCEPTUAL SCHEME

The frame of reference is one in which an *agent* serves (or fails to serve) the *interest* of a *principal*. The agent is a person who has accepted an obligation (as in an employment contract) to act on behalf of his principal in some range of matters and, in doing so, to serve the principal's interest as if it were his own. The principal may be a person or an entity such as an organization or public. In acting on behalf of his principal an agent must exercise some *discretion;* the wider the range (measured in terms of effects on the principal's interest) among which he may choose, the

From *The Journal of Law and Economics*, vol. 18, no. 3 (December, 1975), pp. 587–605. Copyright 1976 by the University of Chicago Press. Reprinted by permission.

broader is his discretion. The situation includes *third parties* (persons or abstract entities) who stand to gain or lose by the action of the agent. There are *rules* (both laws and generally accepted standards of right conduct), violation of which entails some probability of a penalty (cost) being imposed upon the violator. A rule may be more or less indefinite (vague, ambiguous, or both), and there is more or less uncertainty as to whether it will be enforced.

An agent is *personally corrupt* if he knowingly sacrifices his principal's interest to his own, that is, if he betrays his trust. He is *officially corrupt* if, in serving his principal's interest, he knowingly violates a rule, that is, acts illegally or unethically, albeit in his principal's interest.

Agents are in varying degrees *dependable*. The more dependable an agent, the larger the psychic costs to him of a corrupt act and accordingly the higher his reservation price for the performance of the act.

MINIMIZING CORRUPTION: CONSTRAINTS

As a means of showing the relationships among these and other variables it will be useful to imagine a situation every feature of which tends to minimize corruption. In such a situation agents are selected after an elaborate search on the basis of their exceptional dependability and lawabidingness, all of their other qualities being deemed of no importance as compared to these. The agents are given whatever kinds and amounts of incentives will motivate them to loyal service and whatever disincentives (for example, high risk of discovery followed by dismissal with loss of pension rights) will deter them from disloyalty. The principal's interest (ends, objectives, goals, purposes, etc.) is fully explicated, and agents are given discretion no broader than is judged necessary to fully serve that interest. Rules are definite (that is, neither vague nor ambiguous), and it is known whether or not they will be enforced. If an agent's duty requires him to try to attain mutually exclusive or competing ends his dilemma is resolved by his principal. An agent's performance is carefully monitored; if there is any doubt about his loyalty, he is dismissed forthwith. Monitors are themselves carefully monitored.

Obviously all of this implies centralized control: there must be an authority (that of a chief executive or "top management") capable of selecting dependable agents, establishing an effective incentive system, explicating the principal's interest, monitoring monitors, and so on.

It will be seen that the situation just described—one in which everything possible is done to minimize corruption—is in many respects highly unrealistic. The principal may be an abstract entity such as a corporation, labor union, or public, in which case some surrogate—a board of directors or chief executive—must explicate its interest and be a monitor-in-chief who cannot himself (or themselves) be monitored.[1] There may be no central authority capable of doing the things necessary to minimize corruption—for example, to dismiss agents whose loyalty is questionable. The rules may be indefinite,[2] and there may be much uncertainty as to whether they will be enforced.[3]

Perhaps the least realistic feature of the "ideal" situation is the implicit assumption that there exists only one objective—namely to minimize corruption. In the real world there are always competing objectives, a condition that makes it necessary to give up something in terms of one in order to get more in terms of another. Each of the measures that might be taken to reduce corruption entails cost. Getting the information upon which an estimate of the dependability of a prospective agent can be

[1]"Unfortunately," writes Frederick Andrews, "no one has devised a way to impose effective [internal] controls on the very top officers, those whose rank enables them to over-ride controls." *Wall Street Journal*, 12 June 1975, p. 1, col. 5.

[2]The recently enacted federal pension law "requires that fiduciaries follow the 'prudent man rule' which requires them to act 'with the care, skill, prudence and diligence' that a prudent man 'acting in a like capacity and familiar with such matters would use in the conduct of an enterprise of a like character and with like aims.' " *Wall Street Journal*, 14 Feb. 1975, p. 28, col. 1. The bribery conviction of former Senator Daniel B. Brewster was overturned by a U.S. Court of Appeals because the "trial court's instructions did not set forth a clear and comprehensive standard for the jury to make the distinction between receiving bribery payment in return for being influenced in the performance of an official act, receiving illegal gratuities and receiving legal, normal campaign contributions." *Philadelphia Evening Bulletin*, 2 Aug. 1974, p. 1.

[3]The uncertainties regarding enforcement of some rules may be seen from the following: "If the Public Officers Law [of New York State] were enforced, and those who accepted or promised a reward in return for a vote were actually incarcerated, few of the state's legislators would remain outside prison bars . . . the act of paying for a judgeship is, after all, an indictable offense, though never enforced.

"To survive politically, most congressmen must overlook the Corrupt Practices Act, which places a ceiling on campaign expenditures, and pretend ignorance on the subject of where their money originates." Martin Tolchin and Susan Tolchin, *To the Victim—Political Patronage from the Clubhouse to the White House* (New York: Random House, 1971), 94, 146, 246.

based is costly, and to secure the services of one who has this scarce, and hence valuable, quality is also costly—costly not only in money and other resources that are paid out but also, and perhaps chiefly, in terms of opportunities foregone, it being highly unlikely that the candidates standing highest in dependability will also stand highest in all other qualities (for example, intelligence, energy, willingness to take risks, etc.) which may be of value to the organization.[4] Similarly supplying the incentives and the disincentives necessary to secure loyalty is costly in resources and in opportunities foregone.[5] So is the explication of the principal's objective and the negotiation of an agreement between a principal and an agent.[6] So also is monitoring: it entails not only direct costs, such as

[4]Ancient Athens provides an interesting exception. These magistrates, whether chosen by lot or elected, were subjected to an examination not to prove capacity or talent but "concerning the probity of the man." Fustel de Coulanges, *The Ancient City* (Baltimore: Johns Hopkins, 1956), 330. For a close analysis of the costs of policing the agent see Barry Mitnick, "The Theory of Agency: The Concept of Fiduciary Rationality and Some Consequences." (Ph.D. diss., University of Pennsylvania, 1974). With regard to "indirect" costs of such policing, see his "The Theory of Agency: The Policing 'Paradox' " and Regulatory Behavior, *Public Choice* 24 (Winter 1975): 27–42.

[5]Gary S. Becker and George J. Stigler, "Law Enforcement, Malfeasance, and Compensation of Enforcers," in *Capitalism and Freedom, Problems and Prospects*, ed. Richard T. Selden (Charlottesville, VA: University Press of Virginia, 1975), 242. This article first appeared in *Journal of Legal Studies* 3 (1974):1. One of the methods proposed by Becker and Stigler to deter malfeasance or nonfeasance is "to raise the salaries of enforcers (agents) above what they could get elsewhere, by an amount that is inversely related to the probability of detection and directly related to the size of bribes and other benefits from malfeasance." However, malfeasance, Becker and Stigler say, can be eliminated without paying the enforcers lifetime salaries exceeding what they could get elsewhere by requiring them to post a bond which they would forfeit if fired for malfeasance. Ibid., 237.

In some jurisdictions an official who resigns before charges are brought may retain his pension rights. The Knapp Commission remarked on this and another practical difficulty in the way of making the threat of pension rights effective. "The result of the present [New York] forfeiture rule," it said, "has been that the courts on appeal have directed the reinstatement of patently unfit officers because they could not tolerate the injustice involved in the forfeiture of vested pension rights." New York City, *The Knapp Commission Report on Police Corruption* (1972), 26, 228–229 (hereinafter cited as *Knapp Commission*).

[6]"With some recent exceptions, police agencies have tended to keep the policies under which they act ambiguous and unwritten. The reasons may include the

the salaries of monitors,[7] but also indirect ones such as the lowering of morale that may occur when agents feel "spied upon." The same may be said of narrowing agents' discretion: it is costly to form, an estimate of the amount of the corruption that would be prevented by setting this or that limit; moreover, narrowing discretion may injure morale (the exercise of discretion being for many an important nonmonetary reward) and, while preventing the agent from doing (corrupt) things that are slightly injurious to the principal it may at the same time prevent him from doing (non-corrupt) ones that would be very beneficial to him. If simply to prevent corruption an agent is given a narrower discretion than would be optimal if there were no possibility of corruption, whatever losses are occasioned by his having a suboptimal breadth of discretion must be counted as costs of preventing corruption.[8]

It is evident that the costs of eliminating, or controlling, corruption may on occasion be greater than the gains from doing so. One can imagine a firm's spending itself into bankruptcy in an effort to end corruption or a labor union's sacrificing the advantage of its monopoly position by employing an honest but incompetent business agent.[9]

This being so, one might expect the management of an organization to try to discover: (a) what level of corruption is optimal for it—that is, the level at which the marginal cost of anticorruption measures equals the gain from them, and (b) what trade-offs among the variously "priced"

---

fear of articulating clear guidelines because of possible controversy and the difficulties in formulating rules for the varied situations police encounter." Police Foundation, *Toward a New Potential* (1974).

For a homely example see William A. Niskanen's account of his difficulties in inducing a selling agent to maximize his (N's) returns from the sale of a house, *Capitalism and Freedom, Problems and Prospects*, ed. Richard T. Selden (Charlottesville, VA: University Press of Virginia, 1975), 26.

[7] "'If we examined every item over $100,000—which really isn't much for a huge corporation—that would drive our fees sky-high. There's a real cost-benefit problem here,' says . . . a top partner at . . . the largest audit firm." *Wall Street Journal*, 12 June 1975.

[8] For example, the cost to New York City of prohibiting a uniformed patrolman from making a gambling arrest unless a superior officer is present. New York City, *Knapp Commission*, 90.

[9] Cf. Simon Rottenberg, *A Theory of Corruption in Labor Unions*. American Association for the Advancement of Science, Symposia Studies Series, no. 3, 4 (National Institute of Social and Behavioral Science, June 1960). Reprinted in University of Chicago, Industrial Relations Center, Reprint Series, no. 96.

anticorruption measures will yield an optimal set, that is, one in which the marginal return from each measure is the same.

Because of technological factors the substitution possibilities may be severely limited. For example, in certain circumstances it may be impossible to substitute monitoring for dependability (the agent's work may have to be done in absolute secrecy); similarly, in certain circumstances it may be impossible to substitute a narrowing or discretion for dependability (the work may require the exercise of a very broad discretion).[10]

It should be noted also that the nature of corruption puts special difficulties in the way of getting information on the basis of which to make a rational allocation of resources. Thus, for example, a principal cannot know how much a third party may bid to secure a corrupt action from an agent and therefore he cannot know how much it will pay him (the principal) to invest in agent loyalty.

INSTITUTIONAL FORMS: BUSINESS

The concepts and relationships that have been set forth in the abstract will be useful in describing some of the main structural features of a concrete form of organization: the "typical" business (more precisely, the competitive corporation). The reason for discussing the business organization here is to provide a contrasting background, so to say, against which the characteristic features of governmental organization can be more readily seen.

1. In a business organization the principal's interest consists of one— or of a very few—objectives the parameters of which—for example, a satisfactory level of profit and beyond that the maximization of emoluments (including staff and expenses) to managers—are easily ascertained. The goods and services produced by the organization can generally be brought under the measuring rod of money and can be distributed via market competition, thus reducing, and to some extent eliminating, the

[10]Gary S. Becker and George J. Stigler, "Law Enforcement," 243, remark that the role of trust in an employment contract is larger: the less easily and quickly the quality of performance can be ascertained, the more diverse the activities of the enterprise, the more rapidly it is growing or declining, and the more unstable the industries in which it is operating in each case.

need to exercise discretion. If over time the revenues of the firm do not cover its costs it must go out of business.

2. The incentive system of a business organization is based very largely upon personal, material incentives, especially money. Although the employee may find his work intrinsically interesting and may get satisfaction from "associational benefits," money rewards, or the expectation of them, are in the usual case by far the most significant of the inducements which motivate him. Business executives whose attachment to the organization they serve is almost purely pecuniary are probably not at all uncommon.

3. There exists a highly integrated system of control through which a chief executive (sometimes a team: "top management") can: (a) reduce the objectives of the organization to lower levels of generality by defining "targets;" (b) select agents; (c) fix limits on their discretion; (d) give or withhold rewards and punishments; (e) arrange for monitoring the performance of agents (and also of monitors). The chief executive may himself be chosen and monitored by a board of directors.[11] So long as profits are satisfactory he is not likely to be disciplined or removed by his board. Frequently its monitoring of him is *pro forma* since it must depend largely upon him for information (although some boards have independent auditing committees), and he is likely to have selected most of its membership.

4. Since there exists in a business organization an ultimate authority (the chief executive or, in some matters, the board of directors), it is possible, in principle at least, for agents to get authoritative rulings as to the terms on which conflicts among ends should be settled. Questions which lie outside an agent's direction—for example, which of two mutually exclusive criteria of choice should be invoked in a concrete situation— are passed up the hierarchy to one who *has* discretion in the matter.

5. The chief executive of a business organization normally continues in office if profits are "satisfactory" until he reaches retirement age. That is, the one condition he must normally meet in order to maintain his control of the organization is business success.[12] Apart from a poor showing

---

[11] On the changing role of the corporate director see the article by John V. Conti, *Wall Street Journal,* 17 Sept. 1974, p. 1, col. 6.

[12] Union members not uncommonly tolerate corruption on the part of agents who win them advantageous contracts. Philip Taft, "Corruption and Racketeering in the Labor Movement," *New York State School of Industrial & Labor Relations Bulletin* 38 (1958): 6, 28. Similarly, voters sometimes prefer candidates whom they have reason to believe corrupt. Of a sample of 1,059 Boston homeowners

on the balance sheet, the principal danger to his tenure is from a hostile takeover or a merger; this danger exists only as outsiders believe they could operate the corporation enough more profitably to yield them a net gain over the costs of acquiring control—costs which the incumbent chief executive is usually in a position to make discouragingly high.[13]

6. The business organization may do whatever is not prohibited by law or government regulation. (Technically a corporation may do only what its charter allows, but as a practical matter it is usually possible to obtain a charter that allows almost anything lawful.) Within the limits set by law and regulation, the business organization may withdraw from one line of activity and enter upon another; it may hire and fire, reward and punish as it sees fit, and it may purchase what it pleases (including the services of consultants of all sorts) at whatever price it is able and willing to pay. Except as it must make disclosures in accordance with government regulations, its affairs are secret.

SOME IMPLICATIONS

These features of business organization have several implications relevant to a discussion of corruption:

(1) Its principal—perhaps its only—object being profit, the business organization will incur costs to prevent corruption insofar—but *only* insofar—as it expects them to yield marginal returns equal to those that could be had from other investments. Similarly, it will incur costs in order to corrupt (that is, to induce the agents of others to betray their trust) insofar—but again *only* insofar—as it expects them to contribute to profit.

(2) The business organization will invest heavily in search costs and in incentives to assure that its chief executive (a) will not be personally corrupt, and (b) will be officially corrupt insofar as may be necessary to secure the success, or at least avoid the failure, of the business. These qualities, although obviously not sufficient conditions of a good chief executive (ability is probably much more important), are surely necessary

(taken in 1966), 41% agreed that "a mayor who gets things done but takes a little graft is better than a mayor who doesn't get much done but doesn't take any graft." Unpublished data gathered by James Q. Wilson and the writer.
[13]See Henry G. Manne, "Mergers and the Market for Corporate Control," *Journal of Political Economy* 73 (1965): 110. Also, "Company Executives Shore Up Defenses against Take-Overs," *Wall Street Journal*, 21 Oct. 1974, p. 1, col. 6.

ones, and in the effort to secure them the board of directors may choose someone who has close family or friendship ties with a principal (large stockholder) or who has "come up through the organization"—that is, whose "loyalty" has been tested over a long period in positions of successively greater responsibility. In order to identify the interest of the chief executive even more closely with that of his principal the board may give him not only a high salary and generous pension rights but also bonuses in the form of stock or stock options.

(3) The task of the business executive being to optimize rather than to minimize corruption, he may follow a policy of "leniency" in dealing with certain types of personal corruption (for example, petty pilfering of supplies) because to do otherwise might create disaffection among employees who object to being "checked up on." What Dalton calls "unofficial rewards"—in plain language, petty graft—is sometimes a significant element in an incentive system.[14]

(4) In dealing with personal corruption at a high level of hierarchy the chief executive is likely to shun publicity; rather than prosecute an offender he may transfer him, force him to resign, or if there is no other way, arrange for his early retirement. To acknowledge that there was personal corruption in a high place would be "bad for the organization": it might produce unfavorable publicity and, worse, encourage a takeover bid, that sort of corruption being widely regarded as indicative of poor management. (Insofar as the chief executive chooses to ignore or "cover up" such corruption because the revelation of it would reflect on his own work he is of course personally corrupt.) On the other hand, personal corruption at the bottom of the hierarchy in excess of the permitted limit may be dealt with harshly, for the exposure of such corruption is generally taken as a sign of good management.

(5) Obviously to be effective a monitoring system in a business organization must operate selectively; its mission is: (a) to keep corruption at the lower levels within permitted levels and, (b) to inform the chief executive about corruption at the higher levels but in a manner that does not oblige him to let its existence become generally known.

(6) Although the structure of the business organization makes possible the resolution of the dilemma faced by an agent who is required to

---

[14]Melville Dalton, *Men Who Manage* (New York: Wiley, 1959), Chapter 6. See also Alvin W. Gouldner, *Patterns of Industrial Bureaucracy* (Glencoe, IL: Free Press, 1954), 87, 159–62, 176.

attain mutually exclusive objectives, in practice it may fail to do so. When "top management" insists—presumably unwittingly—that agents do what they cannot possibly do without violating rules (that is, without being officially corrupt) the monitoring system is likely to be by-passed, or if it is not, to adapt itself to the situation by "turning a blind eye" on all except the most flagrant rule violations. There will be a tendency also for colleague groups and perhaps monitors as well to define the situation so as to reduce the psychic costs of rule violations (for example, it was "against the rules" but "not really wrong or unethical").[15]

(7) Arrangements which allow for the by-passing of monitors or for selective monitoring may easily become dysfunctional by withholding, sometimes for a monitor's own corrupt purposes, information that a chief executive wants and expects to have. A situation which produces rule violation in the line of duty (that is, official corruption) in effect taxes dependable agents (that is, those for whom rule violation entails psychic costs) and subsidizes the undependable. One would expect that under these circumstances the undependable would eventually replace the dependable.

(8) If its sole purpose is the maximization of profit, the business organization may have an incentive to corrupt the agents of other organizations—competitors, labor unions, government agencies, and the like.[16] Its only disincentives are in the nature of business risks—for example,

---

[15]For an account of a business organization (GE) in which "disjointed authority" led to men being required to do what was illegal see Richard Austin Smith, *Corporations in Crisis* (New York: Doubleday, 1963), Chapters 5–6. Striking parallels are to be found in Joseph S. Berliner, *Factory and Manager in the USSR* (Cambridge, MA: Harvard University Press, 1957), Chapters 11 and 12.

[16]Taft found that racketeering in labor unions tends to appear in industries that are highly competitive and have a highly mobile labor force, and that when these conditions coexist some corruption is "almost inevitable." Philip Taft, "Corruption and Racketeering," 33. The Knapp Commission found the second largest source of police corruption in New York City (the first was organized crime) to be "legitimate business seeking to ease its way through the maze of City ordinances and regulations. Major offenders are construction contractors and subcontractors, liquor licensees, and managers of businesses like trucking firms and parking lots, which are likely to park large numbers of vehicles illegally. If the police were completely honest, it is likely that members of these groups would seek to corrupt them, since most seem to feel that *paying off the police is easier and cheaper than obeying the laws or paying fines and answering summonses.*" (Italics added) New York City, Knapp Commission, 6, 68.

loss of reputation within the trade, unfavorable publicity, and fines and other legal penalties (until recently criminal sanctions have rarely been imposed upon officers of corporations which violated the law). The expected disutility of these will presumably be weighed against the expected gain of successful corruption.

(9) One would expect the tendency to corrupt other organizations to be the strongest among those profit-maximizing businesses which must depend upon a small number of customers or suppliers (whether of capital, labor, or materials) and whose profit margin in the absence of corruption would be nonexistent or nearly so. These might fail if they indulged in any form of "social responsibility."[17] Oligopolistic businesses, having ample "slack," although avoiding the vulgar *quid-pro-quo* forms of corruption, probably tend to be lavish in entertainment, consultants' fees, and other expenditures intended to "create goodwill," "maintain good working relations," and "make friends." Insofar as such expenditures cause corruption, it is likely to be of a kind difficult or impossible to identify clearly as such.

INSTITUTIONAL FORMS: GOVERNMENT

The structural features of "typical" American governmental organization differ strikingly from those of business.

1. Fragmentation of authority both within and among federal, state, and local jurisdictions, a conspicuous and distinctive feature of the American political system, gives incentive to the formation and energetic activity of a multitude of pressure groups. Accordingly, American governmental organizations characteristically have objectives that are numerous, unordered, vague and ambiguous, and mutually antagonistic if not downright contradictory.[18] The product (services generally) of a governmental organization is frequently of such a nature as not to be susceptible to being

---

[17] Arrow has remarked that an ethical code "may be of value to the running of the system as a whole, it may be of advantage to all firms if all firms maintain it, and yet it will be to the advantage of any one firm to cheat—in fact the more so, the more other firms are sticking to it." But he concludes that the value of maintaining the system "may well" be apparent to all and that "no doubt" ways will be found to make ethical codes a positive asset in attracting consumers and workers. Kenneth J. Arrow, "Social Responsibility and Economic Efficiency," *Public Policy* 21 (1973): 315–316, 330.

[18] These points are elaborated in Edward C. Banfield, *Political Influence* (Glencoe, IL: Free Press, 1961), especially Part 2.

priced in a market or perhaps to quantitative measurement of any kind. Almost all government regulation is of this character.[19] Governmental products that might be priced usually are not, and if they are, the price is usually set so low as to subsidize the consumer and (if the supply is short) to create a rationing problem which necessitates exercises of discretion on the part of governmental agents.[20] As Robert C. Brooks wrote early in the present century, "as soon as regulation is undertaken by the state a motive is supplied . . . to break the law or bribe its executors."[21] When, as is often the case, the governmental organization has a monopoly a strong incentive exists for third parties to seek to influence the agent's exercise of discretion by offering a bribe—that is, to pay a monopoly price, the money going not to the government but to the agent.

The governmental organization's existence is not jeopardized by selling its products at prices that are below the cost of production since typically it gets some or all of its revenue by taxation. If there is any threat to its existence, it is likely to arise from its having failed to distribute

[19]These features of governmental organization are illustrated and analyzed by James Q. Wilson, *Varieties of Police Behavior* (Cambridge, MA: Harvard University Press, 1968).

[20]See Arnold J. Meltsner, *The Politics of City Revenue* (Oakland: University of California Press, 1971), 33–35.

[21]Robert C. Brooks, *Corruption in American Politics and Life* (1910; reprint, New York: Arno, 1974), 166. The construction industry provides a case in point. New York City has a 843-page building code; a builder is required to get at least 40–50 permits and licenses (for a very large project as many as 130) from a maze of city departments. "Each stage," John Darnton writes in the *New York Times*, 13 July 1975, p. 5, sec. 4, col. 3, "is an invitation to a payoff. By withholding approval, or concentrating on a minor infraction, or simply not showing up at all, an inspector can cost a builder dearly or delay his recouping a multi-million-dollar investment." In practice, the Knapp Commission found, "most builders don't bother to get all the permits required by law. Instead, they apply for a handful of the more important ones (often making payoff to personnel at the appropriate agency to insure prompt issuance of ther permit). Payments to the police and inspectors from other departments insure that builders won't be hounded for not having other permits." New York City, *Knapp Commission*, 125. Recently two-thirds of the construction inspectors in Manhattan were suspended without pay on bribery charges. None of the charges seems to have resulted from a builder's effort to get around the requirements of the building code. What was being bought and sold, an official said, was time. Robert E. Tomasson, in the *New York Times*, 18 July 1975, p. 5.

enough in subsidies to get the support it needs in the legislature or at the polls.

2. Although the incentive system of the governmental organization is based mainly on money and other personal material incentives, other types of incentives usually bulk larger in it than in the incentive system of business organization. At the lower levels of government hierarchy, job security is an important incentive. At the middle levels the satisfactions of participating in large affairs, "serving in a good cause," and of sharing (albeit perhaps vicariously) in the charisma that attaches to an elite corps (for example, the F.B.I.) or to a leader (for example, J. Edgar Hoover or Robert Moses) are also of importance. At the top level, power and glory are among the principal incentives (for example, Hoover and Moses). It would probably be hard to find in government a very high level official whose attachment to his job is purely pecuniary.[22]

3. From a legal-formal standpoint and often in fact, control of a governmental agency is in many more hands than is control of a business. Normally a chief executive or a small team ("top management") has authority over all of the operations of a business. A governmental agency, by contrast, is usually run by a loose and unstable coalition of individuals each of whom has independent legal-formal authority over some of its operations. Not only is there separation among legislative, judicial, and executive functions, but the executive function is itself divided. In the federal government, for example, there are numerous independent bodies (for example, the Federal Reserve System) which exercise an authority independent of the president's. In state and local government, executive authority is much more widely distributed.[23] The mayor of a moderate-sized city, for example, has no control whatever over at least a dozen bodies whose collaboration is indispensable. The comptroller and district attorney, for example, are usually independently elected, and there are numerous bodies, notably the civil service commission, in which the

---

[22]Moses, Caro writes, found "a hundred ways around" civil service pay and promotion rules. But his men were mainly attached to him because they were "caught up in his sense of purpose" and because they "admired and respected" him. Moses himself, Caro asserts, came to be motivated solely by an insatiable lust for power and glory. Robert A. Caro, *The Power Broker, Robert Moses, and the Fall of New York* (New York: Knopf, 1974), 273.

[23]See Edward C. Banfield and James Q. Wilson, *City Politics* (Cambridge, MA: Harvard University Press, 1963), Chapter 6.

mayor has little or no voice. The governmental chief executive appoints his principal subordinates (subject usually however to their confirmation by a separate authority), but the incentive system is fixed by law and regulation and is as a rule of such a nature as to preclude competition with business organizations for executive talent. Moreover the governmental chief executive and his principal subordinates, having no control over the tenure or pay of lesser ("career") executives, have not much more than nominal authority over the lower levels of bureaucracy. For example, the mayor of New York, according to Sayre, has few levers to move the several hundred bureau chiefs; he works "at the margins of bureau autonomy" by creating and staffing new bureaus and by praising or attacking those who are or are not responsive to him, but "his victories are temporary and touch only a few bureaus."[24]

4. Whereas the business organization may hire, promote or demote, and dismiss salaried employees at will (subject to civil rights and other such laws), the governmental one is severely restricted by civil service regulations. The governmental executive cannot dismiss a career civil servant whom he considers untrustworthy; instead he may prefer formal charges supported by evidence before a trial board from the decision of which the employee may usually appeal. The procedure is so time consuming and dismissals so hard to get that it is invoked only in cases so flagrant as to make it unavoidable.[25]

5. The fragmentation of formal authority in government is overcome to a greater or lesser degree by informal arrangements: officials exchange favors (for example, voting support, jobs, opportunities to make money by legal or other means) with other officials and with interest groups and

---

[24]Wallace S. Sayre, in *Agenda for a City*, eds. Lyle C. Fitch and Annemarie Hauek (Beverly Hills: Sage, 1970), 576.

[25]General Motors' procedure in dismissing salaried employees is in sharp contrast to that of the New York Police Department. When there was reason to believe that some of them were accepting favors and "kickbacks," GM marched salaried employees, some of whom had been with the company for more than twenty-five years, through an "assembly line" for questioning by two company investigators. Forty-three who admitted taking gifts of more than nominal value were fired on the spot. *Wall Street Journal*, 24 April 1975, p. 1, col. 1.

In the New York Police Department, on the other hand, an officer under indictment for a felony may not be suspended; instead he is placed on "modified assignment," retaining his salary, fringe benefits, and gun until final disposition of his case, which may take three or four years. *New York Times*, 19 July 1974, p. 39, col 1.

voters in order to assemble the authority they require to maintain and if possible increase their power.[26] In the extreme case the result is a stable structure in which control is as highly integrated as in the business organization; this is the "machine," the chief executive of which (who may or may not be the chief executive of a governmental organization) is the "boss."[27] One difference between the machine and the business organization requires particular notice: the business executive's control, resting as it does on a solid legal-formal base, is stable as compared to that of the boss, which arises from extralegal, if not illegal, arrangements, is *ad hoc*, and must be continually renewed by "deals" in order to prevent it from collapsing.

6. The agent of a governmental organization is both likely to be required to serve objectives that compete or are mutually exclusive (this is because, as noted, the objective function of a government organization is characteristically vague, ambiguous, and unordered) and unlikely to get the dilemma that he faces resolved by passing it up the hierarchy: there is hardly ever a chief executive with authority over all the matters involved in the conflict, and, if there is and if the conflict is passed up the hierarchy to him, the answer that will come back to the agent will probably be "Maximize *all* objectives."[28] The fundamental fact of the situation is that the electorate or some set of interest groups (that is the principal) *demands* states of affairs that are mutually exclusive.

7. The chief executive of a governmental organization normally serves

---

[26] For an account of the patronage, favors, and "honest graft" distributed in a relatively "clean" medium-sized city (New Haven) see Raymond E. Wolfinger, *The Politics of Progress* (Englewood Cliffs, NJ: Prentice-Hall, 1974), Chapter 4.

[27] On the machine, see Edward C. Banfield and James Q. Wilson, *City Politics*, Chapter 9.

[28] In the Federal bureaucracy, Kaufman writes, subordinates may have to pick and choose among many directives for justification. "This obligation may be thrust upon them by the inescapable ambiguities as well as inconsistencies of the instructions to them." Herbert Kaufman (with the collaboration of Michael Couzens), *Administrative Feedback: Monitoring Subordinates' Behavior* (Washington: Brookings Institute, 1973), 2.

In Oakland, California, the city manager sends a guidance letter to officials involved in the budget process. "The important thing to understand about the manager's letter is that it contains nonoperational guidance, or decision rules that are not decision rules." Arnold J. Meltsner, *Politics of City Revenue*, 165, 171.

a two- or a four-year term after which he must face "takeover" bids first in a primary and then in a general election; in many jurisdictions he may not succeed himself more than once or twice. As compared to his opposite number in business, his tenure is uncertain and brief.[29] Although as an incumbent he has a decided advantage, much of what he does or fails to do in office is with a view to increasing the probability of his reelection.

8. Unlike the shareholder, the citizen cannot easily disassociate himself from a corrupt organization: to escape it he must incur the costs of moving to another city, state, or country. Nevertheless, in the usual case he will have no incentive to invest in its reduction because it, or rather the absence of it, is a "public good" the benefits of which will accrue as much to "free riders" as to others.[30] There are, however, institutions, notably the media and the ambitious district attorney, that stand to gain by searching out and making much of corruption, whether real or seeming.

9. From a formal standpoint, a governmental organization may do *only* what the law expressly authorizes, and it *must* do what the law expressly requires. In fact, when public opinion permits, it sometimes withdraws from tasks assigned by law which expose it to corruption.[31] Its freedom to avoid exposure to corrupting influences is small as compared to that of the business organization however.

10. A governmental organization may have few secrets. To a large and increasing extent, its affairs (for example, the salaries of employees, the number of widgets bought and the price paid, etc.) are matters of public record. Generally, public hearings must be held and public participation secured before a governmental undertaking may get under way. Sometimes there must be a referendum. Meetings at which decisions are to be made must often be open to the public. In many instances officials are required to disclose their property holdings and business connections. As "public figures" they are for all practical purposes unprotected by laws against slander and libel.

[29]On turnover of mayors, see Raymond E. Wolfinger, *The Politics of Progress*, 394–395.

[30]See Mancur Olson, *The Logic of Collective Action* (Cambridge, MA: Harvard University Press, 1965).

[31]The New York Police Department announced that as an anticorruption measure it would not arrest low-level figures in gambling combines or enforce the Sabbath laws (except upon complaint) or the laws pertaining to construction sites (unless pedestrians are endangered or traffic impeded). *New York Times*, 19 Aug. 1972, p. 1.

CONSEQUENCES OF THESE DIFFERENCES

Obviously the structural differences between business and governmental organizations have consequences affecting corruption:

(1) In governmental organization the costs of preventing or reducing corruption are not balanced against gains with a view to finding an optimal investment. Instead, corruption is thought of (when it comes under notice) as something that must be eliminated "no matter what the cost." Even when no deterrent effect could be expected, a governmental organization—the IRS, for example—would not act uncharacteristically in spending, say, $50,000 to uncover and punish a misdeed which could not possibly have cost the government more than, say, $500.

(2) In the absence of central control there is no real (as opposed to nominal) chief executive or "top management" which can make substitutions among anticorruption measures (for example, more investment in dependability and less in monitoring) in an effort to "balance the margins."

(3) In the absence of central control, an agent whose duties are mutually incompatible cannot get a resolution of his dilemma by administrative action; therefore—unless he resigns—he must act corruptly. As Rubinstein remarks in *City Police*, "a policeman cannot escape the contradictions imposed on him by his obligations."[32]

(4) Where authority is highly fragmented, there is no centralized system of monitoring or, more generally, of control. Under these circumstances corruption is likely to occur "because the potential corrupter needs to influence only a segment of the government, and because in a fragmented system there are fewer centralized forces and agencies to enforce honesty."[33] Although there may be officials in roles specialized to perform monitoring or other anticorruption functions—typically an independently elected district attorney—these frequently fail to perform vigorously when doing so would be contrary to their political or other interests; in any case no chief executive can *require* them to—that is, there is no effective provision for monitoring the monitors.

(5) Because of the inflexibility of government pay scales and promotion rules, a government executive is often unable to offer an agent incentives (monetary and nonmonetary) equal to those he could earn elsewhere. One consequence is that the government agent is likely to be

[32]Jonathan Rubinstein, *City Police* (New York: Ballantine, 1973).
[33]Raymond E. Wolfinger, *Politics of Progress*, 114.

less dependable and less able than his business organization counterpart. Of more importance, perhaps, the government agent has less to lose from dismissal.[34]

(6) The nature of governmental activity often precludes precisely stating or narrowly limiting the breadth of an agent's discretion.[35] If the objective is no more definite than to "improve the qualify of life," the agent's decisions must necessarily be on grounds that are highly subjective. The monitoring of such decisions presents obvious difficulties: almost any can be given a plausible rationale.

(7) When output standards cannot be made definite, the organization may try to compensate by making input specifications very detailed. The effect of this, however, is to reduce the number of competitive suppliers, thereby making collusion easier between sellers and corrupt purchasing agents. Sometimes an agent establishes specifications that only one supplier can meet.

(8) Both because of the relatively open (not secret) nature of most government activities and because of the fragmentation of control, monitoring agents from *outside* the organization is probably more common in government than in business. "Unplanned feedback" (that is, information and clues that come to the organization from clients, competitive agencies, the mass media, public prosecutors, civic groups, etc.) may be of special importance in governmental organizations because of the lack (due to fragmentation) of *planned* feedback.[36] Selective monitoring (that is, the

[34]As Becker and Stigler remark, "Law Enforcement," 242, "Trust calls for a salary premium not necessarily because better-quality persons are thereby attracted, but because higher salaries impose a cost on violations of trust." The principle was applied by New York Police Commissioner Murphy who limited the "exposure [to temptations offered by gambling and narcotics interests] to officers of higher rank who presumably have a greater stake in maintaining their reputations." New York City, *Knapp Commission*, 238.

[35]Recently the General Services administrator took steps to monitor some of his own necessarily subjective decisions by delegating the choice of architects to a panel of career civil servants which will rate candidates in writing on the basis of published criteria. Final decisions remain the administrator's responsibility. *Wall Street Journal*, 11 June 1974, p. 7. For an account of the difficulties police administrators have in managing the discretion of policemen see James Q. Wilson, *Varieties of Police Behavior*, 64–65.

[36]Herbert Kaufman, *Administrative Feedback*, 10, writing about the Federal bureaucracy, describes the sources of "unplanned feedback" as follows: "Clientele or customer objections, the normal interactions of organization with each

filtering out of information damaging to certain "higher ups" or posing a threat to the successful functioning of the organization) is difficult to manage when there is much "unplanned feedback."

(9) Governmental organizations are much less likely than business ones to be permissive about petty corruption ("unofficial rewards"). For one thing, the governmental organization, which is not seeking to maximize profit, is willing to accept whatever loss of productivity may result from employees' disaffection at being closely watched. For another, no one in the organization has authority to permit "stealing from the government." Such petty corruption as exists within a governmental organization reflects a failure of management, not, as it might in a business organization, a cost deliberately incurred to avoid a still larger one.

(10) Insofar as government executives are motivated more by nonpecuniary values (for example, power, participation in large affairs, "serving in a good cause," etc.) than are business executives, they are probably less susceptible to pecuniary inducements to corruption. By the same token, they are probably *more* susceptible to nonpecuniary inducements such as "the good of the organization."[37]

(11) Because the opportunity to exercise wide discretion in important matters (that is, to wield power) normally comprises a larger part of the "package" of incentives offered to government executives than of that offered to business executives, close monitoring is probably more disruptive of management in government than in business.

(12) The existence of governmental organizations in which authority is highly fragmented presents an opportunity for a political entrepreneur to "purchase" pieces of authority (that is, to bribe or otherwise influence

---

other, for example, staff agencies such as budget personnel, administrative management, audit, legal counsel, and public relations can be expected to turn up some information for line executives about the behavior of line subordinates. Competitive agencies may reveal a good deal about each other's field activities. Some individuals and organizations thrive on exposing shortcomings in public agencies employing sensational investigatory commissions. Subordinates themselves give clues about their own activities when they seek clarification of policy announcements, overlapping circles of acquaintances, membership in clubs and community groups, gossip and humor." See also, Ibid., 35, 41, 74.

[37] Barnard believed corrupt acts such as falsifying books "for the good of the organization" to be rare in industrial organizations "but undoubtedly have occurred not infrequently in political, governmental, and religious organizations." Chester I. Barnard, *The Functions of the Executive* (Cambridge: Harvard University Press, 1938), 277.

the possessors of authority to use it as the "purchaser" requires) and thus to create a highly integrated system of control (machine).

(13) Where the formal decentralization is not overcome by an informal centralization (that is, a machine) corruption is likely to be widespread, there being no mechanism capable of regulating it. "Prudential considerations restraining corruption," Brooks wrote early in the present century, "are apt to be much more keenly felt by a thoroughly organized machine than in cases where corruption is practiced by disorganized groups and individuals each seeking its own or his own advantage regardless of any common interest."[38]

(14) To the extent that there is a machine, that is, to the extent that the formal fragmentation of authority has been replaced by an informal centralization, the governmental organization will resemble the typical business organization. Its chief executive ("boss") will have reasonably secure tenure, a well-defined objective function (to maintain and enhance the organization), and the ability to exercise control.[39] He will invest heavily in the dependability of his principal subordinates (one "comes up through" a machine by demonstrating loyalty over time),[40] regulate their

[38] Robert C. Brooks, *Corruption*, 107. See also Raymond E. Wolfinger, who writes, *Politics of Progress*, 114: "John A. Gradiner's study of the notoriously corrupt city of Wincanton provides evidence for the proposition that decentralized political systems are more corruptible, because in a fragmented system there are fewer centralized forces and agencies to enforce honesty. The Wincanton political system is formally and informally fragmented; neither parties nor interest groups (including the criminal syndicate) exercise overall coordination. The ample patronage and outright graft in Wincanton are not used as a means of centralization. Indeed, governmental coordination clearly would not be in the interests of the private citizens there who benefit from corruption, nor of the officials who take bribes. Attempts by reformers to stop graft or patronage founder on the cities commission form of government, which is both the apotheosis of local governmental fragmentation and a hospitable environment for machine politics."

[39] Early in the present century an observer found New York state's formal administration "a drifting, amorphous mass, as helpless as a field of seaweed in the ocean" with power dispersed among nearly 170 units and "no head, no manager, no directing will *legally committed* to preside." At the same time the extralegal side—that is, the political parties— had "no loose ends, no irresponsible agents"— that is, "authority is clearly defined, obedience is punctiliously exacted; the hierarchy is closely interlinked, complete, effective." Quoted in Clifton K. Yearley, *The Money Machines* (Albany, NY: State University of New York Press, 1970), 254.

[40] Writing of James Marcus, a high official of the Lindsay administration who was convicted of bribery, Moscow remarks: "Marcus could not have happened in a

breadth of their discretion, maintain an incentive system that motivates machine workers (especially job patronage, legal fees, the purchase of insurance, construction contracts, etc.), and monitor them to check unauthorized corruption.

SOME DYNAMIC FACTORS

In the nature of the case it is impossible to know how much corruption there is at any given time. On theoretical grounds, however, it seems safe to say that for several decades corruption has been increasing in the United States and that it will continue to do so.

Adam Smith remarked that those who trade often with each other find that honesty is the best policy.[41] For traders, mutual adherence to rules constitutes a public good. If the traders are fewer than some critical number, each will find it to his advantage to abide by the rule and even to contribute to its enforcement upon others. But if they exceed the critical number, each may find it to his advantage to violate the rule, and none will voluntarily contribute to its enforcement because each will know that the situation will be essentially the same no matter what he does and that therefore it will pay him to be a "free rider."[42]

It is easy to point out instances in which honesty is still the best policy (the Chicago grain pit is one example and the Chicago political "machine" another). Nevertheless, it seems likely that for several decades the proportion of situations of the large-number type, in which dishonesty is likely to be the best (that is, most profitable) policy, has been increasing. Certainly there are relatively few executives who think of their organization as an entity (for example, a "house") that will exist in essentially the same environment for generations to come. This being so, there is

---

Tammany administration, in which a man of loose morals might have been appointed, but not without full knowledge of his weaknesses." Warren Moscow, *The Last of the Big-Time Bosses* (Briarcliff Manor, NY: Stein & Day, 1971), 202–203.

[41] Quoted from Adam Smith's *Lectures* by Edwin Cannon in his introduction to *The Wealth of Nations* (New York: Modern Library, 1937), xxxii.

[42] For a general discussion see James M. Buchanan, "Ethical Rules, Expected Values, and Large Numbers," *Ethics* 76 (Oct. 1965): 1–13 and his further remarks on The Samaritan's Dilemma, in *Altruism, Morality and Economic Theory*, ed. Edmund F. Phelps (New York: Russell Sage, 1975).

now relatively little incentive to invest in acquiring reputation ("character," to use Smith's word), something which, unlike an "image," can only be had by consistent adherence to the rules over a long period of time.

A second factor tending to increase corruption in the United States has been the dramatic enlargement of the scope and scale of government, local, state, and national. Doubtless this enlargement has been to some extent a response to problems that have arisen because of the inability of traders who interact in large numbers and over short periods to maintain the rules that could be taken for granted when most traders interacted frequently and over long periods of time. As honesty gradually ceased to be the best policy, there were more demands that government offer incentives and disincentives to make it so once again. Moreover, the American passion for equality has always encouraged the replacement of the invisible hand of competition by the visible—and supposedly fairer—one of bureaucracy. (As Tocqueville remarked, "equality singularly facilitates, extends, and secures the influence of a central power" and every central power "worships uniformity [because] it relieves it from inquiry into an infinity of details.").[43] But uniformity was bound to leave many individuals dissatisfied—the infinity of details could not be ignored without cost—and thus to create pressures for special treatment.

Whatever their causes, every extension of government authority has created new opportunities and incentives for corruption. Over the long run this has helped to make it appear normal, tolerable, and even laudable.[44]

Had the growth of government been accompanied by the centralization of control and certain other structural changes, the increase in corruption would doubtless have been less. But the structural changes that occurred were mainly in the "wrong" direction: executive control has been reduced by merit system practices, recognition of public employee unions, civil rights legislation, laws requiring "citizen participation," "sunshine" laws, and the like. At the same time, the extralegal arrangements through which control was informally centralized in a "machine" which, sometimes at least, found it advantageous to moderate and limit corruption

[43] Alexis de Tocqueville, *Democracy in America* Vol. 2 (New York: Knopf, 1945), 295.
[44] Indeed, as has been frequently noted, corruption frequently serves socially desirable functions—for example, it may make a grossly unfair tax more nearly equitable, it may keep the government going in a time of hyperinflation, it may deter a policeman from beating an innocent person, etc.

have in most instances been wiped out or rendered less effectual by "good government" reforms.

A third factor, closely related to the second, has been the imposition upon business organizations of constraints much like those under which government operates. Public opinion (including often that of business-men!) more or less obliges the business organization to subordinate the profit criterion to other objectives—ones which, as in government, are vague and conflicting. Like government organizations, businesses are more and more expected to tolerate, even to encourage, participation in their affairs by outsiders ("public interest groups") and to give the public details of dealings the success of which requires secrecy. Frequently courts and regulatory agencies play leading roles in making business decisions.[45]

If these are indeed the trends, one may well ponder what their outcomes will be. Presumably the culture that is being formed today contains a much smaller stock of dependability than did that formed a generation, or two, or three ago. Substitutes (for example, monitoring) can take the place of much dependability, but they will surely be relatively costly, and there is doubtless some "technological" limit to the amount of substitution that is feasible: it is hard to believe that complex social organizations can exist in the complete absence of dependability.[46] In any case there are more important questions. In a society in which *dishonesty* is the best policy, will not the individual feel contempt for himself and for his fellows, and will he not conclude—rightly perhaps—that he and they are "not worth saving"?

---

[45]Consider, for example, the rules by an administrative law judge of the National Labor Relations Board that a newspaper may not prohibit its reporters and editors from accepting gifts ("freebies") from news sources. *New York Times*, 17 Jan. 1975, p. 43, col. 4. Weidenbaum has written recently of a "second mana-gerial revolution" involving a shift "from the professional management selected by the corporation's board of directors to the vast cadre of government regulators that influences and often controls the key decisions of the typical business firm." Murray L. Weidenbaum, *Government-Mandated Price Increases* (Washington, D.C.: American Enterprise Institute for Public Policy Research, 1975), 98.

[46]Cf. J. S. Mill: "There are countries in Europe, of first-rate industrial capabilities, where the most serious impediment to conducting business concerns on a large scale, is the rarity of persons who are supposed fit to be trusted with the receipt and expenditure of large sums of money." *Principles of Political Economy*, 5th ed. (London: J. W. Parker), 151. See also Edward C. Banfield, *The Moral Basis of a Backward Society* (Glencoe, IL: Free Press, 1958).

ACKNOWLEDGMENTS

The writer is grateful for the encouragement and suggestions given by his colleague, Julius Margolis, and for criticisms by Susan Rose-Ackerman and Barry M. Mitnick. He has also benefited greatly from reading Mitnick's "The Theory of Agency: The Concept of Fiduciary Rationality and Some Consequences" (unpublished Ph.D. diss. University of Pennsylvania, 174).

CHAPTER 9

Ends and Means in Planning

The word *planning* is given a bewildering variety of meanings. To some it means socialism. To others, the layout and design of cities. To still others, regional development schemes like TVA, measures to control the business cycle, or "scientific management" in industry. It would be easy to overemphasize what these activities have in common; their differences are certainly more striking than their similarities. Nevertheless, there may be a method of making decisions which is to some extent common to all these fields and to others as well, and that the logical structure of this method can usefully be elaborated as a theory of planning.

Such an attempt leads at once to the action frame of reference, the means–ends schema, and the usual model of rational choice. An actor (who may be a person or an organization) is considered as being oriented toward the attainment of ends. Planning is the process by which he selects a course of action (a set of means) for the attainment of his ends. It is "good" planning if these means are likely to attain the ends or to maximize the chances of their attainment. It is by the process of rational choice that the best adaptation of means to ends is likely to be achieved.

In this article I propose first to develop these common conceptions sufficiently to provide a simple theory of planning, one which is essentially a definition. It will be descriptive in approach and will deal with how planning would have to be done in order most fully to attain the ends sought, not how it actually is done (this latter would be a theory of the sociology of planning). I shall then argue that the actual practice of organizations does not in fact even roughly approximate those described in

From *The International Social Science Journal*, vol. 11, no. 3 (1959), pp. 361–368. Copyright 1959 by UNESCO. Reprinted by permission.

the theoretical model; this argument will be illustrated with brief reference to a particular case which a colleague and I have described elsewhere.[1] I shall then consider the question of why it is that organizations do so little planning and rational decision making.

I

The concept of rational choice has been expounded with great rigor and subtlety.[2] Here, a much simplified approach will suffice; a rational decision is one made in the following manner: (a) the decision maker lists the opportunities for action open to him; (b) he identifies the consequences that would follow from the adoption of each of the possible actions; and (c) he selects the action that would lead to the preferred set of consequences. According to this definition, no choice can ever be perfectly rational, for there are usually a very great—perhaps an infinite—number of possible actions open to the actor and the consequences of any one of them would ramify *ad infinitum*. No decision maker could have the knowledge (or the time!) to evaluate even a small fraction of the actions open to him. It is possible, however, to be more or less systematic in the canvass of alternatives and probable consequences, so that the conception is not an entirely useless one. For practical purposes, a rational decision is one in which alternatives and consequences are considered as fully as the decision maker, given the time and other resources available to him, can afford to consider them.

A plan (unless we depart very far from customary usage) is a decision with regard to a course of action. A course of action is a sequence of acts which are mutually related as means and are therefore viewed as a unit; it is the unit which is the plan. Planning, then, as defined here, is to be distinguished from what we may call "opportunistic decision making," which is choosing (rationally or not) actions that are *not* mutually related as a single means. The rational selection of a course of action, that is the making of a rational plan, involves essentially the same procedure as any

[1]The conceptual scheme and much of the ensuing argument is set forth more elaborately in Martin Meyerson and E. C. Banfield, *Politics, Planning, and the Public Interest* (Glencoe, Ill.: Free Press, 1955).
[2]For example, Sidney Schoeffler, *The Failures of Economics* (Cambridge, Mass.: Harvard University Press, 1955), Appendix I.

rational choice: every possible course of action must be listed, the consequences of each course must be identified, and that course selected the consequences of which are preferred.

A plan is rationally made by a process which may conveniently be described under four main headings:

Analysis of the Situation

The planner must lay down in prospect every possible course of action which would lead to the attainment of the ends sought. His task is to imagine how the actor may get from where he is to where he wants to be, but his imagination must work within certain conditions fixed by the situation, especially by the resources at his disposal (not merely possessions, of course, but legal and other authority, information, time, executive skill, and so on) and by the obstacles in his way. His opportunity area consists of the courses of action "really" open to him; that is, those which he is not precluded from taking by some limiting condition. It may be, of course, that he has no opportunity area at all—that there is absolutely no way by which the ends sought may be achieved—or that the opportunity area is very restricted.

End Reduction and Elaboration

An end is an image of a future state of affairs toward which action is oriented. The formulation of the end may be extremely vague and diffuse. If so, the end must be reduced to specific or "operational" terms before it can serve as a criterion of choice in the concrete circumstances. The formulation of the end may be elliptical; in this case the planner must clearly explain the meaning in full. An end may be thought of as having both active and contextual elements. The active elements are those features of the future situation which are actively sought; the contextual are those which, while not actively sought, nevertheless cannot be sacrificed without loss. (The man who burned down his house in order to get the rats out of the cellar ignored a contextual end in his effort to achieve an active one.) The planner's task is to identify and clarify the contextual as well as the active components of the ends. If they are not fully consistent, he must also "order" them; that is, he must discover the relative value to be attached to each under the various concrete circumstances envisaged in the courses of action or, as an economist would say, prepare an "indifference map."

The Design of Courses of Action

Courses of action may have a more or a less general character. At the most general level, a development course of action implies a description of the "key" actions to be taken or the commitments to be made. These constitute the premises upon which any less general course of action is based, at the "program" or "operations" levels, for example. In other words, decisions of a less general character represent choices from among those alternatives that are not precluded by the more general decisions already taken. A development course of action may be chosen arbitrarily or capriciously, and a program course of action may then be selected with elaborate consideration of alternatives and consequences; when this happens, there is "functional rationality" but "substantive irrationality."

The Comparative Evaluation of Consequences

If the plan is to be rational, all consequences—not merely those intended by the planner—must be taken into account. To a large extent, then, good planning is an endeavor to anticipate and to provide for unintended consequences. The planner cannot pick and choose among the consequences of a given course of action: he must take them, the unwanted along with the wanted, as a set. Their evaluation must therefore be in terms of the net value attached to each set. If all values could be expressed in terms of a common numerical index (like prices), this would raise no great difficulties. In practice, however, the planner must somehow strike a balance between essentially unlike intangibles. He must decide, for example, whether x amount of damage to a beautiful view is justified by y amount of increase in driving safety.

II

So far the discussion has been intended to make reasonably clear what is meant by *rational planning*. If we now take this definition as a yardstick and apply it to organizational behavior in the real world we are struck by two facts: there is very little planning, and there is even less rationality.

In general, organizations engage in opportunistic decision making rather than in planning: rather than lay out courses of action which will lead all the way to the attainment of their ends, they extemporize, meeting each crisis as it arises. In the United States even the largest industries do not look forward more than five or ten years. In government, the American planning horizon is usually even less distant. Moreover, such plans as are made are not the outcome of a careful consideration of alternative courses of action and their probable consequences. As a rule the most important decisions—those constituting the development course of action—are the result of accident rather than design; they are the unintended outcomes of a social process rather than the conscious products of deliberation and calculation. If there is an element of rationality, it is "functional" rather than "substantive."

A few years ago the writer and a colleague set out to describe how decisions were made by a large and progressive public body, the Chicago Housing Authority.[3] We knew that the housing agency was one of the best administered in the United States, and we therefore assumed that if we observed closely enough we could see how a large organization lays out alternative courses of action, evaluates their probable consequences, and so arrives at what is, in the circumstances, a rational decision. We did not expect to find that the model described was being followed consciously or in detail, of course, but we did suppose that the course followed would roughly approximate it.

What we found was entirely different from what we anticipated. The authority might conceivably have sought to attain its ends by any one of various courses of action. (It might, for example, have given rental subsidies to enable people with low incomes to buy or rent housing in the market. Or it might have built small housing projects for eventual sale. Or again, following the example of the United Kingdom, it might have built new towns in the hinterland beyond the metropolis.) No major alternative to what it was already doing was in fact considered. The development course of action—to build large slum clearance projects—was treated as fixed. This course of action had been arrived at cumulatively, so to speak, from a number of unrelated sources: Congress had made

[3]Martin Meyerson and E. C. Banfield, *Politics, Planning, and the Public Interest.*

certain decisions, the Illinois legislature certain others, the City Council certain others, and so on. Unless the housing authority was to embark upon the unpromising task of persuading all these bodies to change their minds, the development "plan" had to be taken as settled—settled on the basis of decisions made without regard to each other. "The process by which a housing program for Chicago was formulated," we wrote,

> resembled somewhat the parlor game in which each player adds a word to a sentence which is passed around the circle of players: the player acts as if the words that are handed to him express some intention (i.e., as if the sentence that comes to him were planned) and he does his part to sustain the illusion.

The idea of planning, or of rational decision making, assumes a clear and consistent set of ends. The housing authority, we found, had nothing of the kind. The law expressed the objectives of housing policy in terms so general as to be virtually meaningless, and the five unpaid commissioners who exercised supervision over the "general policy" of the organization never asked themselves exactly what they were trying to accomplish. Had they done so they would doubtless have been perplexed, for the law said nothing about where, or in what manner, they were to discover which ends, or whose ends, the agency was to serve.

The agency had an end system of a kind, but its ends were, for the most part, vague, implicit, and fragmentary. Each of the commissioners— the Catholic, the Jew, the Negro, the businessman, the labor leader— had his own idea of them, or of some of them, and the professional staff had still other ideas. There were a good many contradictions among such ends as were generally agreed upon. Some of these contradictions went deep into fundamental questions. For example, the authority wanted to build as much housing as possible for people with low incomes, but it also wanted to avoid furthering the spread of racial segregation. These objectives were in conflict, and there was no way of telling which should be subordinated or to what extent.

Most of the considerations which finally governed the selection of sites and of the type of projects were "political" rather than "technical." A site could not be considered for a project unless it was large enough, unless suitable foundations for high-rise construction could be sunk, and so on. But, once these minimal technical conditions were met, for the most part the remaining considerations were of a very different kind: Was the site in the ward of an alderman who would support the project or oppose it?

III

Unfortunately there does not exist a body of case studies which permits of the comparison that would be interesting—comparison, say, between large organizations and small, public and private, single-purpose and multipurpose, American and other. Despite this lack, some general observations are possible. While the Chicago Housing Authority may be a rather extreme case, there are compelling reasons which militate against planning and rationality on the part of all organizations.

Organizations do not lay out courses of action because the future is highly uncertain. There are very few matters about which reliable predictions can be made for more than five years ahead. City planners, for example, can know very little about certain key variables with which they must deal: How many children, for example, will require schools or playgrounds ten years hence? Recent experience has shown how little even demographic predictions can be trusted. The Chicago Housing Authority could not possibly have anticipated before the war the problems it would have to face after it. Some people, knowing that they cannot anticipate the future but feeling that they ought to try, resolve the conflict by making plans and storing them away where they will be forgotten.

Not only do the conditions within which the planner works change rapidly, but so also do the ends for which he is planning. A public housing program begun for slum clearance may, before the buildings are occupied, be primarily an instrument for the reform of race relations. It need hardly be said that the means most appropriate to one end are not likely to be most appropriate to the other.

When an organization is engaged in a game of strategy with an opponent, the element of change is likely to be of special importance. The opponent tries to force change upon the organization; the organization's actions must then be a series of countermeasures. In the nature of the case these cannot be planned. To a considerable extent all organizations—and not especially those engaged in "competitive" business—are constantly responding to changes which others are endeavoring to impose upon them.

When it is possible to decide upon a course of action well in advance, it is likely to be imprudent to do so, or at least to do so publicly (as, of course, a public agency ordinarily must). For to advertise in advance the actions that are to be taken is to invite opposition to them and to give the opposition a great advantage. This is a principle which many city planners have learned to their cost.

Organizations, especially public ones, do not consider fundamental alternatives because usually there are circumstances which preclude, at least in the short run, their doing anything very different from what they are already doing. Some of these circumstances may be the result of choices the organization has already made; others may be externally imposed. The housing authority, for example, could not cease building its own housing projects and begin giving cash subsidies to private builders: public opinion favored projects rather than subsidies, and the agency had recruited and trained a staff which was project minded and not subsidy minded. An organization's commitments, and often other obstacles as well, may be liquidated over time and a new course of action initiated. But the liquidation is expensive: it may be cheaper to retain for a while an obsolescent course of action than to incur the costs of instituting a new one. If the organization could see far enough into the future it might liquidate its commitments gradually, thus making an economical transition to a new course of action. As a rule, however, it cannot anticipate very clearly or surely what it will want to do a few years hence. Moreover, if it acknowledges its doubt about the wisdom of what it is presently doing it risks giving aid and comfort to its enemies and damaging its own morale.

Organizations have a decided preference for present rather than future effects. One might think that public organizations, at least, would be more willing than are persons to postpone satisfactions—that, in the language of economics, they would discount the future less heavily. They do not seem to, however, and this is another reason why they do not plan ahead.

The reason they discount the future so heavily is, perhaps, that they must continually be preoccupied with the present necessity of maintaining what Barnard has called the *economy of incentives*. That is to say, the heads of the organization are constantly under the necessity of devising a scheme of incentives by means of which they can elicit the contributions of activity required to keep the organization going. Any scheme of incentives is inherently unstable. It must be continually rebuilt according the needs of the moment. "Indeed, it is so delicate and complex," says Barnard, "that rarely, if ever, is the scheme of incentives determinable in advance of application."[4]

The end of organizational maintenance—of keeping the organization going for the sake of keeping it going—is usually more important than any substantive end. The salmon perishes in order to give birth to its young. Organizations, however, are not like salmon; they much prefer sterility to death. Given the supremacy of the end of organizational maintenance, opportunistic decision making rather than planning is called for. Indeed, from the standpoint of maintenance, the organization may do well to make as few long-term commitments as possible. Advantage may lie in flexibility.

The end system of an organization is rarely, if ever, a clear and coherent picture of a desirable future toward which action is to be directed. Usually a set of vague platitudes and pious cant is used to justify the existence of the organization in the eyes of its members and of outsiders. The stated ends are propaganda, not criteria for guiding action. What John Dewey said in *Human Nature and Conduct* of an individual applies as well to an organization: it does not shoot in order to hit a target; it sets up a target in order to facilitate the act of shooting.

It follows that serious reflection on the ends of the organization, and especially any attempt to state ends in precise and realistic terms, is likely to be destructive of the organization. To unify and to arouse spirit the ends must be stated in vague and high-sounding terms. When they are given definite meaning they lose their magic, and, worse, they become controversial. Had it tried to formulate a set of ends in relation to racial policy, the Chicago Housing Authority would certainly have destroyed itself at once.

It follows also that organizations do not as a rule seek to maximize the attainment of their ends or (to say the same thing in different words) to use resources efficiently. If the ultimate end is the maintenance of the organization, how indeed is "maximization" possible? The organization may endeavor to store up the largest possible quantity of reserves of a kind which may be used for its maintenance at a later time (to accumulate "good will," or the wherewithal to procure it, for example, in advance of

[4]Chester I. Barnard, *The Functions of the Executive* (Cambridge, Mass.: Harvard University Press, 1938), 158.

need). In this case a *quantity*—utility—is being maximized. But if substantive ends are regarded, Herbert A. Simon is right in saying that organizations "satisfice" (that is, look for a course of action that is satisfactory or good enough) rather than maximize.[5]

Laying out courses of action, clarifying ends, and evaluating alternatives take time and money and cannot be done without the active participation of the chief executives. However great may be the resulting gain to the organization, full attention to the present crisis—assuming the supreme importance of organizational maintenance—is likely to result in far greater gain. Paradoxical as it may seem, if all costs are taken into account, it may be rational to devote very little attention to alternatives and their consequences.[6]

Rationality, as defined before, is less likely to be found in public than in private organizations. One reason for this is that the public agency's ends often reflect compromise among essentially incompatible interests. This is not an accidental or occasional feature of public organization in a democracy. Where conflict exists and every conflicting element has to be given its due, it is almost inevitable that there be an end system which "rides madly off in all directions."

Whether or not conflict is built into the end system, the end systems of public organizations are vastly more complex than those of private ones. Contextual ends, in particular, are far more numerous. A private builder, for example, does not concern himself with the effect of highrise construction on birth rates and family life, but a public one must. The more complex the end system of the organization, the harder to devise courses of action, the more consequences must be evaluated, and the greater the likelihood that some ends will be sacrificed in the endeavor to attain others. That rationality, in the sense of the definition, becomes more

[5]See the discussion of this in the preface to the second edition of *Administrative Behavior* (New York: Macmillan, 1957), xxiv–xxv and the references given there to Simon's more technical writings. Simon says (p. xxiv) that human beings "satisfice" because "they do not have the wit to maximize." This does not seem to be quite the right way of putting it. If the trouble is merely that they lack wit to make the necessary calculation, then they are trying to maximize and failing or, in another view of the matter, succeeding given their limitations. The point being made here, at any rate, is not that organizations lack wit but that they lack will to maximize; in other words, it is the nature of their end systems, not their ability to compute, which is here in question.
[6]See Sidney Schoeffler, *The Failures of Economies.*

difficult to achieve is not, of course, an argument against public enterprise: perhaps private enterprise does not take *enough* ends into consideration.

IV

The reader may by now have come to the conclusion that because organizations are so little given to the rational adaptation of means to ends, nothing is to be gained from constructing such a model of planning as that set forth here.

Certainly this would be the case if one's interest were mainly sociological. For the study of how organizations actually behave, an altogether different conceptual scheme would probably be more rewarding.

But if the interest is normative—if it is in describing how organizations would have to act in order to be in some sense more effective or efficient—it is hard to see how reference to such a model can be avoided or, indeed, why its lack of realism should be considered a defect. And students of administration are, after all, chiefly interested in describing organization in order that they may improve it. Their problem, then, is to find a theoretical model which, without being so far removed from reality as to be a mere plaything, is yet far enough removed to suggest how organizations may be made to function better.

It would be a contribution to the development of a suitable theory if there were a body of detailed case studies, all of them built on a common conceptual scheme so as to allow of significant analytical comparisons. It would be particularly helpful to have a full account of the workings of an organization which is so placed as to be able to encourage the fullest development of planning and rational choice: one, let us say, with a few clearly defined purposes, free of political and other conflict, blessed with a large opportunity area, and headed by persons who try to be rational. How fully and clearly would such an organization explain and define its ends? How often and how elaborately would it consider alternative courses of action at the various levels of generality? How exhaustively would it inquire into probable consequences, the unintended as well as the intended? Would it perhaps carry planning and rationality beyond the point where marginal cost equals marginal return? And would it "maximize" or "satisfice"?

CHAPTER **10**

The Training of the Executive

What has to be learned is not an abstract idea, or a set of tricks, not even a ritual, but a concrete, coherent manner of living in all its intricateness.

MICHAEL OAKESHOTT[1]

The postwar popularity of executive development programs raises in slightly new form the old question of what should be the training of the executive. An executive development program is a conference, course, or seminar lasting from one or two days to a month. Executives, drawn usually from "middle management," are brought together, usually under the auspices of an academic institution or a trade association (the American Management Association is one of the chief operators in the field), to hear talks by "experts," usually professors of administration or executives of large organizations, to engage in discussions, and in some cases to engage in "simulation" of administrative situations.[2] Many businesses every year

From *Public Policy*, vol. 10, eds. Carl J. Friedrich and Seymour E. Harris. A Yearbook of the Graduate School of Public Administration (Cambridge, MA: Harvard University Press, 1960), 16–43. Copyright 1960 by John Wiley & Sons, Inc. Reprinted by permission.

[1]Michael Oakeshott, "Political Education," in Peter Laslett, ed., *Philosophy, Politics and Society* (New York: Macmillan, 1956), 17.

[2]See National Industrial Conference Board, *Executive Development Courses in Universities*, Studies in Personnel Policy no. 160 (New York, 1957); Ward Stewart, "Executive Development Programs in Collegiate Schools of Business," *Higher Education*, May and September, 1958 and April, 1959; Louis J. Rago, "Executive Training Programs," *Advanced Management* 22 (December, 1957): 22–23; Harland Fox, *Leadership and Executive Development; A Bibliography*, Industrial Relations Center Bulletin 14 (University of Minnesota Press, 1954), and L. V. Murphy, "A Select Bibliography of Employee Training and Development," *Personnel Management* 19 (September–October, 1956): 62–68.

send several of their "coming men" to these institutes at company expense. Until recently federal employees were not so fortunate; the agencies were not permitted to use their appropriations to send employees to "outside" training institutions. Many federal employees took development courses, but they did so at their own expense or on grants from foundations. The Government Employees Training Act of 1958 has given the agencies more freedom, however, and no doubt the demand for executive development will increase accordingly.

City and state executives are being "developed" in large numbers. In the spring of 1959, for example, a foundation arranged to send the state commissioners of mental health of the United States to the University of Chicago for a two-week course on decision making. Recently, the Ford Foundation made a large grant to enable the New York University and the New York City Department of Personnel to establish a "pilot program" of executive development for New York City and the New York metropolitan area. Under this arrangement, deputy commissioners and bureau chiefs of the city government will attend ninety hours of lectures, problem-solving conferences, and workshops.

The approach taken to the training of the executive varies greatly from program to program. Some programs emphasize practical training with "on-the-job application." Others offer "broad perspectives and new cultural horizons." Still others have it both ways: their courses are "down to earth" and "oriented toward creative leadership." Generally the executive is encouraged to think that he will be shown how to use some expertise that academic authorities have produced by research or leading practitioners have distilled out of experience. Not infrequently he is led to suppose that the methods and results of social science can be useful to him. Sometimes the jargon of social science—"decision making," "role systems," "identification," and so on—gives the course a scientific aura.

The brochure published by New York University to describe its new Ford Foundation-sponsored program of executive development is not unrepresentative in the claims that it makes for the value of technical knowledge in administration:

> The field of management is an ever expanding body of knowledge, procedure and tested theory. Its contribution to intelligent leadership in government and industry is well recognized today. The Executive Program for the City of New York and metropolitan area has been established to enable top government executives to benefit in the performance of their duties from this essential body of information.

POVERTY OF THE LITERATURE

When one turns from such brochures to the literatures of administration one is struck forcibly by the disparity between claims and
achievements. March and Simon have recently combed the literature
thoroughly, and they observe that although bits, pieces, and snatches of
theory and empirical data can be assembled from a wide range of sources
(in particular the reminiscences of executives; the products of the scientific
management movement; sociology, especially that influenced by Weber's
writings on bureaucracy; social psychology; the writings of political scientists on governmental efficiency, and parts of economics), "the literature
leaves one with the impression that after all not a great deal has been
said about organizations, but it has been said over and over in a variety
of languages."[3] There is also, they add, a great disparity between the
hypotheses in the literature and the evidence to support them.

> Much of what we know or believe about organizations is distilled
> from common sense and from the practical experience of executives.
> The great bulk of this wisdom and lore has never been subjected to
> the rigorous scrutiny of scientific method. The literature contains
> many assertions, but little evidence to determine—by the usual stan-
> dards of public testability—whether these assertions really hold up
> in the world of fact.[4]

Although they take a dim view of the present state of scientific
knowledge about organization, March and Simon have high hopes for its
future state. Their labors will be more than repaid, they say at the end
of their book, "if they encourage others to join in the task of replacing
fancy with fact in understanding the human mind and human behavior
in an organizational setting."[5] Presumably they think there is nothing to
prevent the replacement of fancy with fact if the task is approached in
the right way.

This optimism is hard to understand. There are very grave methodological difficulties in the way of studying the important aspects of
organization scientifically. As Simon explained in the first edition of
Administrative Behavior, it is an "indispensable condition" of scientific
analysis "that sufficient experimental control be exercised to make possible

[3]James G. March and Herbert A. Simon, *Organizations* (New York: Wiley, 1958).
[4]Ibid.
[5]Ibid.

the isolation of the particular effect under study from other disturbing factors that might be operating on the organization at the same time."[6] In the study of the important aspects of organization this degree of experimental control can seldom if ever be established; unless the effect under study is one that is frequently repeated, "disturbing factors" are sure to appear.[7] That there has so far been virtually no rigorous testing of hypotheses about organizations is not to be explained simply by lack of methodological sophistication or by dullness on the part of researchers.

The poverty of the literature should be enough, one would think, to raise doubts about the worth of such executive development programs as claim to impart knowledge abut organization that is scientific or, in any except a very loose sense, "tested." There are, however, other and more fundamental questions that should be raised about the mentality that produces such courses and makes them popular.

THE SUBJECT MATTER OF A SCIENCE

Before entering upon these questions, it is necessary to clarify what is meant by the science of organization and what it is that an executive does. Obviously any conclusions that may be reached about the place of the science of organization in the training of the executive will depend largely upon the understanding one has of both these matters.

It is by no means clear from the literature what the science of organization is supposed to be about. March and Simon say, although somewhat indirectly, that it is "about organizations" and about "human

[6]Herbert A. Simon, *Administrative Behavior* (New York: Macmillan, 1945), 42.
[7]The writer made this point, along with others, in a review of the second edition of *Administrative Behavior*. The point was illustrated by reference to Simon's own early study on workloads in a public welfare agency. See E. C. Banfield, "The Decision-Making Schema," *Public Administration Review* XVII (Autumn, 1957), especially p. 280. Simon's reply is to be found in the next issue of the same journal; see especially p. 61.
The difficulties are only somewhat less grave if one is willing to settle for what Helmer and Rescher have called "quasi-laws," that is, historical laws that obtain not always but "as a rule," there being forthcoming adequate explanations demonstrating the exceptional characteristics of apparent exceptions. Quasi-laws, however, are dangerously close to what Simon derides as "proverbs." See Olaf Helmer and Nicholas Rescher, "On the Epistemology of the Inexact Sciences," *Management Science* 6 (October, 1959): 29.

behavior in organizations."[8] Whether these are the same thing is problematical, to say the least.[9] In any case, to say that the science of organization is concerned with behavior in organization is circular: it defines
"organization" in terms of "organization." And apart from this, what is
meant by behavior "in" organization? Apparently not "behavior in a role
formally defined as organizational," for this would leave outside the pale
of the theory behavior which is not in an organizational role but is in
some way crucial to the maintenance of, or the success of, an organization[10]
as well as those aspects of the behavior of role incumbents which are
unrelated to their formal roles.

For present purposes these difficulties may be left unresolved. It
will suffice here to divide "behavior in organizations," whatever it may
be, into two analytical categories. The first consists of behavior aimed at
establishing or maintaining the relationships, formal or informal, by which
action is consciously concerted for the attainment of a purpose. This
category of behavior will be called *administrative*. The other category is
a residual one: it includes everything else done *in* organization. Much of

[8]March and Simon, *Administrative Behavior*, 1–11. They say also (p. 2) that they
are interested, from the point of view of the social psychologist, in influences
that impinge upon the individual from his environment and in his response to
them. Organizations, they point out, are an important part of that envioroonment.
   In *Administrative Behavior*, Simon said in one place (p. xlvi), that "decision-
making is the heart of administration" and in another (p. xvi), that the term
organization means "the complex pattern of communications and other relations
in a group of human beings . . . a 'role system'."

[9]Analytically there seem to be at least three distinct subject matters:
   (a) Behavior in organizational roles. This would be an aspect of social structure,
on the same level as behavior in friendship or kinship roles.
   (b) Organizations as actors. This would view the organization as an entity and
would explain its behavior, including its behavior vis-à-vis other actors.
   (c) Behavior concerted for the attainment of purpose. This would explain
behavior (without regard to role), from the standpoint of its significance for a
system of activity being concerted.
   All such subject matters are of course to be sharply distinguished from "behavior likely to have practical consequences for most organizations." A theory telling
typists how to type faster by holding their fingers differently would be of this
sort. It would be very useful to organizations, but it would not belong to the
science of organization.

[10]Barnard regards customers, voters, and others who do not deliberately serve
the organization as "contributors" to the "system of activity" all the same. See
Organization and Management (Cambridge: Harvard University Press, 1948),
Chapter 5.

this other behavior, which will be called *nonadministrative*, has to do with the achievement of the substantive ends of the organization. This will be called *substantive* behavior.

By way of illustration, let us think of an executive in the Department of Agriculture whose job is to estimate corn yields. When this man concerns himself with the relationships by which activity is concerted—for example, when he divides up work with his assistant—he is engaged in administration. His other activities are nonadministrative. Those having to do with estimating yields of corn are substantive.

It will be useful to subdivide the administrative category into two classes. The first of these consists of behavior which, although it may be immediately directed at a particular feature of the system of relationships by which activity is concerted, nevertheless takes into account the state of the system as a whole and indeed deals with the particular feature only in order to bring about a change in the state of the system. This class of behavior will be called *administrative management*. The other class consists of behavior which operates upon particular features of the system without taking the system as such into account. This will be called called performance of *administrative tasks*. When the Department of Agriculture executive acts with a view to "improving our operations" (i.e., changing the state of a subsystem), he is engaged in administrative management. When he fills out a payroll form, he performs an administrative task.

It may be helpful to compare the organization to a system of pipes through which some substances flow—a refinery, for example. In the operation of the refinery some people are mainly concerned with the substances inside the pipes and others are mainly concerned with the arrangement and proper functioning of the pipes themselves. The first category of people is, in the language used here, performing nonadministrative (substantive) acts. The second category is engaged in administration. Those who work on particular pipes and valves without regard to the operation of the whole system of pipes (e.g., those who paint the pipes to keep them from rusting) perform administrative tasks. Those who concern themselves with the pipes as a system are engaged in administrative management.

The organization, however, is both a system of relationships and a system of concrete activities. Behavior having to do with the state of this more inclusive system—that is, with the system of relationships *and* the concrete activities being carried on by virtue of them—will be called *organizational management*. Thus, one who is concerned with both the pipes *and* the substances inside of them—that is, with the operation of the refinery as a business enterprise—is engaged in organizational management.

Employing these distinctions, it is safe to say that most "behavior in organizations" is substantive, or at least nonadministrative, in character. Every employee engages in some administrative behavior, of course. Certain employees work full-time at administrative tasks like personnel classification, the preparation of procedure manuals, and auditing. A few, usually a very few, are almost entirely occupied with administrative management. Usually, however, administrative management is done by the same people who manage the organization; for them, administrative management is only one activity among many and often the least time-consuming or urgent.

The Department of Agriculture man is an *executive* in the usual meaning of the term, perhaps, but he is not mainly, or even largely, concerned with administrative or organizational management. The same is usually true, although perhaps to a lesser extent, of the personnel officer and the budget officer. They supervise the performance of administrative tasks. These may not have any more to do with administrative or organizational management than does estimating corn yields.

The science of organization deals, presumably, with administrative management, and not with the performance of administrative tasks (e.g., personnel classification) except as these may impinge upon, or enter into, administrative management. It does not deal, presumably, with the management of the organization.[11]

Whatever the value of the science may be, then, there is not much to be gained by teaching it to the corn estimator, the personnel classifier, and others who have little or nothing to do with administrative management. To teach these people the science of organization on the grounds that they are "executives" makes about as much sense as would teaching them the theory of kinship systems on the grounds that they have cousins.

[11]Cf. the distinctions among (a) "methods of managerial problem-solving," (b) "organization theory" (the scientific study of "how human beings function in organizations"), (c) "management principles" (generalizations about the best of current management practices), and (d) "human relations" (either rules for dealing with employees or relevant portions of psychology and sociology), in R. A. Gordon and J. E. Howell, *Higher Education for Business* (New York: Columbia University Press, 1959), 179–183.

LIMITS ON THE USE OF SCIENCE

The science of organization, if it will benefit anyone, will benefit the relatively few persons who are largely engaged in administrative management. Its value to these few, however, is very much limited by at least four important circumstances.

First, one who does administrative management must usually deal with a unique case, whereas the science of organization describes regularities that have appeared in large numbers of cases. It is one thing (and, as was pointed out before, usually an impossible thing) to find a large number of cases which are alike with respect to all important factors. It is another thing (assuming the first to have been done) to find a problem situation which corresponds reasonably well to the experimental one. With regard to most problems of administrative management, the science will either have nothing whatever to say (the problem never before having arisen), or it will be very doubtful what significance is to be attached to what it has to say. Even in matters that have been elaborately studied, the situation in which the results are to be applied almost always involves the introduction of factors which were not present when the research was done. After painstaking (and of course futile) efforts to find ten submarine crews alike in all significant respects, certain generalizations were made about "leader behavior."[12] What bearing, if any, do these generalizations have upon the crews of minesweepers? Of destroyers? Of an eleventh submarine? Or, indeed, of the same ten submarines one month later (perhaps after the outbreak of war!)? These are all "new" questions, which must either be made the subject of fresh research or else decided unscientifically—that is, by common sense or guesswork.

As a practical matter, there will be very few cases in which common sense will not recognize glaring disparities between the situation in which the research was done and the one in which it is to be applied. Even when the organizations in question are superficially similar—of about the same size, employing personnel with the same class and ethnic characteristics, organized on the same structural principles, doing the same kind of work, and so on—the differences will almost always be profound. Indeed, the same organization is likely to differ greatly from year to year and even

[12]See Donald T. Campbell, *Leadership and its Effects Upon the Group*, Research Monograph no. 83, Bureau of Business Research, Ohio State University, 1956, esp. pp. 15 and 71.

from month to month. To paraphrase the philosopher, "No organization ever steps into the same river twice."

Second, the task of the executive is to deal with "difficulties."A difficulty is different from a "puzzle," which is deliberately made up and can usually be solved in only one way. It is also different from a "problem," which, although not deliberately made up, can be solved as if it were a puzzle.[13] A difficulty is "coped with," not "solved." The science of organization consists of, or can be formulated as, solutions to puzzles in administrative management. Other puzzles, those having to do with the substantive matters that the organization deals with, are "science which may be useful in dealing with an organization's difficulties."[14] This has nothing to do with "the science of organization."

There is, of course, no *a priori* reason for supposing that a difficulty will turn out to be a problem. It is quite possible that the executive will have to deal with his difficulties not by finding solutions to them but by avoiding them, ignoring them, or getting out of them. Indeed, if he approaches his difficulties "scientifically"—that is, on the assumption that they will turn out upon close examination to be problems—he may incapacitate himself for dealing effectively with them when in fact they cannot be solved but must be got around.

A difficulty may consist of elements some of which are susceptible to solution as problems and some of which are not. If the science of organization (or some other science) is used to deal with those elements of the difficulty which are problems, the difficulty, it may be argued, is thereby made more manageable. Suppose, for example, that a difficulty of administrative management has twenty elements and that the science of organization has something conclusive to say about two or three of them. This is a gain, it may appear, even though not a very large one: the amount of guesswork necessary has been reduced. But whether this

[13]The distinctions and some of the use that is made of them are taken from T. D. Weldon, *The Vocabulary of Politics* (London: Pelican, 1953), 75–83.

[14]This is what operations research is, namely, a collection of rigorous tools of analysis (e.g., statistical decision theory, game theory, linear and dynamic programming, information and communication theory, probability theory, value theory, etc.), that have been found useful in identifying the "puzzles" in what present themselves as concrete "difficulties." It is a method of finding and solving puzzles and thus of dealing with difficulties, not a way of "contributing to science"; in most applications it has little or nothing to do with the "science of organization."

seeming gain is a real one is open to question. Having scientific answers to two or three of the twenty elements of a difficulty may be about as useful as having two or three windows securely fastened in a house which is without doors and lacking an outside wall.

Third, the relevant science may be descriptive rather than recommendative. That is, it may merely describe behavior, not tell what behavior would be necessary to achieve the ends of the organization. Descriptive science can often be turned into recommendative science, but it cannot *always* be: the actual behavior is sometimes so far from the desirable that nothing about the desirable can be learned from it. The executive who wants to know how to budget will not be helped by generalizations, however powerful and well tested, about budgeting that is done stupidly or capriciously.

That most organizations do not have clear and well-ordered objectives makes this limitation all the more severe. In the first edition of *Administrative Behavior*, Simon remarked that it is an "indispensable condition" of scientific analysis (there were two such conditions) "that the objectives of the administrative organization under study be defined in concrete terms so that results, expressed in terms of these objectives may be accurately measured."[15] In practice (as Simon acknowledged in the second edition), it is usually impossible to determine which of two administrative situations is better, because without clearly defined and ordered goals there is no way of knowing how the various diagnostic criteria should be weighted. To show that submarine officers who delegate authority and participate socially in group activities are best liked by their crews is not very helpful unless it is known what difference it makes to the effectiveness of the ship whether the officers are liked or not. This raises the question: What precisely is "effectiveness" in the operation of a submarine? What is the relative weighting to be given the elements that comprise it? If these questions cannot be answered, it is impossible to turn the descriptive proposition into a recommendative one.[16]

[15] Simon, *Administrative Behavior*, 42. See also p. xxxiv and the review and reply referred to in footnote 7.

[16] The study by Campbell cited before begins by trying to relate leadership behavior to ship effectiveness. Effectiveness proves next to impossible to define, however, and so the focus of the study shifts first to "ship effectiveness and morale" (p. 16), and then to "morale," by which is meant "a syndrome of attitudinal contentment" (p. 29). The question of what, if anything, "attitudinal contentment" has to do with ship effectiveness is never raised.

Fourth, as Knight has remarked, "scientific explanations of what is demonstrates that it is inevitable under the given conditions."[17] These conditions, often enough, cannot be changed, or cannot be changed according to plan. Even when they can be, the executive may not be able to bring about the change according to *his* plan. Most of economic theory, for example, is useless to a businessman: it explains what happens, but it does not enable him to make things happen differently. If it is useful in a practical way at all, it is useful to those—legislators, for example— who are in a position to change the structure of the situation. The same is true of scientific knowledge about organizations or about "behavior in organizations." Knowing the laws of his behavior may on occasion be about as useful to an executive as knowing the laws of *his* would be to a man in prison.

WHAT THE EXECUTIVE DOES

Even if these limitations did not exist, the science of organization could tell the executive only a small part of what he needs to know. For, as was said previously, he is usually occupied mainly with other matters than administrative management. The system of relationships by which activity is concerted is only one feature, and often an inconspicuous one, of the whole situation in which he operates. To return to the analogy used before, he is concerned not with the pipes but with the refinery as a business enterprise. At best, then, the science of organization is only one of the sciences he needs, and even if it were as well developed as other sciences there would often be no more reason to train him in it than in some other science.

There is, however, no science at all, either of organization or of anything else, which will help the executive much in performing his most essential and characteristic functions.[18] These are the following:

1. *Making value judgments.* The executive decides what state of affairs the organization will seek to bring about (what its "mission" is), he reduces its ends from higher to lower levels of generality (translating

---

[17]Frank H. Knight, *On the History and Method of Economics* (Chicago: University of Chicago Press, 1956), 143.
[18]Cf. the discussion of the nature of business competence in R. A. Gordon and J. E. Howell, *Higher Education*, Chapter 6.

"purpose" into "goals" and "goals" into "objectives" and "targets"), and he decides the terms upon which competing ends are to be compromised (what is to be given up when something must be given up).

2. *Making moral judgments.* He decides what it is the "duty" or "obligation" of the organization to do, what its "institutional character" ought to be, and what is "fair" and "just" in particular cases. Here he is guided not by the preference system internal to the organization but by the "moral order," that is, what the society thinks "right" or "obligatory."[19]

3. *Making estimates of probability.* He estimates how the future will affect the organization and what the organization should do now in the light of the probable future. In particular, he estimates how opinion (of the public, legislators, customers, foreign statesmen) will respond to this eventuality or that; what changes are likely to occur in the state of the arts; what the strategy of opponents is likely to be; what are the capacities of personnel, either individuals or abstract groups; and which way of doing something is most likely to "work" or to prove efficient. In addition, he must, as Knight has pointed out, estimate the value of his estimates.[20]

4. *Conceiving possibilities.* By the exercise of imaginative insight or intelligence, he gets ideas for the organization; he perceives dangers, devises courses of action, and sees how to deal with difficulties.

The first three of these activities are carried on by anyone who manages an organization. The fourth is not essential, at least in the short run. For convenience here, however, it will be assumed that all belong to the same concrete process, which will be called *judgment.*[21]

[19] For an explication of the difference between a value and a moral judgment, see W. D. Lamont, *The Value Judgement* (Edinburgh: University of Edinburgh, 1956), 11. For the concepts "mission" and "character" of organization, see Philip Selznick, *Leadership in Administration* (Evanston: Northwestern, 1957), 38–42, and Chapter 3.

[20] "A man," Knight says, "may act upon an estimate of the chance that his estimate of the chance of an event is a correct estimate." Frank H. Knight, *Risk, Uncertainty, and Profit* (Boston: Kelley, 1921), 227. Knight's treatment of the psychology of probability estimates is extremely apposite.

[21] It is a serious limitation of this discussion that it deals only with the cognitive aspects of what the executive does. Other aspects may be even more important. Barnard stresses (in *Organization and Management*) the importance of a high level of vitality, of decisiveness, and of responsibility. To these may be added the ability to act on the basis of organizational rather than personal or other inappropriate ends and the ability to inspire morale. In some of these matters training may make a difference.

To a large extent, executives make judgments not about concrete matters but about the capacity of subordinates to deal with these matters. "The fundamental fact of organized activity," Knight has written,

> is the tendency to transform the uncertainties of human opinion and action into measurable probabilities by forming an approximate evaluation of the judgment and capacity of the man. The ability to judge men in relation to the problems they are to deal with, and the power to "inspire" them to efficiency in judging other men and things, are the essential characteristics of the executive.[22]

The executive, it will be seen, has much in common with the politician or statesman. If politics is, as Oakeshott says, "the activity of attending to the general arrangements of a set of people whom chance or choice has brought together,"[23] then the management of an organization is the activity of attending to the general arrangements of a set of people whose behavior is consciously concerted for the attainment of a purpose. The executive is a little statesman—a statesman whose "nation" is the organization. His specialty, exercising good judgment in attending to its general arrangements, is from an analytical standpoint the same as the statesman's. In the case of large organizations, or perhaps rather of organizations in which large numbers of people are pervasively and profoundly involved, it may be practically very similar as well. Accordingly, much of what may be said about the training of the executive may be said also about that of the statesman, and vice versa.

TECHNICAL KNOWLEDGE VERSUS ART

The science of organization, and science in general, has nothing to contribute to what have been called the most essential and characteristic functions of the executive. Science cannot tell him how to make a value judgment or a moral judgment; there is not even a logic of casuistry by which he can derive particular valuations from axiological principles. In matters that are subject to statistical treatment—that is, that involve many instances of the same thing—science has a great deal to say about probability. But, alas, the matters the executive is called upon to decide are almost never of this kind. He must make probability judgments about unique events, and in this science has no help to offer.

[22]Knight, *Risk, Uncertainty, and Profit*, 311.
[23]Oakeshott, "Political Education," 2.

There are many ways in which research can stimulate and guide the imagination of the executive. But there is no science or technique that can take the place of the flash of insight by which he gets an idea or sees the solution to a problem. Nor can any objective test take the place of subjective ones in estimating the capacity of a man. As Knight has written,

> We form our opinions of the value of men's opinions and powers through an intuitive faculty of judging personality, with relatively little reference to observation of their actual performance in dealing with the kind of problems we are to set them at. Of course we use this sort of direct evidence as far as possible, but that is usually not very far. The final decision comes as near to intuition as we can well imagine; it constitutes an immediate perception of relations, as mysterious as reading another person's thoughts or emotions from subtle changes in the lines of his face.[24]

It cannot be concluded, however, that judgment is not based upon knowledge of some kind, or that men cannot be trained to exercise good judgment or, at any rate, better judgment.

To discuss these matters usefully, it is necessary to distinguish radically between two kinds of knowledge. *Technical knowledge* can be communicated in the form of propositions. *Art* (or "skill") cannot. Art can be communicated—*imparted* is a better word—but not by propositions or, more precisely, not by what propositions are expressly meant to convey. Art appears as "knack," "flair," "feel," or "innate ability"; it is learned by doing, and especially by imitation, not from books and lectures.[25]

The distinction is analytical. A concrete piece of knowledge may be a mixture of technical knowledge and art. Swimming, for example, is mostly art; there is a little technical knowledge that will help the swimmer ("kick your legs hard while holding your feet this way"), but no one ever learned how to swim from it alone. The same is true of riding a bicycle. Good cooking is also largely a matter of art, although cookbooks are very useful. Teaching is to a great extent art, and so is research, including research in the exact sciences.[26]

Our society exhibits a general bias toward treating every problem as if its solution were to be supplied mainly, or entirely, by technical

[24]Knight, *Risk, Uncertainty, and Profit*, 293. See also Knight's discussion of the mental processes by which decisions are made. Ibid., 211.

[25]For an elaborate inquiry into these matters, see M. Polanyi, *Personal Knowledge* (Chicago: University of Chicago Press, 1958), especially Chapter 4.

[26]Polanyi's remarks on the place of "personal knowledge" in exact science are particularly interesting.

knowledge rather than by art. We have how-to-do-it books on everything (including how to win friends, how to be a successful parent, and how to be happy), and we are apt to derogate, or to regard with suspicion, any individual or profession that admits reliance on intuition, skill, or "feel of the situation." Most of us would have more confidence in a gardener who had taken a course in horticulture than in one with a "green thumb," more in a mother with a Ph.D. in child psychology than in one with a maternal instinct, and more in diagnosis by an electronic computer than by a skillful physician.[27]

Our distrust of art is to be explained in part, perhaps, by ample evidence that "wisdom," meaning here the account that practitioners of an art give of how they achieve their results, proves to be a tissue of nonsense and of untruths when subjected to rigorous analysis and experiment. (When the politician, for example, tries to explain the basis of his decisions he makes a very poor showing compared to the engineer or planner.) Proving "wisdom" to be wrong does not, however, prove anything at all about the art it claims to explain. For it is in the nature of the art that no one, not even the skillful practitioner, can explain how it is done, in the sense of offering rules which are sufficient for the doing of it.

This misunderstanding of the difference between art and wisdom does not entirely account, however, for our persistent search for ways of doing by technical knowledge what has hitherto been done by art. Obviously there are enormous advantages in using technical knowledge. An activity which is solely a matter of technical knowledge (assuming for a moment there are such) can be taught relatively quickly to anyone who can understand language. Art, on the other hand, is always the monopoly of a few.[28] Those who possess the technical knowledge may, moreover,

[27]The British do not share our bias in this, perhaps because their elite has been educated in the classics. Recently the *Times* commented: "There is no doubt that many of the problems of tomorrow in both industry and political life will involve scientific considerations. There will have to be men with the ability to assess them and to give advice on the consequences that will follow one course of action or another. But when it comes to the taking of ultimate decisions then quite a different quality is called for. That is wisdom." "In the final analysis," the *Times* concluded in a tone very characteristically British, "cleverness will never count for so much as character." 23 Oct. 1959.
[28]The American preference for the technician to the man of judgment is to be understood partly as an expression of our general taste for decentralization. Leaving matters to be decided in a mysterious, nonpublic way by men of

be treated as interchangeable parts: any one of them can do as well as any other, and so their performance is predictable and subject to standardization.

Clearly there is much to be said for trying hard to replace art with technical knowledge. In organizational management, for example, it is advantageous to rely as little as possible upon the executive's art of judgment and as much as possible upon technical information. Judgment is a scarce commodity—one of the scarcest—and, moreover, it is at best highly uncertain in its operation.

It must be acknowledged, however, that there are dangers in this. We may push too hard. We may make the mistake of treating as the same, or as equivalent, things that are really very different. If what the psychologist tests is what the person with a "knack for judging men" senses (or if, perchance, it is something still more relevant) then obviously we ought by all means to employ the test rather than the knack. The danger is that the function performed by technical knowledge may without our realizing it be different than and inferior to that performed by art. One can get a "do-it-yourself" painting kit and by following instructions—"use paint A on all spaces marked A"—paint a picture. But it would be ridiculous to suppose that "painting" done in this fashion has anything in common with "painting" as done by an artist. It is no less ridiculous to suppose that an application of technical knowledge in organization is the equivalent of art merely because it deals with what by the conventions of language is called the same problem.

AN ILLUSTRATIVE CASE

Before reaching some conclusions, it may be well to draw together and illustrate the principal points that have been made so far. This may conveniently be done with a brief case study—one which, appropriately enough, was put before the writer when he asked a class of state commissioners of mental health and superintendents of state hospitals for the mentally ill (they were undergoing executive development at his hands!)

experience and superior gifts is "undemocratic," whereas deciding everything on the basis of "facts" and "logic" is "democratic." George Kennan in an article on the organization of external relations remarks that "the present system is based, throughout, on what appears to be a conscious striving for maximum fragmentation and diffusion of power." "America's Administrative Response to Its World Problems," *Daedalus* (Spring 1958): 13.

for examples of the difficulties they would face when they returned to their offices.

The case is as follows:

A certain state hospital was built many years ago on such a plan that the employees lived within its walls in close proximity to the patients. Not only were the employees housed at state expense but their family laundry was done in the institution's plant without charge. A few years ago the superintendent of the hospital became convinced that it would be best for all if the employees lived outside the institution in normal community surroundings. He was able to compensate them for the loss of the housing, and the change was made with no great difficulty. Recently he decided that the laundry privilege should also be ended: patients in dirty clothes were washing the laundry of the employees, and this, he felt, was not only unjust but gave both the employees and the patients the impression (a very bad one for the mental health of the patients, he felt) that the institution was run for the employees rather than for the patients. Unfortunately it was not possible to compensate the employees for the loss of the laundry privilege. The superintendent knew that they would complain bitterly—and with justice—that free laundry had been one of the terms of their employment. Some might quit, and others would "lie down on the job." There might even be a strike. Whatever happened, the mood of the employees would probably have some effect upon the health of the patients.

The superintendent had to decide what to do. Should he risk creating bad feeling among the employees, and thus perhaps worsening the health of the patients, by withdrawing the laundry privilege? If he withdrew it, how should he do it? What should be the timing of his announcement? What explanation should he give the employees? And if there was a strike, how should he deal with it?

This is exactly the kind of thing which makes up the daily life of a top executive. It should therefore offer a fair test of the application of the argument that has been made here.

Clearly the science of organization could not, even if it were much more highly developed than it is, tell the superintendent much that would help him. His difficulties—and it should be noted that they were *difficulties*, not *problems*—were not entirely in the field of administrative management: maintaining a system of relationships was only one aspect of the larger problem of securing a successful result (i.e., the improvement in the health of the patients). But even in those aspects of the matter involving other "behavior in organization," the science of organization could say little. There was no theory, tested or otherwise, that applied to the situation. Even if there had been a large body of rigorously tested knowledge about how mental hospital employees behave when laundry privileges are withdrawn (an unlikely possibility, to say the least), the superintendent would have had to decide—by guesswork, of course— whether particular personalities and other special circumstances made his institution a special case. Very likely he would decide that they did.

Even if he decided that his institution was perfectly typical, he would be left with most of his difficulties. He would still have to make crucial value and moral judgments (how much weight should be given to the "rights of employees" as against the "welfare of patients"?) and probability judgments (how much improvement in the health of the patients would accrue from changing the character of the hospital?). If he was to think of a way of getting around his difficulties (or, for that matter, of solving them, if it be assumed that they constituted a problem), it would have to be by "getting an idea," not by systematic fact gathering or by the application of a technique.

Some will suppose that the superintendent would benefit from knowing such social science as has to do with small-group behavior, leadership, power structure, and so on. He might, for example, learn from such studies that "democratic" leaders are most successful, and on this principle "encourage" the employees to elect a study committee to analyze the problem and make recommendations to him.[29]

If the superintendent were a man from Mars—someone with no previous experience whatever of our society and its ways and accordingly having no basis whatever for forming judgments—such textbook rules would indeed be helpful. There may in fact be some executives who are not much better equipped than men from Mars to exercise judgment. Most executives, however, do well to rely heavily upon judgment. For them the textbook rules are more of a hazard than a help. If their confidence in the use of judgment is undermined, they may throw up a smokescreen of "research" under cover of which to escape their responsibilities, and this is likely to be disastrous.

---

[29]This was the suggestion of a behavioral scientist to whom the superintendent's problem was put. For an account of what happened when this advice was acted upon in very similar circumstances, see E. C. Banfield, *Government Project* (Glencoe: Free Press, 1951), esp. 178–81.

TRAINING IN THE ART

The key question, then, is: By what training can the executive learn the art of sound judgment?

It is not certain that he can learn it at all. Like other aspects of intelligence, sound judgment may be one of the givens of the individual's personality and perhaps even of his biological and cultural heredity. Assuming, however, that some improvement is possible, it seems likely that once the individual has left childhood the range within which it can take place is narrow and fixed.

A necessary but not a sufficient condition of sound judgment is possession of a suitable stock of *background* knowledge. This may include scientific and other organized knowledge, but it is likely to be largely—for many executives mainly—what Hayek has called "knowledge of particular circumstances of time and place,"[30] that is, knowledge inherently unsusceptible of being centralized and therefore not found in books. Whatever the nature of its component elements, background information is, in the words of Helmer and Rescher (from whom the concept is borrowed), vague in extent, indefinite in content, and deficient in explicit articulation.

> One is unable to set down in sentential form everything that would have to be included in a full characterization of one's knowledge about a familiar room; and the same applies equally, if not more so, to a political expert's attempt to state all he knows that might be relevant to a question such as, for example, that of U.S. recognition of Communist China.[31]

Possession of such knowledge, indispensable as it is to good judgment, is no guarantee of it. The executive must have also a mysterious faculty—there is no other way of describing it—which enables him to draw correct conclusions more often than does the "average" person. To the extent that this skill can be learned or developed, it must be learned or developed as other skills are—by doing, and especially by imitation. The executive, if he is to learn at all, must learn in much the same way as the swimmer learns to swim and the cyclist learns to ride. He must

[30]F. A. Hayek, *Individualism and the Economic Order* (Chicago: University of Chicago Press, 1948), 77–91.
[31]Helmer and Rescher, "On the Epistemology," 39.

be put in a position where he has to make judgments and to bear responsibility for them. He will no doubt benefit from serving an apprenticeship in intimate association with a master craftsman in the art of judgment, but the benefit will come not from learning the craftsman's wisdom (or at least not from the wisdom taken at its face value, for that is likely to be wrong) but from acquiring his "feel of the situation." To do this he must have respect for the master's skill (not for his wisdom; politeness will suffice for that), and he must have sufficient humility to accept the discipline, incomprehensible as it may be, which the master's way imposes upon him.

The exercise of good judgment seems to involve both generalized skills (a manner of thinking) and specialized ones (a manner of thinking about certain subject matter). One may have the generalized skill and yet radically lack the needed specialized one. Thus, for example, a businessman who enters government service leaves behind him his specialized skill for making judgments in business matters while retaining a generalized skill in making judgments; if he remains in government service long enough and if he subjects himself to suitable training, he will, presumably, acquire a set of specialized skills appropriate to his new environment.

What is needed in the training of judgment, this suggests, is not only background knowledge and experience (the two are not the same) of affairs *in general* but also background knowledge and experience in the *specific realm of affairs in which judgment is to be exercised.* To the extent that this realm is a technical one, the executive will of course require technical knowledge. It is essential that the head of a missiles program know some physics. It would be easy, however, to overestimate the amount of it that he should know. All that is really necessary is that he be able to assess correctly the technical competence of his subordinates. Even if he were better qualified than they to make technical decisions, he might not make them anyway.

The specialized subject matter in which the executive most characteristically requires skill is not technical. Rather it is, in Oakeshott's phrase, "a concrete manner of behavior."[32] The executive must have profound knowledge of the ways of his particular tribe—its language (not merely its vocabulary and rules of grammar but also, and especially, its

[32]Oakeshott, "Political Education," 16–21.

nuances of gesture and intonation), customs, modes of thought, and mystique. To know well the concrete manner of behavior of his tribe, he may,
of course, have to have scientific or other specialized training, for this
may be an integral part of that manner of behavior.[33] This is very different,
however, from being trained to use technical knowledge as the basis of
decision.

It would be foolish to assert, with regard to those fields of activity
in which scientific or technical knowledge plays no great part, that the
executive can get nothing from formal training. History, biography, jour-
nalism, novels—these are all ways of learning the ways of one's tribe,
even when, as is almost always the case, they are about the ways of other
tribes. But it is clear that for the most part one acquires skill in a concrete
manner of behaving by associating intimately with others who behave in
that manner and by being subject to their discipline while behaving in
that manner oneself. And it is even clearer that formal training, especially
when it is in subject matter alien to the concrete manner of behaving (as
social science, even at its best, is alien to concrete behavior "in" organization), may only succeed in instilling a "trained incapacity" for suitable
behavior.[34]

---

[33]Liberal education is sometimes recommended for the executive. There is no
denying that he ought to read the great books, but he ought to read them for
the same reason that others ought to read them, namely that intellectual life is
intrinsically worthwhile, not to make himself a better executive.

Very likely some business support of liberal education for executives is a
public relations gesture intended to give the organization an air of elegance and
opulence. But it may also be that the conduct of some businesses is coming to
depend more and more upon their executives' understanding of the "concrete
manner of behavior" of people who are liberally educated. In other words, the
executive must understand the ways of this particular tribe. For suggestions of
this, See Peter E. Siegle, *New Directions in Liberal Education for Executives*
(Chicago: Center for the Study of Liberal Education for Adults, 1958).

By the same token, if—God forbid!—the language of social science, or of the
theory of organization, entered into the ways of thinking of people "in" organizations, then executives would have to learn that language in order to understand the others and make themselves understood.

[34]The view of Chester I. Barnard, an executive of great experience who is himself
one of the principal theorists of organization, are of special interest in this
connection. He remarks that intellectual training "by itself" tends to check
propensities indispensable to leadership (e.g., decisiveness). *Organization and
Management*, 102–106.

THE TRAINING OF THE EXECUTIVE

TYPES OF RESEARCH AND WRITING

It is convenient to distinguish four kinds of research and writing about organization, each of which has characteristic uses and limitations in the training of executives:

1. The theoretical science of organization (whether of "behavior in organizations" or "behavior by which action is concerted for the attainment of a purpose"); the object of science is to enable the making of more definite and certain predictive statements.
2. Operations research, a method of "dealing with difficulties" rather than of "contributing to knowledge"; by the use of rigorous and usually quantitative methods, it identifies puzzles in what appear to the organization as difficulties, and then solves the puzzles.
3. Principles of management, which may be accounts of how typical difficulties have been dealt with on a common sense basis and low-level generalizations, or quasi-laws, based on common sense observation; the scope of a "principle" is not clearly formulated, and no explanation is given to account for the cases in which it will not work.
4. Common sense descriptive accounts of the structure and functioning of organizations as concrete entities. These set forth the rules, formal, and informal, by which behavior in organization is regulated and often show the interaction between the form of the organization, and the substance of its policy or work. The case study, usually an account of the way an organization has met a particular difficulty or made a decision, belongs to this type.

These kinds of research and writing are addressed to three types of audience: social scientists, laymen who want general information about organization because it is an interesting aspect of the world they live in, and the executive.

Naturally a writer finds it tempting to try to do more than one thing, and address more than one audience at a time. The temptation should be resisted, however; except in rare instances, nothing is gained and something is lost by not specializing.

The theoretical science of organization can contribute little to common sense understanding of "behavior in organization," which is what both layman and executive mainly want. Science must employ categories radically foreign to common sense, whereas "understanding" must employ the categories used by those whose behavior is to be understood, namely, the categories of common sense.[35]

Sociologists do not expect books on role patterning within the family to be read as handbooks by parents. Neither should the theorist of organization expect his book to be read by the executive (unless, perchance, his "theory" is not theory at all but only common sense—or common nonsense—dressed up in jargon to look like it).

The science of organization, it may be added, can as a rule learn nothing from accounts of organization that employ common sense, rather than scientific, concepts and theories.

Operations research cannot usefully be combined with the other kinds of research and writing listed before because (among other reasons) it is a way of coping with particular difficulties rather than of organizing knowledge for general use. To the extent that the science of organization succeeds, it will offer solutions to certain puzzles, and so will be among the operations analyst's stock of ready-made solutions. An executive needs to know enough about operations research to judge whether an activity presents a difficulty (it may be there without his seeing it, of course) and, if it does, to judge whether the difficulty is susceptible to being solved as a problem by the methods of operations research. This, of course, does not require that he have even an amateur's knowledge of the methods themselves.

Principles of management are not likely to be derived from any of the other three kinds of research or to be improved by being brought in contact with them. Principles, in this sense, are of no interest to the scientist, and it may be doubted whether they are of any value to the executive. Helmer and Rescher suggest that an expert, that is, one who has a large amount of background knowledge, can use quasi-laws to make relatively good predictive judgments.[36] At the lowest level of generality, however, where the scope of the principle is least well defined, there is probably not much value in writing down "tips," "helpful hints," and "rules of thumb"; these may best be "picked up" through association with other practitioners. At the highest level of generality, on the other hand, where

---

[35]See F. A. Hayek, *The Counter Revolution of Science* (Glencoe: Free Press, 1952), 50.

[36]Helmer and Rescher, "On the Epistemology," 43.

quasi-laws most nearly approach true ones, there is the same danger that attends the use of true science: the executive may accord the principle more respect than it deserves, or he may even substitute the application of principles for the exercise of judgment.

Descriptive studies of organization are doubtless of great value in conveying "general information" to the layman. They are not, however, of much value to the executive. He may get from them some background knowledge, perhaps, but background knowledge *about organizations* is usually a small part of what he needs. So far as he needs knowledge about organization at all, he needs knowledge about his particular organization. And of this he needs much more than could possibly be provided in a work for general distribution. Even in a very large organization, the knowledge he needs can most economically be provided by word-of-mouth instructions from fellow employees and by manuals prepared in the organization for training purposes.[37]

It may be doubted whether any kind of writing about organization will help the executive get the knowledge of a "concrete manner of behavior" that he needs for skill in judgment. It may be that vivid, intimate description of the informal aspects of organization—how decisions are "really" made—will give him some of the "feel" that he needs. If so, case studies are probably the most apppropriate form. These should deal as nearly as possible with the special subject matter and milieu of the organization in which the executive is to act. Case studies written from within the organization by persons who know it well would be best. Needless to say, as a practical matter organizations are not likely to assign valuable employees to writing such studies, or even to permit the telling "within the family" of the whole truth. Doing so would destroy illusions necessary

[37]A case in point is Herbert Kaufman, *The Forest Ranger* (Baltimore: John Hopkins, 1959). Although this has the subtitle, "A Study in Administrative Behavior" and includes references to several sociological studies, it is not, by any stretch of the imagination, a contribution to the scientific literature. Moreover, the author's predilection for the jargon of social science leads him on the false scent of "administrative behavior" and away from a common sense treatment of what to the lay reader is the most interesting question: namely How can such a fantastic degree of centralization be made to work and what are its real costs?

The book can be of little or no use in training forest rangers. Vastly more detail would be required on every point. As Kaufman notes, the Forest Service's procedure manual runs to seven volumes.

to the maintenance of the organization. And it would be wasteful as well, for the amount executives could learn from the cases would be small in proportion to the cost of preparing them.[38]

At best the case study method cannot train the executive in the art of judgment. For, even assuming that the case writer concentrates on the behavior of skillful actors, how can he convey what it is that the skill consists of? When he displays a detail or an anecdote, the case writer in effect says "this was an element contributing to the rightness of the judgment." In this there is implicit a rule for judging soundly. But art or skill cannot be specified, not even by indirection. The case writer does not know of what the skill consists. Nor does the skillful actor himself. How, then, is the case writer to select details that "show," even if they do not "tell," how to act skillfully?

The kind of research and writing which may be most useful in the training of executives is not about organization at all, and so was not listed among the categories mentioned previously. This is that which provides background knowledge. What sort of background knowledge the executive needs and therefore how practicable it may be to provide it by formal instruction depends, of course, upon the nature of the substantive matter with which he deals.

It is worth noting in passing, perhaps, that individual scholars, research and teaching institutions, and journals treating the subject of organization have in most cases sorted themselves out into one or another of the categories listed.[39] This has occurred, one suspects, not so much because scholars have appreciated the fundamental differences in what they are doing as by unconscious response to pressures arising from the inherent logic of these differences.

---

[38]For an example of case study writing of a kind that might be valuable in the training of one class of executives, see the volumes by George S. Kennan on American-Soviet relations at the end of the first World War. These give the reader a "feel of the situation." But note that the situation is one tht existed 40 years ago: possibly the feel of that situation is the wrong one for the executive who acts in the same realm of affairs today. Note also that men of Kennan's experience and capacity seldom have leisure to write such books.

See the discussion of the uses and limitations of the case study method, with many bibliographical references, in R. A. Gordon and J. E. Howell, *Higher Education for Business*, 368–72.

[39]The Graduate School of Public Administration at Harvard, for example, specializes in providing background knowledge pertinent to policy problems.

DANGERS OF UNDERVALUING ART

Some may suppose that in time the advance of science and technology, and particularly the use of electronic computers, will displace the executive and the arts of judgment almost altogether. This possibility need not be taken seriously. To be sure, many decisions now made on the basis of judgment will in the near future be made by the application of technical rules. But this, far from leaving the executive technologically unemployed or underemployed, will surely require the making of more and more judgments. For just as the building of freeways to relieve traffic congestion in a city has the long-run effect of enlarging the city and of encouraging more people to enter it, so the removal of some matters from executive judgment will open the way for a larger number of other matters to come before him.

In the general bias against art and in favor of technical knowledge, a bias which expresses itself in distrust of those, like the executive and the politician, who depend upon intuition and "feel of the situation" rather than upon rigorous or scientific procedures, there lie several grave dangers. First, there is the danger of loss of efficiency from the use of technical knowledge as a substitute for art in matters in which it is not really a fit substitute. If, for example, some men have a remarkable talent for judging the capacities of other men, it is a costly mistake to rely upon poor objective psychological tests rather than upon this talent. Losses of this kind are likely to go unsuspected, for no one knows how well an undertaking would have turned out if art rather than technical knowledge *had* been employed, and it is easy to take for granted, our enthusiasm for science being what it is, that the least bit of technical knowledge is better than any amount of art.

A greater danger, perhaps, is loss of confidence in the art of judgment, loss of confidence on the part of those who exercise judgment and also on the part of those who depend upon its exercise. This loss occurs not in some few particular spheres (where, perhaps, confidence was never justified at all or where art had long since been superseded by an appropriate use of technical knowledge) but in all spheres at once: in those where technical knowledge is least able to supply the deficiency as well as those in which it is most able. This leads inevitably to the growth of wasteful and dishonest rituals by which those who have responsibility for judgment conceal from themselves and from their publics the inevitability of relying on judgment. In these circumstances matters are likely to be decided by default because the decision makers, having no confidence themselves in their judgment or knowing that their publics respect only technical knowledge, let matters turn upon considerations that have nothing to recommend them as a basis for decision save that they are susceptible to technical treatment.

There is danger also that undue confidence in the possibility of deciding matters upon technical grounds will lead to an endless elaboration of bureaucratic resources for fact gathering and analysis, an elaboration which may of itself render the exercise of judgment impossible.[40]

We need to have more respect for the art of judgment and more willingness to rely on it. It is not a very good way of deciding matters, but, where it is appropriate at all, it is the best that we have. One of the great needs of our day is to form a correct collective judgment (and it will have to be a judgment!) of the uses and limitations of technical knowledge and of art, both in general and in particular types of situations and subject matters. Another great need of our day is to train executives so that they can distinguish what can be decided on technical grounds from what must be decided by the exercise of judgment. Still another great need is to create in all our institutions a workable association between the elements which assemble and use technical knowledge and those which, making what use they can of the first, employ the arts of judgment.

[40]Kennan has questioned whether "massive organization can be useful at all" in the essentially intellectual process of synthesis and evaluation of information available to the government in the field of external relations; "—whether, in fact, the attempt to solve problems in this field by large-scale organization does not rest on certain basic misunderstandings as to what can and cannot be accomplished by the working together of many people." *Daedalus*, "America's Administrative Response," 22.

PART **IV**

The Capacities of Local Government

This section views the performance of local political systems from a variety of perspectives. The first two essays try to show how the effectiveness of local government and the rate and character of urban growth are affected by the decentralization of the American political system. The remaining essays offer different and, some readers may think mutually irreconcilable accounts of the role that local government, and indeed government in general, can play in alleviating the so-called urban crisis.

Although in the past decade the proportion of Americans living in urban areas has leveled off at about 75 percent and the largest metropolitan areas of the Northeast and Upper Midwest have either lost population or gained very little, most American cities will continue to grow and—depending upon immigration—some of them may grow very fast. Will local governments be able to cope with the stresses and strains that growth entails? The argument of the first essay is that the answer depends not on numbers of people but rather upon the ratio between the burdens they place upon the political system and the capacity of the system to bear those burdens. Comparison of American with British experience suggests that the ratios are becoming more nearly the same in the two countries, although from opposite causes: in Britain because burdens are increasing relative to capacity and in this country because capacity is increasing relative to burdens.

"The City and the Revolutionary Tradition" was one of a series of bicentennial lectures given in 1976 under the auspices of the American Enterprise Institute for Public Policy Research. Here comparison with (nonrevolutionary) Canada is used to support the contention that the tradition inspired by the Revolution, one of direct democracy and free

209

enterprise, was largely responsible for the extraordinary growth of American cities. What Woodrow Wilson considered a defect of the American system, namely that the public thought that it should have a hand in everything, was a source of enormous and unprecedented energy.

The third essay of this section, "A Critical View of the Urban Crisis," contends that the "urban crisis" is not specifically urban and that there is very little that government can do about it. The social pathologies by which "the urban crisis" is usually defined—multiproblem families, functional illiteracy, crime and drug abuse, and so forth—exist in rural areas as well and therefore cannot be explained in terms of factors, such as overcrowding and the flight of the middle class to the suburbs, which are peculiar to the cities. The crisis, it is maintained, is the result of attitudinal changes, especially with respect to authority, self, hedonism, and egalitarianism, which are pervasive in American culture. These changes, which were set in motion by the ideas of philosophers of the God-is-dead variety, are not likely to be rendered nugatory by exhortations to return to old ways or by government programs intended to maintain social order by the giving of grants or the making of regulations.

The final essay finds government, local and other, to be one of the principal contributing causes of the presence in the inner cities of a large and growing class of disadvantaged persons. The main disadvantage of many of these people, it is argued, consists of barriers to upward mobility that have been placed in their way by government regulations imposed in response to the pressures of coalitions of self-serving interests and well-meaning but mistaken reformers. The minimum wage, for example, is dear to the hearts of labor unions (whose well-paid members produce the things that are substitutes for low-skilled, low-value labor) and by many people who suppose that the law will benefit the poor. In fact the minimum wage is a cause of unemployment among the lowest-skill workers, especially teenage blacks. The Reagan administration's proposed "enterprise zones" would not improve matters significantly in the inner cities, largely because it is politically impossible to repeal the many laws and regulations that in the name of worker protection constitute handicaps.

In this context one sees that the mistaken simplicity of the reform-minded middle class may create a grave and enduring political problem for the society. If, as seems likely, the barriers to upward mobility by the disadvantaged are built ever higher, the inner cities will eventually contain a permanent underclass of such a size and character as to menace the consensual basis of democratic society.

CHAPTER 11

The Political Implications of Metropolitan Growth

The rapid growth of the metropolitan populations will not necessarily have much political effect. To be sure, many new facilities, especially schools, highways, and water supply and sewage disposal systems, will have to be built and much private activity will have to be regulated. But such things do not necessarily have anything to do with politics: the laying of a sewer pipe by a "public" body may involve the same kinds of behavior as the manufacture of the pipe by a "private" one. Difficulties that are "political" arise (and they may arise in "private" as well as in "public" undertakings) only insofar as there is conflict—conflict over what the common good requires or between what it requires and what private interests want. The general political situation is affected, therefore, not by changes in population density or in the number and complexity of the needs that government serves ("persons," the human organisms whose noses are counted by census-takers, are not necessarily "political actors") but rather by actions which increase conflict in matters of public importance or make the management of it more difficult. In what follows, such actions will be called "burdens" upon the political system.

In judging how a political system will work over time, increases and decreases in the burdens upon it are obviously extremely relevant. They are not all that must be considered, however. Changes in the "capability"

From *Daedalus*, Journal of the American Academy of Arts and Sciences, vol. 90 (Winter 1960), pp. 61–78. Copyright 1960 by the American Academy of Arts and Sciences. Reprinted by permission.

of a system, that is, in its ability to manage conflict and to impose settlements, are equally relevant. The "effectiveness" of a political system is a
ratio between burdens and capability. Even though the burdens upon it
increase, the effectiveness of a system will also increase if there is a
sufficient accompanying increase in its capability. Similarly, even though
there is an increase in capability, the effectiveness of a system will decrease
if there is a more than commensurate increase in burdens.

In this article an impressionistic account will be given with respect
to two contrasting political systems, the British and the American, of the
burdens metropolitan affairs place upon them and of their changing capabilities. Naturally, the focus of attention will be upon *ratios* of burdens
to capabilities and upon the significance of these ratios for metropolitan
affairs.

THE TASKS OF BRITISH LOCAL GOVERNMENT

Until recently British local government (meaning not only government that is locally controlled but all government that deals with local
affairs) had, by American standards, very little to do. Until three or four
years ago there was little traffic regulation in Britain because there were
few cars (the first few parking meters, all set for two hours, were installed
in London in the summer of 1958). Now all of a sudden there are 5,500,000
cars—more per mile of road than in any other country—and the number
is increasing by a net of 1,500 per day; by 1975 there are expected to be
13,500,000. Obviously, the need for roads and parking places will be
enormous. But the automobile will create other and graver problems for
local government. When there are enough cars and highways, there will
doubtless be a "flight to the suburbs." The central business districts will
be damaged, and so will mass transit (94 percent of those who now enter
London do so by public transportation) and the green belts.[1]

[1]Dame Evelyn Sharp, permanent secretary, Ministry of Housing and Local Government, recently pointed out that the expected population increase in England
and Wales in the next 15 years (nearly three million) is almost double the increase
on which plans have been based. The number of separate households, moreover,
is growing faster than the number of people. Much of the demand for new
housing, she said, is demand for better and more spacious housing. All this has
increased the pressure on land, especially on the green belt, and particularly
around London. The Government policy, she said, was to encourage the building

Law enforcement has been relatively easy in Britain up to now. The British have not been culturally disposed toward violence or toward the kinds of vice that lead to major crimes. (There are only 450 dope addicts in all of Britain, whereas in Chicago alone there are from 12,000 to 15,000.) British opinion, moreover, has not demanded that some forms of vice be made illegal, much less that vice in general be suppressed. In England adultery is not illegal, and neither is prostitution, although it is illegal to create a nuisance by soliciting. Physicians in England may prescribe dope to addicts. (In the United States, where this is illegal, black-market prices prevail, and the addict must usually resort to crime to support his habit. In Chicago a week's supply of heroin costs at least $105; to realize this much, the addict must steal goods worth about $315. According to the estimate of a criminal court judge, about $50 million worth of goods is shoplifted every year in the central business district of Chicago by addicts.[2]) Never having tried to suppress drinking, gambling, or prostitution, the British have no organized crime.

The task of law enforcement is also becoming more difficult, however. Dope addiction, and consequently crimes of violence, will increase with the number of West Indians and others who are not culturally at home in England. In the past year the horde of London prostitutes has been driven underground, where they may prove a powerful force tending toward the corruption of the police.[3] As traffic fines increase in number and amount, the bribery of the police by motorists will also increase. "All Britain's big cities," an *Observer* writer recently said, "now have enclaves of crime where the major masculine trades appear to be pimping and dealing in dubious secondhand cars."[4]

---

of houses for owner occupation, and how to follow this without wrecking the effort to preserve the green belt was one of the most difficult problems facing the planning authorities. She said there were also increasing demands on land by industry, for great new roads, car parking and garaging, and for power. *The Times,* 23 October 1959.

[2] These facts were supplied by Dr. Arnold Abrams of Chicago in a private communication.

[3] The Wolfenden Committee considered this possibility and concluded that the measures it proposed (chiefly to make it easier for police officers to establish "annoyance") justified the risk. Its measures, the committee said, were not "likely to result in markedly increased corruption. There are other fields of crime where the temptation to the police to succumb to bribery is, and will continue to be, much stronger than it is here." *Report of the Committee on Homosexual Offenses and Prostitution,* Cmnd. 247 (September 1957): 96.

[4] "Table Talk," *The Observer,* 15 May 1960.

Even if motorists, dope addicts, and prostitutes do not seriously corrupt it, the police force is bound to deteriorate. The British have had extraordinarily fine policemen, partly because their social system has hitherto offered the working class few better opportunities. As it becomes easier to rise out of the working class, the police force will have to get along with less desirable types. It is significant that the Metropolitan Police are now 3,000 men short.

State-supported schooling, one of the heaviest tasks of local government in the United States, has been a comparatively easy one in Britain. Four out of five British children leave school before the age of sixteen. The British, it is said, are not likely to develop a taste for mass education.[5] They are demanding more and more state-supported schools, however, and no doubt the government will have to do more in this field.

It would be wrong to infer that because of these changes the burden upon the British political system will henceforth be comparable to that upon our own or, indeed, that it will increase at all. Conceivably, the new tasks of local government will have no more political significance than would, say, a doubling of the volume of mail to be carried by the post office. One can imagine, for example, two opposite treatments of the London traffic problem, one of which would solve the problem without creating any burden upon the political system and the other of which would leave the problem unsolved while creating a considerable burden.

Possibility 1. The Ministry of Transport takes jurisdiction over London traffic. Acting on the recommendations of a Royal Commision, the minister declares that the central city will be closed to private automobiles. His decision is acclaimed as wise and fair—"the only thing to do"— by everyone who matters.

Possibility 2. The boroughs retain their control over traffic because the minister is mindful of organized motorists. People feel that it is an outrageous infringement of the rights of Englishmen to charge for parking on the Queen's highway or to fine a motorist without having first served a summons upon him in the traditional manner. Traffic is unregulated, and everyone complains bitterly.

As this suggests, "governmental tasks" are "political burdens" only if public opinion makes them so. What would be an overwhelming burden in one society may not be any burden at all in another. What would not

[5]Sir Geoffrey Crowther, "English and American Education," *The Atlantic* (April 1960).

be a burden upon a particular political system at one time may become one at another. It is essential to inquire, therefore, what changes are occurring in the way such matters are usually viewed in Great Britain and in the United States. The factors that are particularly relevant in this connection include the intensity with which ends are held and asserted; the willingness of actors to make concessions, to subordinate private to public interests, and to accept arbitration; and, finally, the readiness of the voters to back the government in imposing settlements.

THE RELATION OF CITIZEN TO GOVERNMENT

The British have a very different idea from ours of the proper relation between government and citizens. They believe that it is the business of the government to govern. The voter may control the government by giving or withholding consent, but he may not participate in its affairs. The leader of the majority in the London County Council, for example, has ample power to carry into effect what he and his policy committee decide upon; it is taken for granted that he will make use of his power (no one will call him a boss for doing so) and that he will not take advice or tolerate interference from outsiders.

Locally as well as nationally, British government has been in the hands of middle and upper classes. Civil servants, drawn of course entirely from the middle class, have played leading and sometimes dominant roles. Most elected representatives have been middle or upper class. The lower class has not demanded, and apparently has not wanted, to be governed by its own kind or to have what in the United States is called "recognition." Although Labour has controlled the London Country Council since 1934, there have never been in the council any such gaudy representatives of the gutter as, for example, Alderman "Paddy" Bauler of Chicago. The unions have kept people with lower-class attributes, and sometimes people of lower-class origins as well, off the ballot. They would not have done so, of course, if the lower class had had a powerful itch to have its own kind in office. (In that case the unions would themselves have been taken over by the lower class.) As Bagehot said in explaining "deferential democracy," "the numerical majority is ready, is eager to delegate its power of choosing its ruler to a certain select minority."[6]

[6]Walter Bagehot, *The English Constitution*, in *Bagehot's Works* (Hartford, Conn., Travellers Insurance Co., 1889), Vol. 4, pp. 267–268.

The ordinary man's contact with government inspires him with awe and respect. (Is government respected because it pertains to the upper classes, or does causality run the other way, the upper classes being respected because of their association with government?) "The English workingman," an Englishman who read an earlier draft of this article said, "seems to think that the assumption of governmental responsibilities calls for the solemnest of blue suits. They tend to be so overawed by their position as to be silenced by it."

The ethos of governing bodies, then, has been middle or upper class, even when most of their members have been lower class. So has that of the ordinary citizen when, literally or figuratively, he has put on his blue suit to discharge his "governmental responsibilities" at the polls.

Consequently the standards of government have been exclusively those of the middle and upper classes. There has been great concern for fair play, great respect for civil rights, and great attention to public amenities—all matters dear to middle- and upper-class hearts. At the same time there has been entire disregard for the convenience and tastes of the working man. London pubs, for example, are required by law to close from two until six in the afternoon, not, presumably, because no one gets thirsty between those hours or because drinking then creates a special social problem, but merely because the convenience of pub keepers (who would have to remain open if competition were allowed to operate) is placed above that of their customers. Similarly, trains and buses do not leave the center of London after eleven at night, not, presumably, because no one wants to go home later, but because the people who make the rules deem it best for those who cannot afford taxis to get to bed early.

It is not simply class prejudice that accounts for these things. By common consent of the whole society the tastes of the individual count for little against prescriptive rights. When these rights pertain to the body politic—to the Crown, in the mystique—then the tastes of the individual may be disregarded entirely. Public convenience becomes everything, private convenience nothing.

As heirs of this tradition, the British town planners are in a fortunate position. They do not have to justify their schemes by consumers' preferences. It is enough for them to show that "public values" are served, for by common consent any gain in a public value, however small, outweighs any loss of consumers' satisfaction, however large. Millions of acres of land outside of London were taken to make a green belt without anyone's pointing out that workingmen are thus prevented from having small places in the country and that rents in the central city are forced up by the reduction in the supply of land. It is enough that a public amenity is being created (an amenity, incidentally, which can be enjoyed only by those having time and money to go out of London). The planning authorities of the London County Council, to cite another example of the general disregard for consumers' tastes, consider the following questions, among others, when they pass upon an application to erect a structure more than 100 feet in height:

> Would it spoil the skyline of architectural groups or landscapes? Would it have a positive visual or civic significance? Would it relate satisfactorily to open spaces and the Thames? Would its illuminations at night detract from London's night scene?

It is safe to say that the planners do not weigh the value of a gain in "visual significance" against the value of a loss in "consumer satisfaction." In all probability they do not try to discover what preferences the consumer actually has in the matter. Certainly they do not make elaborate market analyses such as are customarily used in the United States in planning not only shopping places but even public buildings.

Green belts and the control of the use of land are only part of a plan of development which includes the creation of a dozen satellite towns, "decanting" the population of the metropolis, and much else. Where these sweeping plans have not been realized, it has not been because of political opposition. There has been virtually no opposition to any of these undertakings. The real estate, mercantile, banking, taxpayer, and labor union interests, which in an American city would kill such schemes before they were started, have not even made gestures of protest. The reason is not that none of them is adversely affected. It is that opposition would be futile.[7]

---

[7] An English friend comments: "I think you underestimate the sensitivity of central government to local or even private pressures. Parliamentary questions and debates, M.P.'s correspondence, lobbying, etc., provide plenty of opportunity for needling Ministers. The difference [between American and British practice] is, I think, that in Britain the government is not necessarily deflected by the pressures although it does its best to placate them. It does *not* ride rough-shod over protests; it lumbers on, writhing under the criticism and dispensing half-baked compromises."

THE DIRECTION OF CHANGE

Obviously, a political system that can do these things can do much else besides. If the relation between government and citizen in the next half century is as it has been in the past, the "governmental tasks" that were spoken of will not prove to be "political burdens" of much weight. One can hardly doubt, for example, which of the two ways of handling London traffic would, on this assumption, be more probable.

There is reason to think, however, that fundamental changes are occurring in the relations between government and citizen. Ordinary people in Britain are entering more into politics, and public opinion is becoming more ebullient, restive, and assertive. The lower class no longer feels exaggerated respect for its betters,[8] and if, as seems reasonable to assume, respect for public institutions and for political things has been in some way causally connected with respect for the governing classes, the ordinary man's attachment to his society may be changing in a very fundamental way. British democracy is still deferential, but it is less so than a generation ago, and before long it may be very little so.

It would not be surprising if the lower class were soon to begin wanting to have its own kind in office. Lower-class leaders would not necessarily be less mindful of the common good and of the principles of fair play than are the present middle- and upper-class ones, however. The ethos of the British lower class may not be as different from that of the other classes as we in America, judging others by ourselves, are likely to imagine.

There is in Britain a tendency to bring the citizen closer to the process of government. Witness, for example, a novel experiment (as the *Times* described it) tried recently by an urban district council. At the conclusion of its monthly meeting, the council invited the members of the public present (there were about twenty) to ask questions. According to the *Times*:[9]

---

[8]Such an incident as the following, which is supposed to have occurred about the time of the First World War, would be inconceivable today: Hulme [the poet] was making water in Soho Square in broad daylight when a policeman came up. "You can't do that here." Hulme: "Do you realize you're addressing a member of the middle class?" at which the policeman murmured, "Beg pardon, sir," and went on his beat. Christopher Hassall, *Edward Marsh, Patron of the Arts: A Biography* (London: Longmans, Green & Co., 1959), 187.

[9]The *Times*, 24 November 1959.

The Council, having decided to cast themselves into the arms of the electorate, had obviously given some thought to how they could extricate themselves if the hug became an uncomfortable squeeze. The chairman, after expressing the hope that the experiment would be successful, suggested a few rules. It was undesirable, he said, that such a meeting should become an ordinary debate with members of the public debating with members of the council and perhaps members of the council debating with each other. He decreed that the public should be restricted to questions on policy or factual information. He finished the preliminaries by saying that if things got out of hand he would rise and would then expect all further discussion to cease.

This last precaution proved to be unnecessary. The public were pertinent, probing, and shrewd in their questions, but content to observe the proprieties. The more vexed of domestic questions of Nantwich (the demolition of old property, road repairs, housing, and the like) were thrown down quickly and in every case received reasoned replies. The atmosphere of the chamber continued to be one of high good humor.

Carried far enough, this kind of thing would lead to the radical weakening of government. (There is no use giving people information unless you are going to listen to their opinions. And if you do that, you are in trouble, for their opinions are not likely to be on public grounds, and they are virtually certain to conflict.) The British are not likely to develop a taste for what in American cant is called "grass-roots democracy," however; the habit of leaving things to the government and of holding the government responsible is too deeply ingrained for that. What the public wants it not the privilege of participating in the process of government but, as the Franks Committee said, "openness, fairness, and impartiality" in official proceedings.[10]

The tastes of the ordinary man (consumers' preferences) will be taken more into account in the future than they have been in the past, not because the ordinary man will demand it (he may in time, but he is far from doing so now) but because the ruling elite—an elite that will be more sophisticated in such things than formerly—will think it necessary and desirable. The efforts of the Conservative government to let the market allocate housing are a case in point. These have been motivated, not by desire to deprive the workingman of advantages he has had for half a century (that would be out of the question), but by awareness that

---

[10]*Report of the Committee on Administrative Tribunals and Enquiries*, Cmnd. 218 (July 1958).

people's tastes may be best served in a market. The cherished green belts are now being scrutinized by people who are aware of consumer demand for living space, and some planners are even beginning to wonder if there is not something to be said for the American system of zoning. It is not beyond the bounds of possibility that the British will exchange their system of controls of the use of land, which as it stands allows the planner to impose a positive conception, for something resembling ours, which permits the user of land to do as he pleases so long as he does not violate a rule of law.

The conclusion seems warranted that twenty or thirty years from now, when today's children have become political figures, governmental tasks which would not place much of a burden on the political system may then place a considerable one on it. Governmental tasks like traffic regulation will be more burdensome politically both because there will be insistent pressure to take a wider range of views and interests into account, but also, and perhaps primarily, because the ruling group will have become convinced that the preferences of ordinary people ought to count for a great deal even when "public values" are involved. It is not impossible that the elite may come to attach more importance to the preferences of ordinary people than will the ordinary people themselves.

THE CONTRASTING AMERICAN TRADITION

Local government in the United States presents a sharply contrasting picture. It has been required to do a great deal, and the nature of American institutions and culture has made almost all of its tasks into political burdens.

Although there have always been among us believers in strong central government, our governmental system, as compared ot the British, has been extraordinarily weak and decentralized. This has been particularly true of state and local government. The general idea seems to have been that no one should govern, or failing that, that everyone should govern together. The principle of checks and balances and the division of power, mitigated in the federal government by the great powers of the presidency, were carried to extreme lengths in the cities and states. As little as fifty years ago, most cities were governed by large councils, some of them bicameral, and by mayors who could do little but preside over the councils. There was no such thing as a state administration. Governors were ceremonial figures only, and state governments were mere congeries of independent boards and commissions. Before anything could be done, there had to occur a most elaborate process of give-and-take (often, alas, in the most literal sense) by which bits and pieces of power were gathered up temporarily, almost momentarily.

It was taken for granted that the ordinary citizen had a right—indeed, a sacred duty—to interfere in the day-to-day conduct of public affairs. Whereas in Britain the press and public have been excluded from the deliberations of official bodies, in the United States it has been common practice to require by law that all deliberations take place in meetings open to the public. Whereas in Britain the electorate is never given an opportunity to pass upon particular projects by vote, in the United States is usually is. In Los Angeles, according to James Q. Wilson,

> The strategy of political conflict is more often than not based upon the assumption that the crucial decision will be made not by the City Council of Los Angeles, the Board of Supervisors of the county, or the legislature of the state, but by the voters in a referendum election."[11]

Los Angeles is an extreme case, but the general practice of American cities, a practice required by law in many of them, is to get the voters' approval of major expenditures. The New York City government, one of the strongest, is now having to choose between building schools and making other necessary capital expenditures; it cannot do both because the voters of the state have refused to lift the constitutional limit on debt. Such a thing could not happen in London; there all such decisions are made by the authorities, *none of whom is elected at large.*

The government of American cities has for a century been almost entirely in the hands of the working class.[12] This class, moreover, has had as its conception of a desirable political system one in which people are "taken care of" with jobs, favors, and protection, and in which class and

---

[11]James Q. Wilson, *A Report on Politics in Los Angeles* (Cambridge: Joint Center for Urban Studies of Massachusetts Institute of Technology and Harvard University, 1959), 1–13.

[12]A couple of generations ago politics was literally the principal form of mass entertainment. See Mayor Curley's account of the Piano-Smashing Contest, Peg-leg Russell, the greased-pig snatch and other such goings-on at Caledonian Grove. When the working class could pay more than twenty-five cents for its all-day family outing, it went to Fenway Park and baseball pushed politics into second place. James M. Curley, *I'd Do It Again!* (New York: Prentice-Hall, 1957), 54–55.

ethnic attributes get "recognition." The idea that there are values, such
as efficiency, which pertain to the community as a whole and to which
the private interests of individuals ought to be subordinated has never
impressed the working-class voter.

The right of the citizen to have his wishes, whether for favors,
"recognition," or something else, served by local government, has been
an aspect of the generally privileged position of the consumer. If the
British theory has been that any gain in public amenity, however small,
is worth any cost in consumer satisfaction, however large, ours has been
the opposite: with us, any gain to the consumer is worth any cost to the
public. What the consumer is not willing to pay for is not much of value
in our eyes. Probably most Americans believe that if the consumer prefers
his automobile to public transportation his taste ought to be respected,
even if it means the destruction of the cities.

We have, indeed, gone far beyond the ideal of admitting everyone
to participation in government and of serving everyone's tastes. We have
made public affairs a game which anyone may play by acting "as if" he
has something at stake, and these make-believe interests become subjects
of political struggle just as if they were real. "The great game of politics"
has for many people a significance of the same sort as, say, the game of
business or the game of social mobility. All, in fact, are parts of one big
game. The local community, as Norton E. Long has maintained in a
brilliant article, may be viewed as an ecology of games: the games serve
certain social functions (they provide determinate goals and calculable
strategies, for example, and this gives an element of coordination to what
would otherwise be a chaotic pull and haul), but the real satisfaction is
in "playing the game."[13]

Since the American political arena is more a playground than a
forum, it is not surprising that, despite the expenditure of vast amounts
of energy, problems often remain unsolved—after all, what is really wanted
is not solutions but the fun of the game. Still less is it surprising that
those in authority seldom try to make or impose comprehensive solutions.
They mayor of an American city does not think it appropriate for him to
do much more than ratify agreements reached by competing interest
groups. For example, the mayor of Minneapolis does not, according to a
recent report,

[13]Norton E. Long, "The Local Community as an Ecology of Games," *American
Journal of Sociology*, 64(1958): 252.

actively sponsor anything. He waits for private groups to agree on a project. If he he likes it, he endorses it. Since he has no formal power with which to pressure the Council himself, he feels that the private groups must take the responsibility for getting their plan accepted.[14]

American cities, accordingly, seldom make and never carry out comprehensive plans. Plan making is with us an idle exercise, for we neither agree upon the content of a "public interest" that ought to override private ones nor permit the centralization of authority needed to carry a plan into effect if one were made. There is much talk of the need for metropolitan-area planning, but the talk can lead to nothing practical because there is no possibility of agreement on what the "general interest" of such an area requires concretely (whether, for example, it requires keeping the Negroes concentrated in the central city or spreading them out in the suburbs) and because, anyway, there does not exist in any area a government that could carry such plans into effect.[15]

CHANGE IN THE UNITED STATES

The relation of the citizen to the government is changing in the United States as it is in Britain. But the direction of our development is opposite to that of the British: whereas their government is becoming more responsive to popular opinion and therefore weaker, ours is becoming less responsive and therefore stronger. In state and local government this trend has been under way for more than a generation, and it has carried far. Two-thirds of our smaller cities are now run by professional managers, who, in routine matters at least, act without much interference. In the large central cities, mayors have wider spheres of authority than they did a generation ago, much more and much better staff assistance

---

[14]Alan Altshuler, *A Report on Politics in Minneapolis* (Cambridge: Joint Center for Urban Studies of Massachusetts Institute of Technology and Harvard University, 1959), 11–14. The writer has described the posture of Mayor Daley of Chicago, the undisputed boss of a powerful machine, in similar terms. This suggests that it is not lack of power so much as a sense of what is seemly that prevents American mayors from taking a strong line. See E. C. Banfield, *Political Influence* (Chicago: The Free Press, 1961), Chapter 9.

[15]See E. C. Banfield and M. Grodzins, *Government and Housing in Metropolitan Areas*, (New York: McGraw-Hill, 1958), esp. chs. 3 and 4.

(most of them have deputies for administrative management), and greater freedom from the electorate. These gains are in most cases partly offset, and in some perhaps more than partly, by the decay of party machines, which could turn graft, patronage, and other "gravy" into political power, albeit power that was seldom used to public advantage.

Reformers in America have struggled persistently to strengthen government by overcoming the fragmentation of formal authority which has afflicted it from the beginning. The council manager system, the executive budget, metropolitan area organization—these have been intended more to increase the ability of government to get things done (its capability, in the terminology used previously) than to make it less costly or less corrupt.[16]

One of the devices by which power has been centralized and the capability of government increased is the special-function district or authority. We now commonly use authorities to build and manage turnpikes, airports and ports, redevelopment projects, and much else. They generally come into being because the jurisdictions of existing general-purpose governments do not coincide with the areas for which particular functions must be administered. But if this reason for them did not exist, they would have to be created anyway, for they provide a way of escaping to a considerable extent the controls and interferences under which government normally labors. The authority, as a rule, does not go before the electorate or even the legislature; it is exempt from the usual civil-service requirements, budget controls, and auditing, and it is privileged to conduct its affairs out of sight of the public.

The success of all these measures to strengthen government is to be explained by the changing class character of the urban electorate. The lower-class ideal of government, which recognized no community larger than the ward and measured advantages only in favors, "gravy," and nationality "recognition," has almost everywhere gone out of fashion. To be a Protestant and a Yankee is still a political handicap in every large Northern city, but to be thought honest, public spirited, and in some degree statesmanlike is now essential. (John E. Powers, the candidate expected by everyone to win the 1959 Boston mayoralty election, lost apparently because he fitted too well an image of the Irish politician that the Irish electorate found embarrassing and wanted to repudiate.) Many voters still want "nationality recognition," it has been remarked, but they

---

[16]See Don K. Price, "The Promotion of the City Manager Plan," *Public Opinion Quarterly* (Winter 1941): 563–78.

want a kind that is flattering.[17] It appears to follow from this that the nationality-minded voter prefers a candidate who has the attributes of his group but has them in association with those of the admired Anglo-Saxon model. The perfect candidate is of Irish, Polish, Italian, or Jewish extraction, but has the speech, dress, and manner and also the public virtues (honesty, impartiality, devotion to the public good) that belong in the public mind to the upper-class Anglo-Saxon.

The ascendant middle-class ideal of government emphasizes "public values," especially impartiality, consistency, and efficiency. The spread of the council-manager system and of nonpartisanship, the short ballot, at-large voting, and the merit system testify to the change.

Middle-class insistence upon honesty and efficiency has raised the influence and prestige of professionals in the civil service and in civic associations. These are is a position nowadays to give or withhold a good government "seal of approval" which the politician must display on his product.

The impartial expert who "gets things done" in spite of "politicians" and "pressure groups" has become a familiar figure on the urban scence and even something of a folk hero, especially among the builders, contractors, realtors, and bankers who fatten from vast construction projects.[18] Robert Moses is the outstanding example, but there are many

[17]In a study of politics in Worcester, Massachusetts, Robert H. Binstock has written: "Israel Katz, like Casdin, is a Jewish Democrat now serving his fourth term on the Worcester City Council. Although he is much more identifiably Jewish than Casdin, he gets little ethnic support at the polls; there is a lack of rapport between him and the Jewish voter. The voter apparently wants to transcend many features of his ethnic identification and therefore rejects candidates who fit the stereotype of the Jew too well. Casdin is an assimilated Jew in Ivy-League clothes; Katz, by contrast, is old world rather than new, clannish rather than civic-minded, and penny-pinching rather than liberal. Non-Jews call Katz a 'character,' Casdin a 'leader.' It is not too much to say that the Jews, like other minorities, want a flattering, not an unflattering, mirror held up to them." (Robert H. Binstock, A Report on the Politics of Worcester [Cambridge: Joint Center for Urban Studies of Massachusetts Institute of Technology and Harvard University, 1960], Section II, B, 2.)

[18]"In our political or business or labor organizations," Robert E. Sherwood observes in his account of Roosevelt and Hopkins, "we are comforted by the knowledge that at the top is a Big Boss whom we are free to revere or to hate and upon whom we can depend for quick decisions when the going gets tough. The same is true of our Boy Scout troops and our criminal gangs. It is most conspicously

others in smaller bailiwicks. The special function district or authority is, of course, their natural habitat; without the protection it affords from the electorate they could not survive.

The professionals, of course, favor higher levels of spending for public amenities. Their enlarged influence might in itself lead to improvements in the quality and quantity of goods and services provided publicly. But the same public opinion that has elevated the professional has also elevated the importance of these publicly supplied goods and services. It is the upper middle- and the lower-class voters who support public expenditure proposals (the upper middle-class voters because they are mindful of "the good of the community" and the lower-class ones because they have everything to gain and nothing to lose by public expenditures); lower middle-class voters, who are worried about mortgage payments, hostile toward the lower class (which threatens to engulf them physically and otherwise), and indifferent to community-regarding values, constitute most of the opposition to public improvements of all kinds.

Thus it happens that as Britain begins to entertain doubts about green belts, about controls of the use of land that make much depend upon the taste of planners, and about treating public amenity as everything and consumer satisfaction as nothing, we are moving in the opposite direction. There is a lively demand in the United States for green belts (the *New York Times* recently called "self-evident truth" the astonishing statement of an economist that "it is greatly to be doubted if any unit of government under any circumstances has ever bought or can ever buy too much recreation land");[19] the courts are finding that zoning to secure

---

true of our passion for competitive sport. We are trained from childhood to look to the coach for authority in emergencies. The masterminding coach who can send in substitutes with instructions whenever he feels like it—or even send in an entirely new team—is a purely American phenomenon. In British football the team must play through the game with the same eleven men with which it started and with no orders from the sidelines; if a man is injured and forced to leave the field the team goes on playing with only ten men. In British sport, there are no Knute Rocknes or Connie Macks, whereas in American sport the mastermind is considered as an essential in the relentless pursuit of superiority." Robert E. Sherwood, *Roosevelt and Hopkins, An Intimate History* (New York: Harper & Brothers, 1948), 39.

[19]*New York Times*, editorial, 11 April 1960. The economist was Dr. Marion Clawson of Resources for the Future, whose statement appeared in a report sponsored by the New York Metropolitan Regional Council and the New York Regional Plan Association.

aesthetic values is a justifiable exercise of the police power; performance zoning, which leaves a great deal to the discretion of the planner, is becoming fashionable, and J. K. Galbraith has made it a part of conventional wisdom to believe that much more of the national income should be spent for public amenities.

Perhaps in the next twenty or thirty years municipal affairs will pass entirely into the hands of honest, impartial, and nonpolitical "experts"; at any rate, this seems to be the logical fulfillment of the middle-class ideal. If the ideal is achieved, the voters will accept, from a sense of duty to the common good, whatever the experts say is required. We may see in the present willingness of business and civic leaders to take at face value the proposals being made by professionals for master planning, metropolitan organization, and the like, and, in the exalted position of Robert Moses of New York, portents of what is to come.

The presence in the central cities of large numbers of Negroes, Puerto Ricans, and white hillbillies creates a crosscurrent of some importance. For a generation, at least, these newcomers will prefer the old style politics of the ward boss and his "gravy train." How this anomaly will fit into the larger pattern of middle-class politics is hard to imagine. Possibility the lower class will simply be denied representation. And possibly the rate of increase of per capita income being what it is, the assimilation of these people into the middle class will take place faster than anyone now imagines.

SUMMARY AND CONCLUSIONS

It has been argued in this paper that the tasks a government must perform (the number and complexity of goods and services it must supply) have no necessary relation to political matters. Tasks may increase without accompanying increase in the burden placed upon a political system. The important questions for political analysis, therefore, concern not population density or other indicators of the demand for goods and services, but rather the amount and intensity of conflict and the capacity of the government for managing it. Looked at from this standpoint, it appears that the effectiveness of British government in matters of local concern will probably decrease somewhat over the long run. The demands that will be made upon it in the next generation will be vastly more burdensome than those of the recent past (although also vastly less burdensome than the same demands would be in America), and the capacity of the government will be somewhat less. The effectiveness of local government in the United States, on the other hand, will probably increase somewhat. Local government has had more tasks to perform here than in Britain, and these have imposed enormously greater burdens. The tasks of local government will doubtless increase here too in the next generation, but the burdens they impose will probably decline. American local government is becoming stronger and readier to assert the paramountcy of the public interest, real or alleged.

Although each system has moved a considerable distance in the direction of the other, they remain far apart, and each retains its original character. The British, although more sensitive to public opinion, still believe that the government should govern. And we, although acknowledging that the development of metropolitan areas should be planned, still believe that everyone has a right to "get in on the act" and to make his influence felt. Obviously, the differences are crucial, and although the trend seems to be toward greater effectivness here and toward reduced effectiveness in Britain, there can be no doubt that in absolute terms the effectiveness of the British system is and will remain far greater than that of ours. Despite the increase in the tasks it must perform, the burden upon it will remain low by American standards, and its capability will remain high. Matters which would cause great political difficulty here will probably be easily settled here.

The basic dynamic principle in both systems has not been change in population density but rather change in class structure. It is the relaxation of the bonds of status that has caused the British workingman to enter more into politics, that has made his tastes and views count for more, and that has raised questions about the right of an elite to decide matters. In America the assimilation of the lower class to the middle class and the consequent spread of an ideal of government which stresses honesty, impartiality, efficiency, and regard for public as well as private interest have encouraged the general strengthening of government.

The mere absence of dispute, acrimony, unworkable compromise, and stalemate (this, after all, is essentially what the concept *effectiveness* refers to in this connection) ought not, of course, to be taken as constituting a "good" political order. Arrogant officials may ignore the needs and wishes of ordinary citizens, and the ordinary citizens may respectfully acquiesce in their doing so, either because they think (as the British lower class does) that the gentleman knows best or (as the American middle class does) that the expert knows best. In such cases there may be great efffectiveness—no dispute, no acrimony, no unworkable compromise, no stalemate—but far from signifying that the general welfare is being served, such a state of affairs signifies instead that the needs and wishes with which welfare under ordinary circumstances, especially in matters of local concern, is largely concerned are not being taken into account. To say, then, that our system is becoming somewhat more and the British system somewhat less effective does not by any means imply "improvement" for us and the opposite for them. It is quite conceivable that dispute, acrimony, unworkable compromise, and stalemate may be conspicuous features of any situation that approximates the idea of general welfare.

Such conclusions, resting as they do on rough and, at best, commonsense assessments, amply illustrate the difficulty of prediction, and—since the causal principles lie deep in social structure and in culture—the utter impossibility within a free society of a foresighted control of such matters.

CHAPTER 12

The City and the Revolutionary Tradition

It would be very pleasant on such an occasion as this to say that the American city has been and is a unique and unqualified success—and to be able to show that its successes all derive from adherence to principles established and given institutional form in the American Revolution, whose bicentennial we are here to commemorate.

Unfortunately, it is all too evident that even if this were the Fourth of July I would not have license for that sort of oratory. In many important respects the American city is a great success, but there are certainly many things about it that are thoroughly unpleasant, and some that are—or ought to be—intolerable. Moreover, it is obvious that in most important respects—the good and the bad alike—the American city differs more in degree than in kind from cities elsewhere. What we have to be proud of and what we have to worry about are, for the most part, features of modernity and not of anything specifically American.

If we limit ourselves, as this occasion requires, to those features of the city that have been distinctively American over a long period of time, we shall nevertheless have a rather long and varied list. I shall begin by offering *my* list. Then I shall try to account for the items on it with a simple explanatory principle. In the hope of making this explanation more convincing, I shall draw a contrast—necessarily based on fragmentary and impressionistic evidence—between urban development in the United

States and Canada—having chosen Canada because it was a British colony which did not revolt and to whose development my explanatory principle applies, so to say, in reverse. Finally, I shall point to what I consider one of the great ironies of history—that the Founding Fathers created a political system whose essential character turned out to be the very opposite of what most of them intended.

I

My list of features which have distinguished the American city over time will be more manageable if I break it down into three categories. The first I shall call growth and material welfare, the second civility, and the third government. I hope that no attention will be paid to the order of the listings, or to the fact that some items would fit about as well in one category as in another.

Growth and Material Welfare

It should not be necessary to remind a Philadelphia audience how astonishingly fast was the growth and spread of cities in this country. Philadelphia, which in 1775 had a population of 44,000, was the world's eighth largest city a little more than a century later. Of the nine cities in the world with more than a million population in 1890, three were American, and there were then 351 others in the United States of more than 10,000 population.

The cities were built by that often ludicrous and sometimes contemptible fellow—the Worshipper of the Almighty Dollar, the Go-Getter, the Businessman-Booster-Speculator—an upstart, a nobody, but shrewd, his eye on the main chance, always ready to risk his own and (preferably) someone else's money. "Americans," Thomas Low Nichols wrote in 1864,

> are sanguine, and hope to succeed in the wildest speculations; but if they do not, they have little scruple about repudiation. A man cares little for being ruined, and as little about ruining others. But then, ruin there is not like ruin in older countries. Where a man can fail a dozen times, and still go ahead and get credit again, ruin does not amount to much.[1]

[1] Thomas Low Nichols, M.D., *Forty Years of American Life, 1821–1861* (reprinted, New York: Stackpole, 1937), 58; (first published 1864).

In search of the dollar, the American has been constantly on the move. The historian, Stephan Thernstrom, has estimated that, over the past 170 years, probably only 40 to 60 percent of the adult males in most cities at any point in time were in the same city ten years later.[2] "A migratory race" Tocqueville called us, "which, having reached the Pacific Ocean, will retrace its steps to disturb and destroy the social communities which it will have formed and left behind."[3]

The ethnic diversity of our cities has been unparalleled. As early as 1890, one-third of the residents of cities of over one hundred thousand population were foreign born. Ten million foreign born were counted by the 1970 census, and their median family income, it is interesting to note, was not appreciably lower than that of all U.S. families.

The American city has always provided a high level of living for the great majority of its residents. (It was because of what he saw in Europe that Thomas Jefferson came to loathe the city.) The American city dweller has always had more and better schooling, housing (in 1900 one-fourth of the families in most large cities owned their own homes), sanitation, and transportation than city dwellers elsewhere.

Civility

Organized philanthropy has always been conspicuous in the American city. Museums, libraries, symphony orchestras, asylums, hospitals, colleges, parks, and playgrounds—the number and variety of such institutions begun and supported in whole or part by "service" clubs, foundations, and other private efforts is impressive and, I believe, peculiarly American (a point which Tocqueville also made).

Most of these achievements are largely to the credit of the Go-Getter. But he must also be mentioned as a doer-of-evil—as one who, to get things done, has been ready to go to any lengths. Politicians took bribes, Lincoln Steffens remarked, because businessmen paid bribes, and so it was they, the businessmen, who were the real corrupters.

The extent of corruption in American city government has long been the wonder of the civilized world. Some have tried to account for it by

---

[2]Stephan Thernstrom, *The Other Bostonians, Poverty and Progress in the American Metropolis, 1880–1970* (Cambridge: Harvard University Press, 1973), 225.

[3]M. Gustave de Beaumont, ed., *Memoir, Letters, and Remains of Alexis de Tocqueville* (Boston: Ticknor and Fields, 1862), vol. 1, p. 154.

pointing to the masses of poor and politically inexperienced immigrants, but this is surely only a partial explanation. Boss Tweed and his "Forty Thieves" (there were then forty New York City councilmen) were in business before a great many immigrants had arrived. Frank J. Goodnow, writing at the turn of the century in one of the first textbooks on city government, stated the puzzling facts:

> Philadelphia, with a large native-born and home-owning and a small tenement-house population, with a charter which is largely based on what is considered to be advanced ideas on the subject of municipal government, is said to be both corrupt and contented.[4]

The experience of cities like Philadelpha, he concluded, encourages the belief that "there must be something in the moral character of the particular populations."

Moreover, if corruption was common in American cities, so was violent crime. As far back as records go (as much as 100 years in only two cities) the homicide rate has been extraordinarily high by the standards of other countries.

Class differences have, of course, existed in all countries. In America, however, where there has probably been more upward mobility than anywhere else, to be socially defined as "no account" has been crushing in a way that it could not be where everyone knew that rising in the world was out of the question. Perhaps because most have expected to rise, if not themselves then through their children, the American city, unlike cities in most countries, has never produced a radical working-class movement of importance. Perhaps because some have been demoralized by their failure to rise in a society where rising is supposed to be easy, the American city has had a *lumpenproletariat*, a lower as distinguished from a working class—one more conspicuous and possibly more resistant to absortion into normal society than the lower class of other countries.

If the openness of American urban society has produced total alienation in some, it has created disaffection in many more. In a society preoccupied with getting and spending, those who have not managed to get as much as others with whom they compare themselves are likely to feel poor and perhaps to blame themselves and the society for their being relatively badly off even if they are in absolute terms reasonably well off.

---

[4]Frank J. Goodnow, *City Government in the United States* (New York: The Century Co., 1904), 304–305.

This is no new thing. Josiah Strong in his book *Our Country*, written in 1858, observed that

> within a century there has been a great multiplication of the comforts of life among the masses; but the question is *whether that increase has kept pace with the multiplication of wants*. The mechanic of today who has much, may be poorer than his grandfather, who had little. A rich man may be poor, and a poor man may be rich. Poverty is something relative.[5]

Nichols, from whose book (written at about the same time as Strong's) I have already quoted, pointed out wider implications of this "struggling upward."

> There is no such thing in America as being contented with one's position or condition. The poor struggle to be rich, the rich to be richer. Every one is tugging, trying, scheming to advance—to get ahead. It is a great scramble, in which all are troubled and none are satisfied. . . . Every other ragged little boy dreams of being President or millionaire. The dream may be a pleasant one while it lasts, but what of the disappointing reality? What of the excited, restless, feverish life spent in pursuit of phantoms?[6]

Government

What is perhaps most conspicuous to the foreigner is the localism of our politics—localism in two senses: First, every city, even every village, has, by the standards of other countries, an extraordinary degree of independence in dealing with a wide range of matters, including police and schools. (Where else could the voters of a small town decide not to permit the construction of a $600 million oil refinery?)[7] Second, in America city politics turns on local, often neighborhood, concerns, not on national issues or on ideologies.

Our cities have been, and still are, run—to the extent that they can be said to be run at all—by politicians (meaning persons whose talent is for managing conflict), not by career civil servants or planners (meaning persons whose talent is for laying out consistent courses of action to attain agreed-upon goals). To be sure, thousands of documents called *plans* have

[5]Josiah Strong, *Our Country*, ed. Jurgen Herbst (Cambridge: The Belknap Press of Harvard University, 1963), 147.
[6]Nichols, *Forty Years of American Life*, 195.
[7]*New York Times*, 8 March 1974.

been made under the auspices of American local governments. It would be hard to find one that has been carried into effect, however, unless perhaps by an accident of politics.

The "problem of metropolitan organization" exists in this country in a form that may be unique. Actually, it is really two quite different problems. One comes from the multiplicity of more or less overlapping jurisdictions within a single metropolitan area, and the other from the absence, in any such area, of a general-purpose government having jurisdiction over the whole of the area. It is a peculiarly American practice to refer a great many matters to the electorate—not only the choice of mayors and councilmen (and, in many places, of judges) but decisions about capital expenditures, zoning, and governmental structure as well.

Finally, it is remarkably easy for a small number of persons, especially if they are organized, to prevent an American local government from carrying out undertakings which are alleged to be—and which may in fact be—in the interest of the large majority. Ours is, in David Riesman's phrase, a system of *veto groups*.

II

This has been a sketchy listing of what I take to be the distinctive features that American cities have exhibited over time. I turn now to what I regard as the "key" difference—the one which, better than any other, accounts for or "explains" the items on the list. This "key" difference is the extreme fragmentation of authority in the federal system, especially in state and local government. Our constitutions and charters divided authority into a great many small pieces and distributed the pieces widely. The fragmentation, great to begin with, was further increased in the half-century from 1830 to 1880; governors and mayors were mainly for show, and the executive function was carried on by a multitude of separately elected boards and commissions, most of them subject to constant interference by legislatures, courts, and electorates. In recent decades there has been a considerable amount of centralization, but even now ours is, by the standards of other countries, an extraordinarily fragmented system.

How does this explain the features of the American city that I have held to be distinctive? Let me begin with the governmental category. Fragmentation of authority explains why the cities have been run by people adept at managing conflict—the "politicians"—and not by people adept at devising comprehensive and internally consistent courses of action—the "planners." It also explains both sorts of localism. The wide distribution of authority has meant that in order to exercise power on the state or national scene one had to have a local base. Political parties in the United States are not really national organizations; rather they are shifting coalitions of those who, by winning elections or otherwise, have assembled enough pieces of local authority to count.

Because there is power at stake locally, able and ambitious men and women exert themselves to get it. They have always been able to afford to offer the voter (enough voters to make a difference) inducements more substantial than mere ideology—jobs, favors, ethnic recognition. Politics in the American city has been serious business—that is, the politician has been a sort of businessman and the businessman a sort of politician. Obviously this would have been impossible if power had been centralized.

The fragmentation of authority has not only permitted but also encouraged its informal centralization by means—notably the machine and the boss—that were corrupt. If, as Steffens said, businessmen gave bribes because they had to—because it was impossible to operate a street railway without doing so—it is also true that politicians took them because they had to—because, to centralize enough power to get things done, they had in one way or another to "purchase" pieces of authority from voters and others. Without this easy access to power on the local scene, the Go-Getter would not have had the opportunity to "go get." As it was, he could extend the grids of nonexistent cities into the hinterland confident that he could induce some public body to build the canal, railroad, highway, arsenal, or whatever that would send land values up. Even the new immigrant's ethnic ties had a political value that could be converted into the small amount of capital he needed to get started.

These incentives released prodigious amounts of energy. The freedom—near anarchy in places— of the politician-businessman-entrepreneur was a necessary condition of the great scramble to advance which, Thomas Low Nichols said, left all troubled and none satisfied. (In Europe, Nichols wrote, in a part of the passage that I did not quote, as a rule the poor man knows that he must remain poor, and he submits to his lot. "Most men live and die in the position to which they are born.") Also, where laws were made and unmade by majority vote and enforced or not depending upon who paid how much to whom, the consequence must have been not only general disrespect for law but also for the persons and institutions that claimed to act under its authority. The same conditions that made the Go-Getter also helped to make the conman and the gunslinger.

That the system produced a high and ever-rising material level of living for most city dwellers must not blind us to the fact that those who did not know how to work the system, or who for one reason or another were prevented from working it, fared badly. Those who took "favors" from the machine and its boss made a very poor bargain, judged at least by middle-class standards. As Jane Addams remarked in *Twenty Years at Hull-House* (1916):

> The policy of the public authorities of never taking an initiative, and always waiting to be urged to do their duty, is obviously fatal in a neighborhood where there is little initiative among the citizens. The idea underlying self-government breaks down in such a ward. The streets are inexpressibly dirty, the number of schools inadequate, sanitary legislation unenforced, the street lighting bad, the paving miserable, and altogether lacking in the alleys and smaller streets, and the stables foul beyond description.[8]

III

The explanation that I have offered to account for the distinctive features of the American city would be more convincing if I could show that in another country an opposite principle produced opposite results. I believe I can. The history of urban development in Canada provides such a test, for the Canadian political system has been the opposite of ours in what for me is the crucial respect. I am not, of course, implying that the Canadians are less attached to democracy than we are. Rather, my point is that their idea of it is essentially different from ours. In Canada the British tradition has never been interrupted: the duty of government has always been to govern—not, as in the United States, to preside over a competition of interests. Canadians, writes Professor Tom Truman of McMaster University, "insist on strong stable executive government, which, once it has made up its mind on what the public interest requires,

---

[8] Quoted by Louis Wirth, *The Ghetto* (Chicago: University of Chicago Press, 1956), p. 196.

should take the necessary action quickly and with determination to see it through completely."[9]

It goes without saying that the comparison with Canadian experience cannot provide a wholly satisfactory test of my argument, for there are manifestly many differences between the two countries that may account for much of what I am trying to explain. Although Canada is larger in area than the United States, its great natural resources have been, especially in the nineteenth century, much less accessible. It has always had an important French-speaking minority. And it has always been profoundly affected by events in this country. The influence of these and other circumstances on urban development has certainly been great. I believe, however, that the centralized structure of political authority in Canada accounts—better than any other single principle—for the differences between Canadian and American cities in the features I have listed.

A detailed account of Canadian experience is obviously out of the question here, but let me call your attention to a few relevant facts:

1. The growth of cities in Canada was slow. As late as 1911 Canada had only six cities of fifty thousand or more population, of which only two (Montreal and Toronto) had more than three hundred thousand.

2. The Go-Getter-Businessman-Booster-Speculator has been (until recently) conspicuous by his absence. Horatio Alger heroes, it seems, have never been popular in Canada.[10] It may be indicative of the difference in business ethos that there are about twice as many lawyers per capita in the United States as in Canada: in 1955, one lawyer in private practice per 868 persons here compared to one per 1,630 there.[11]

3. Immigration into Canada was, until well into this century, mainly from the British Isles. British immigrants were long favored by law. By American standards, assimilation of non-British and non-French-speaking immigrants was slow: not until this century, I understand, was one elected to public office.

[9]Tom Truman, "A Critique of Seymour M. Lipset's Article, Value Differences, Absolute or Relative: The English-speaking Democracies," *Canadian Journal of Political Science*, 4, no. 4 (1971): 513.

[10]Seymour M. Lipset, *The First New Nation: The United States in Historical and Comparative Perspective* (New York: Basic Books, 1963), 251.

[11]Ibid., p. 264.

4. Generally speaking, the level of public services has been low by American standards.

5. Organized philanthropy began late—about World War I, an import from the United States.[12]

6. Large-scale corruption has never been a feature of city life.

7. There has been very little violent crime.

8. Social mobility has been less than in the United States.

9. Although radical working-class movements (the Canadian Commonwealth Federation and the National Democratic Party) have been able to form governments only on the prairies, they have had more supporters in the urban areas than among the wheat farmers.

10. Urban Canada does not seem to have had a *lumpenproletariat* on anything like the American scale.

11. "The incessant exercise of voting power," Lord Bryce remarked, "has never possessed any special fascination for the Canadian."[13]

12. Toronto has a metropolitan government—one much admired by American reformers. It was created in 1953, over the objections of most of the local governments concerned, by the Provincial Government on recommendation of the Ontario Municipal Board, a quasi-judicial body. The possibility of a referendum was never seriously discussed.[14]

Can these features of Canadian development be accounted for in large part by the centralized structure of government? I do not have time to develop evidence in support of this claim, but I must quote one of many pertinent passages in a work by the Canadian sociologist S. D. Clark. He writes in *The Developing Canadian Community:*

> A force of Royal Engineers put an end to lawlessness in the mining camps of British Columbia. Settlement of the western prairies and the gold rush to the Klondike took place under the close control of the North West Mounted Police. Even in Canadian cities, serious threats to law and order have been met by the decisive use of force.

[12]Aileen D. Ross, "Organized Philanthropy in an Urban Community," *Canadian Journal of Economics and Political Science* 18, no. 4 (1952): 474–75.

[13]James Bryce, *Modern Democracies* (London: Macmillan & Co., 1921), vol. 1, pp. 553–554.

[14]Harold Kaplan, *Urban Political Systems: A Functional Analysis of Metro Toronto* (New York: Columbia University Press, 1967).

The result was to establish a tradition of respect for the institutions of law and order. The population generally did not feel the
need of taking the law into its own hands through mob action or the
organization of vigilantes. There was lacking that intense jealousy of
local rights which in the United States made it difficult for federal
forces to intervene. The way in which the North West Mounted Police
came into being was in striking contrast with that of the Texas Rangers. In the United States the frontier bred a spirit of liberty which
often opposed efforts to maintain order. In Canada, order was maintained at the price of weakening that spirit.[15]

IV

One of the great ironies of history is to be found in these developments, for it was a centralized system like the Canadian, not a fragmented one like the American, that the principal figures among the
Founding Fathers thought they were creating.

The Revolution, John Adams wrote in a letter in 1818, was effected
before the war; it was "in the minds and hearts of the people; a change
in their religious sentiments of their duties and obligations." So long as
the king and all in authority under him were believed to govern according
to the laws and constitution derived to them by their ancestors, the
colonists thought themselves bound to pray for them as "ministers of God
ordained for their good." However,

> when they saw those powers renouncing all the principles of authority
> and bent upon the destruction of their lives, liberties and properties,
> they thought it their duty to pray for the continental congress and
> the thirteen state congresses.[16]

On this view, the intention of the revolutionaries was to bring about a
change of regime, not of political principles. Rulers who would not act
as ministers ordained by God were to be replaced by others who would.

There is nothing to contradict this in the Declaration of Independence. Jefferson, in writing that governments "derive their just powers
from the consent of the governed," did not assert something novel. Since
1689 British monarchs had needed the consent of the House of Commons

[15]S. D. Clark, *The Developing Canadian Community* (Toronto: The University of
Toronto Press, 1962), 191–92.
[16]Adrienne Koch and William Peden, eds., *The Selected Writings of John Quincy
Adams* (New York: Alfred A. Knopf, 1946), 203.

in order to raise revenue. And, as Martin Diamond pointed out in his lecture, the Declaration says that consent is required to institute or establish a government, not for the conduct of its affairs. The unchallenged principle was that the conduct of affairs belonged in the hands of those authorized to govern.[17]

Adams wanted not only to follow the principles of the British system but, so far as American conditions allowed, to recreate its forms as well.[18] That the executive authority was to be in the hands of one chosen by election did not seem to him or most others to constitute a fundamental change. It had long been understood that in Britain almost all real, as opposed to nominal, authority was in the hands of ministers, not of the king. As Gouverneur Morris put it later when addressing the Constitutional Convention, "Our President will be the British minister."[19]

It was in that convention that the distinctively American political arrangements were worked out. They represented neither the reestablishment of the essential principles of the British system nor the assertion of contrary principles. They were a compromise—that is, the acceptance of contradictory principles. Expediency prevailed, and the result was not a plan but an accident.

Hamilton and Madison acknowledged that "the deliberate sense of the community" should govern the conduct of those in office, but they added that this did not require "an unqualified complaisance" to every transient impulse of the people. "When occasions present themselves in which the interests of the people are at variance with their inclinations, it is the duty of the persons whom they have appointed to be the guardians of those interests, to withstand the temporary delusion." The humors of the legislature did not require unqualified complaisance either: "It is certainly desirable that the Executive should be in a situation to dare to

[17]Jefferson's view, according to Harvey C. Mansfield, Jr., was that government "derives from the people, where it is 'deposited,' and yet acts on the people to keep them independent by making them republican." He was, Mansfield says, "willing to trust the people, not to govern, but to choose their governors." See his essay, "Thomas Jefferson," in Morton J. Frisch and Richard G. Stevens, eds., *American Political Thought* (New York: Charles Scribner's Sons, 1971), 38–39.

[18]Bernard Bailyn, *The Ideological Origins of the American Revolution* (Cambridge: Harvard University Press, 1967), 290 (footnote).

[19]Max Farrand, ed., *Records of the Federal Convention*, 3 vols. (New Haven: Yale University Press, 1937), vol. 2, p. 104.

act his own opinion with vigor and decision." Also: "It is one thing to be subordinate to the laws, and another to be dependent on the legislative body."[20]

In his farewell address Washington warned that "all combinations and associations, under whatever plausible character, with the real design to direct, control, counteract, or awe the regular deliberation and action of the constituted authorities," are "of fatal tendency."

It is fair to say that until John Quincy Adams left the White House in 1829 there had been no revolution, so far as any of the presidents were concerned, if by revolution is meant fundamental change of political principles. One might even say that there was an effort at counterrevolution— a return to the established principles of the British constitution which were, as A. V. Dicey has said, supremacy of law and "the omnipotence or undisputed supremacy throughout the whole country of the central Government."[21] Nevertheless, there were signs, before the second Adams left the White House, that the government of the United States would never be the "monarchical republic" that his father and some of the others had intended it to be and imagined that it was.

Immediately before and during the revolutionary war public opinion turned against all things British, including the idea that there ought to be a ruler—a minister ordained of God to act for the common good. The expansion of the frontier and the increase in the number and prosperity of tradesmen and craftsmen in the towns and cities gave the *local Demagogues,* as Gouverneur Morris called them, an unassailable power. In its first years the national government was without physical force to support its measures (the army consisted of a few hundred men), and then, almost at once, the War of 1812 absorbed all its resources. Under the circumstances the executive could not as a practical matter exercise the power that it claimed in principle. Washington meant to sell the public lands gradually and in a way that would encourage compact settlement (this had long been the British policy), but his plan could not be carried out: the minimum price of public land, set at $2.00 per acre in 1796, was reduced under pressure from frontiersmen and speculators to $1.20 in 1820 and, a few years later, again cut by almost half.[22] The comprehensive

---

[20]*The Federalist* 71.
[21]A. V. Dicey, *Law of the Constitution* (London: Macmillan & Co., 1902), p. 179.
[22]V. Webster Johnson and Raleigh Barlowe, *Land Problems and Policies* (New York: McGraw-Hill, 1954), 35–36.

plan for internal improvements put forward by Jefferson's secretary of the treasury, Gallatin, became, after long delay, a pork barrel for the states which Madison vetoed the day before he left office.[23]

The same forces that prevented the national executive from establishing its mastery led to the development of political parties on a local rather than (as in Canada) a national basis. The parties were coalitions within each state of local interests which, every four years, formed loose federations to nominate and elect a president.

By the 1830s the American political system had assumed its characteristic and lasting form. The president was indeed an "elective monarch," but only in matters in which he was willing to invest the whole force and energy of his office; in the nature of things, there could be few such matters at any one time. In other matters the system functioned to accommodate competing and more or less parochial interests, not to deliberate about (much less enforce) an idea of the common good. State and local governments were organized in imitation of the much-revered national one, but the imitations did not extend to the feature the Founding Fathers had considered crucial: a strong executive—a minister ordained of God for the people's good. Governors and mayors, as I have said, were little more than ceremonial figures. In state and local government, the principle of interest balancing prevailed.

Those with a taste for irony will relish the fact that by the time the American Revolution had worked itself out to this conclusion, the British system—whose corruption in the eighteenth century had set the American events in motion—had somehow reformed itself and was operating on the principles that most of the Founding Fathers unqualifiedly admired and had meant to copy.

As I said at the outset, this is not a Fourth of July oration. But I do not wish to leave the impression that I consider the American Revolution to have been a mistake. Even if I were sure that a strong central government, operating with consent and under law, would produce effects that are on the whole preferable to those produced by a system of interest balancing, I would not think that the Revolution was a mistake. For there is no doubt that without the American example before them, other nations, including the British and the Canadian, would not have succeeded as well—perhaps not at all—with their brand of democracy.

[23]Carter Goodrich, "National Planning of Internal Improvements," *Political Science Quarterly*, vol. 63, no. 1 (1948).

That a people could, by a deliberate process, accomplish what has always been regarded as the highest and noblest of all tasks—the creation of a political order that assures to them and their posterity the blessings of life, liberty, and the pursuit of happiness—has had, not only for us but for the whole world, a significance no other event could possibly have had. But if there is great reason for pride in this achievement, there is also reason for apprehension—certainly for pondering such questions as those asked by Thomas Low Nichols in the book from which I have several times quoted:

> If the only source of power is the will of the people expressed by the votes of a majority, what are the institutions that may not be overthrown?—what are the institutions that may not be established? The whole people owns the whole property; what shall hinder them from doing with it as they will? So the people are above their institutions, and may frame, modify, or abolish them according to their sovereign will and pleasure. Right is a matter of opinion, and to be determined by a majority. Justice is what that majority chooses. Apparently expediency is the only rule of conduct.[24]

Plainly Nichols thought justice is not what the majority chooses, and expediency is not the only rule. And so do I.

[24]Nichols, *Forty Years of American Life*, 244.

CHAPTER 13

A Critical View of the Urban Crisis

From the farmhouse in Vermont where this is written, it is several miles to the nearest city, the population of which is about ten thousand, but one can find here most of the "big problems" that together are usually thought of as constituting "the urban crisis." There are multiproblem families; welfare mothers—some of them visited, it is alleged, by mates; substandard housing; functional illiteracy—some of it on the part of high school graduates; hard-core unemployment—meaning able-bodied men who will not work; sexual promiscuity; drug abuse—a helicopter hovers overhead as this is written, searching for plantings of marijuana; pornography—this eventing, a local movie advertises, one may take an "unforgettable journey into a new erotic world" of "way-out German sex practices"; venereal disease—Operation Venus has recently been locally organized; and crime—mostly vandalism and stealing by juveniles, but occasionally violence by adults.

How is all this to be reconciled with the account usually given of the causes of the "urban crisis?" The problems of the cities, it is commonly held, arise principally from the following causes: the mass movement of a rapidly increasing population into metropolitan areas, producing congestion; the flight of the middle class, as well as of much manufacturing and commerce, to the suburbs, resulting in a financial crisis for the cities; the

From the *Annals of the American Academy of Political and Social Science*, vol. 405 (January 1973): 8–14. Copyright 1973 by the American Academy of Political and Social Science. Reprinted by permission.

spread in the central cities and in the older, larger suburbs of slums and blight and, more generally, the prevalence of poverty and a physical environment that makes a decent style of life impossible for many people; "white racism," which keeps Negroes and some other minority groups— notably Puerto Ricans in New York and in a few other cities—subject to various all-but-insuperable handicaps; and finally, a local governmental structure so chaotic and fragmented as to render it ineffective if not entirely futile.

It is obvious that these cannot be the causes of the problems—the *same* problems—that manifest themselves in rural Vermont. The density of population in my township is sixty persons to the square mile; there is only one Negro family—referred to politely by its neighbor as "complected"—and if the local government is ineffective, it is not because of fragmentation, although, to be sure, the Board of Selectmen, the Board of Listers (tax assessors), the Road Commissioner, the School Board, and the Volunteer Fire Department are all separate and independent.

Two putative causes of the "urban crisis" *do* exist there: there are some houses that would be called *slums* in the city, and the income of some families is certainly low. It must be added, however, that the correlation between poverty and the social problems to which it is supposed to give rise is very weak if it exists at all.

Enough has been said, perhaps, to prepare the way for my main contention, which is that urban problems, so-called, arise in the main not so much from environmental conditions—certainly not from ones peculiar to cities—as from ideas, or to put it more generally, from states of the public mind. This argument will be developed by a critical examination of the several factors widely held to be the principal causes of the "urban crisis."

METROPOLITAN GROWTH

No one has produced evidence supporting the notion that "overcrowding" in cities is harmful. Nevertheless, much discussion proceeds on the assumption that it is and also that it is widespread and increasing. These assumptions seem to get support from figures constantly being issued in official reports—figures that, taken out of context and read hastily, appear to justify alarm: for example, the recent estimate that by the year 2000 some 85 percent of all Americans will live in metropolitan areas.[1]

Unlike a country or a continent, a metropolitan area is merely a statistical convenience. The Bureau of the Budget tried, apparently with little success, to alert the unwary to this by terming its contrivance *Standard Metropolitan Statistical Area* (SMSA). An SMSA is, to simplify slightly, a central city of fifty thousand or more inhabitants, or two or more contiguous cities having that population in total, together with such adjacent counties as are "functionally integrated" with the central city. That SMSA does not mean *metropolis*—a word found in dictionaries—is evident from the fact that the Meriden, Connecticut, area, population fifty-six thousand, is an SMSA. That measures of "metropolitanization" are artifacts of the definition is evident from the fact that some six million persons became "metropolitan" in 1970 without—so far as anyone knows— anything having changed except some details of the definition.

If the census and other suppliers and users of data used words like *metropolis, city,* and *urban* in the senses in which they are commonly understood, there might be a good deal less concern about "congestion." By many measures it is decreasing. The percentage of the population living in cities of one hundred thousand or more (incidentally, these figures have also been affected by redefinitions, the reader should be warned) reached a peak of 30 percent in 1930 and has dropped gradually ever since. Of those who live outside of central cities—but within SMSAs— two-thirds live in places of less than five thousand. To be sure, the definition assures that these little places are within counties adjacent to and "functionally integrated" with central cities, but it should not be assumed from this that all or even most who live in them have much contact with the city. That the most recent New York State Development Plan sets as a goal for 1990 that no "urban citizen" shall have to travel more than fifteen miles to reach an "urban center" is indicative.

[1]Most of the figures in this section are to be found in three recent official publications, none of which is open to the criticisms made here of the frequent misuse of such figures. They are The White House, *Report on National Growth 1972* (Washington, D.C.: U.S. Government Printing Office, 1972); Commission on Population Growth and the American Future, *Population and the American Future* (Washington, D.C.: U.S. Government Printing Office, 1972); and Allen R. Bird, "Migration and its Effect on Agriculture and Rural Development Potential" (Paper presented at a national symposium at Muscle Shaols, Ala., June 22– 23, 1972), mimeographed by the Economic Development Division, Economic Research Service, U.S. Department of Agriculture, Washington, D.C.

It is an error to assume, as many do, that practically all economic and other growth is occurring within SMSAs. Patterns differ by region. In the South the same percentage of places in each size class increased in population whether they were more or less than fifty miles from a central city. In Pennsylvania, manufacturing employment grew faster between 1960 and 1966 in towns of less than twenty-five thousand than in SMSAs. Small, relatively isolated places may have more of a future than sweeping generalizations based upon aggregate data would lead one to expect.

The problems of the core areas of the large cities cannot be caused by increasing densities; these have been declining since at least the turn of the century, and they are still doing do. In the past decade, more central cities in every size class lost population than gained it and—leaving out of account changes due to annexations—the aggregate gain in population by central cities amounted to only about one percentage point.

Taking the SMSAs as a whole, there has apparently been a continuing decline in densities:

        1950   5,408 persons per square mile
        1960   3,752
        1970   3,376

("Apparently" is used because changes of definition, as well as of income and technology, make such comparisons hazardous. Income and technology are relevant because presumably density is measured as an indicator of accessibility; more and better transportation may actually have increased accessibility despite the drop in density.)

The only sort of congestion that common sense tells us may plausibly have a direct bearing upon health and psychic well-being—not to mention convenience—is that which can be measured in persons per room. This, too, has been decreasing. In 1950 nearly 17 percent of units were occupied at more than one person per room, but in 1970 the figure was less than 7 percent.

THE "FLIGHT" TO THE SUBURBS

If congestion is not a principal cause of the "urban crisis," neither is "the flight of the middle class to the suburbs." Families have always tended to move, when they could afford to, from the relatively old and crowded inner parts of the cities to newer and less crowded neighborhoods at the outskirts of the city or beyond them. They have done so—and do so still—because they can get more of what they want for their money by building on vacant land which, because it is not close to centers of accessibility, is not in great demand for competing uses and is therefore relatively cheap. What such families want is more spacious and up-to-date housing, easy parking, better schools, and a "nicer" neighborhood. Factories and offices have followed them partly because they depend upon their labor, but also because of the inducement of cheap land and easy access to (heavily subsidized) truck and air routes.

To speak of those who move to the suburbs as "fleeing" the central city makes about as much sense as it would to say of one who trades an old Chevrolet for a new Buick that he is "fleeing" used cars. Of course there are districts in all large cities from which many people move, out of fears—perhaps not entirely justified by the facts—of violence or of the loss of lifetime savings from a sudden drop in property values. Such moves are properly called *flights*. They are not, however, typical in most cities.

The argument, so often heard, that the departure of the well-off from the central cities has created a "financial crisis" for them by reducing their tax bases is unsound. In the first place, very few central cities face anything that can reasonably be called a financial crisis. With few exceptions, the credit ratings of the large cities are as high or higher than they were thirty years ago; what the central cities have lost in tax base has been partly made up for in state and federal grants—state and federal aid to local governments increased from $8.8 billion in 1963 to an estimated $43.5 billion in fiscal 1973, and most, but not all, central cities get a larger per capita share of this than do their suburbs;[2] suburbanites viewed collectively are not escaping financial responsibility for central cities, though the individual—who would have to pay the state and federal taxes if he remained in the central city—does have a tax incentive to move. In any event, we have the authority of Richard Musgrave for asserting that under the circumstances that are likely to obtain, cities will come within $5 billion of meeting their revenue needs in 1975, an amount well within their capacity to raise by tax increases.

[2]Charles L. Schultze *et al.*, *Setting National Priorities, the 1973 Budget* (Washington, D.C.: The Brookings Institution, 1972), 291 and Table 9–1, 294–95. For documentation of other statements in this paragraph see E. C. Banfield, "Revenue Sharing in Theory and Practice," *The Public Interest*, no. 23 (Spring 1971): 33–45.

THE PHYSICAL ENVIRONMENT

Twenty years ago there were those who thought that a wide array of social problems could be solved, or much relieved, simply by providing "a proper environment." The Pruitt–Igoe Public Housing Project in St. Louis, designed by some of the world's most gifted architects, represented such an effort on a grand scale—twenty-eight thousand units on fifty-seven acres built in 1954 at a cost to the government of $36 million. Today, having been for years the nation's most conspicuous slum—one to which respectable people could not be attracted even by rents of $20 a month—only six hundred units are occupied, and about $39 million is being spent to tear down two buildings, lower others, and in general redesign the project.

The contemporary expression of the idea underlying Pruitt–Igoe is the "new town"—a city built from the ground up as a technologically advanced "life support system" for a large and diverse population. "You are what you live in" could have been the slogan of public housing; it would serve as well today for the new town. The ills of the cities cannot be understood in such terms, however.

The material requisites of a wholesome and satisfying life are within the reach of almost all. It would cost little—in relation to welfare costs—to make the elements of a nutritionally adequate diet as free as air; the trouble with the idea is, of course, that powdered milk, dried peas and beans, cereals, and the like, would not be acceptable to most of the twenty million Americans who are said to go to bed hungry. Studies of the effects of housing quality on family health have shown them to be minimal.[3] In any event, the amount and quality of housing in American cities is, on the whole, good and getting better. From 1950 to 1970, real per capita consumption of housing in the United States more than doubled, and we are now producing about two units of housing for every new household.[4] The poor and the black have shared in this progress; between 1960 and 1968, the percentage of housing occupied by Negroes in central cities and described by the census as "not meeting specified conditions" fell

[3]D. M. Wilner, *The Housing Environment and Family Life* (Baltimore, Md.: Johns Hopkins Press, 1962).

[4]Henry J. Aaron, *Shelter and Subsidies* (Washington, D.C.: The Brookings Institution, 1972), 28. See also the address by George Romney on March 27, 1972, reprinted in *HUD News* (Housing and Urban Development).

from 25 percent to 9 percent. Air pollution, although a health problem in all large and some not-large cities, is not specifically an *urban* problem—a study of death certificates in New York State showed more deaths from emphysema, the leading respiratory disease, in rural than in urban areas.[5] Moreover, as far as anyone knows, it is not a contributing cause of the other problems that are generally thought of as constituting the "urban crisis."

"WHITE RACISM"

The essential basis of urban problems today, some assert, is that the high visibility of the Negro has marked him inescapably as a victim of "white racism." This, they think, makes the present situation fundamentally different from any earlier one. Whereas other disadvantaged groups coming to the city could expect to be assimilated into the "mainstream," the Negro, because he is indelibly marked, cannot.

It is of course impossible to disprove such statements. In evaluating them, however, several things should be kept in mind. One is that physical characteristics making for high visibility have not prevented Orientals from moving close to the top of the income distribution and from being accepted socially.

Another is that the attitudes and the behavior of whites toward blacks have changed dramatically since World War II. This is shown by numerous public opinion polls—notably those taken between 1940 and 1970 by the National Opinion Research Center, the most recent of which found that "the trend has been distinctly toward increasing approval of integration"[6]—and more persuasively (since actions speak louder than words) by the gains Negroes have made in the past decade in schooling, income, and health and by the successes of many at the polls. The indications are that, if present trends continue, Negroes, despite their initial handicap, will overtake whites in income, schooling, and other crucial respects much sooner than one would have thought possible a decade or two ago.[7]

[5] *New York Times*, 30 October 1970, 32.
[6] Andrew M. Greeley and Paul B. Sheatsley, "Attitudes toward Racial Integration," *Scientific American* 225, no. 6 (December 1971): 13.
[7] See Michael J. Flax, *Study in Comparative Urban Indicators: Conditions in 18 Large Metropolitan Areas* (Washington, D.C.: The Urban Institute, 1972).

Much of what is perceived as the effect of *presently* existing white prejudice is, in fact, largely or even entirely the effect of factors that owe their existence to prejudice *in the past*. Differences in arrest rates and in earnings, for example, either disappear or are sharply reduced when account is taken of such factors as age, education, place of origin—whether urban or rural, Southern or non-Southern—and class culture.[8]

It is to be expected that, as their incomes rise, Negroes will leave the central cities in increasing numbers for the newer suburbs. In the 1960s the black suburban population grew by 762,000, or 42 percent, a signficant increase over the previous decade. In the decade to come, the increases will be larger still. Despite these movements, huge areas of the largest cities, and also many of the suburbs into which Negroes are moving, will be predominantly or even exclusively black. In and of itself this need not be a problem. What matters is not segregation as such, but *involuntary* segregation, something that, strange as it may seem, no one has attempted to measure.

METROPOLITAN ORGANIZATION

The average SMSA has within it about a hundred local governments, and no SMSA has a general purpose government with jurisdiction over the whole area. Both circumstances, but especially the latter, are frequently listed among the principal causes of the "urban crisis," the theory being that since proper governmental organization would solve, or at least alleviate, problems, the want of it must itself be counted as a major problem. Thus on this view, the continued concentration of the poor and the black in the central cities results largely from the absence of a metropolitan government which could sweep aside local zoning and other barriers.

This reasoning would be valid if it could be assumed that a metropolitan-area government would indeed exercise its powers in these ways. But this amounts to assuming that the local political forces now

[8]Edward Green, "Race, Social Status, and Criminal Arrest," *American Sociological Review* 35, no. 3 (June 1970); James Gwartney, "Discrimination and Income Differentials," *The American Economic Review* 60, no. 3 (June 1970); and Nathan Glazer, "Blacks and Ethnic Groups; The Difference and the Political Difference It Makes," *Social Problems* 18, no. 4 (Spring 1971).

pressing in the contrary directions would disappear or lose their influence. The implausibility of such an assumption can be seen from the example of a large city—London—which *does* have a metropolitan government, but which nevertheless has been signally unsuccessful in moving the working poor from the inner city to the outlying suburbs.[9] It can also be seen, at much closer range, in the recent decision of the New York State Urban Development Corporation, a body amply endowed with authority and funds to enable it to ignore local zoning ordinances, to defer the building of low-income housing in rural Westchester County.[10]

THE STATE OF THE PUBLIC MIND

If these are not principal causes of the "urban crisis," what are? My answer is that they are mainly changes in the way things are perceived, judged, and valued, and in the expectations that are formed accordingly—in a phrase, changes in the states of the public mind.

Changes in the world of physical things and of technology, for example, population density or techniques for making poisonous substances, influence the public mind, of course. In the main, however, changes occur in other ways: partially by the gradual unfolding of potentialities immanent in the culture, that is, in the collective experience that is passed from one generation to the next; partially by the influence of one group upon another—culture, or subculture, contact; and partially (in the long run most importantly) by the ideas of philosophers—of whom there may be two or three in the world per century. Philosophical ideas somehow make their way directly and via the intuitions of great literary and other artists to lesser lights—professors, journalists, and others—whose writings convey them to politicians, lawyers, businessmen, and other managers of affairs, until finally the ideas (by this time much diluted and otherwise altered) become widely, even generally, accepted as the views and standards—when verbalized, the cant—of the middle class.

[9]See "London Fight with Suburbs Predicted," news story by Malcolm Stuart, *The Manchester Guardian*, 5 January 1972.

[10]*New York Times*, 5 August 1972, 38. For an account of tactics used to foil a Massachusetts law designed to open the suburbs to the poor, see *Wall Street Journal*, 17 October 1972, 1.

This is not the place to try to describe the changes that have been occurring in the state of the public mind. The best that can be done is to incorporate by reference, as lawyers say, some writers that *do* describe them.[11] These stress (a) decline of authority in all its forms and manifestations; (b) preoccupation with self and its fulfillment; (c) rationalism which justifies any act expected to benefit *ego* without perceptible injury to *alter*; (d) hedonism which takes plentitude rather than scarcity to be the fundamental fact; (e) egalitarianism which asserts not that equals should be treated equally, but that all should be treated as if they were the same; and, finally, (f) moral fervor directed not against the sins of individuals, but against those of institutions—business firms, churches, and governments ("consumerism").

How these changes in the state of the public mind cause the social problems that have been under discussion cannot be explained here for lack of space, but an illustration may be useful.[12] Jurors in at least one large city are said to tend to take for granted the "unfairness," even the untruth, of whatever the prosecution charges. If this is the case, the attitude would seem to reflect (a) loss of respect for authority; (b) belief that it is the juror's right as a "person" to ignore the rules of judicial procedure if he "feels like it"; (c) awareness that adherence to or violation of a rule in a particular case will not in certain circumstances contribute perceptibly, let alone decisively, either to the maintenance or to the undermining of the principle that rules must be obeyed; (d) faith that the crime could be controlled without inflicting punishment if only we "work to find the way"; (e) belief that it is inequitable that only jurors and not, say, courtroom spectators as well, render verdicts; and (f) feeling that whatever the accused may have done, he is a "victim," and "society," which is "really" to blame, should be put on trial.

These changes in state of mind proceed from the city—from one or two of the largest, in fact. Operation Venus came into being in rural

[11]Particularly apposite are Robert A. Nisbet, "The Twilight of Authority," *The Public Interest* 15 (Spring 1969): 3–9; Marvin Zetterbaum, "Self and Political Order," *Interpretation* 2 (Winter 1970): 233–46; Edward A. Shils, "Plentitude and Scarcity," *Encounter* (May 1969); John Passmore, "Paradise Now, the Logic of the New Mysticism," *Encounter* (November 1970); and a brief letter from Paul Craig Roberts in *Science* 169, no. 3948 (August 28, 1970): 816.
[12]The point could be as well illustrated by describing effects that are socially desirable—for example, increased concern about racial and other injustice. This article, however, happens to deal with "problems."

Vermont because two high school girls read in a national magazine that it was a way of "helping others." Very likely some of the others needed this help because of things that *they* had read in national magazines or had seen in the movies—"way-out German sex practices" perhaps. That the city is the place from which the changed standards are disseminated is, however, an incidental circumstances, not a cause of the changes.

THE FUTILITY OF PLANNING

If the view taken here is correct, the "urban crisis" is not to be solved or alleviated by government programs, however massive. Planners and other managers may put their fingers in dikes, but their doing so will not make any perceptible difference in the long run. Nor, unfortunately, will suggestions such as those recently made with hesitation—for fear of "horrifying many readers"—by a professor of political science writing in the *New York Times* that educators teach the values of honesty and truthfulness, trustworthiness, work well done, kindness and compassion, the courage to admit mistakes, racial tolerance, respect for law, nonviolence in pursuit of one's goals, and respect for democratic rights.[13] If teachers *would* teach these values and pupils *would* learn them, a professor would not find it necessary to make such a suggestion hesitantly or otherwise. That these values are no longer valued is the *cause* of the problem; his "solution" assumes away the very fact with which we must cope.

[13]"Reo M. Christenson, "The Old Values are the Best Values," *New York Times,* 3 June 1972.

CHAPTER 14

The Zoning of Enterprise

This chapter seeks to make two principal points. The first is that upward mobility on the part of disadvantaged persons in the cities has been, is being, and doubtlessly will be, hampered by laws and regulations the manifest purpose of which is to make them better off. The second is that as our society becomes more sensitive to social injustices (real and imagined) it thereby becomes less capable of coping with certain of its problems; indeed, it increasingly confronts the dilemma that a good society, if it is to remain one, must sometimes do things that are incompatible with its goodness.

The word *disadvantaged* as used here refers to persons who are unskilled or low skilled, whose command of English is poor or nonexistent, who are subject to discrimination because of race or class, and who live in an urban enclave consisting prdominantly or entirely of the disadvantaged. As used here, then, the word is not synonymous with *poor*, although for obvious reasons disadvantaged persons are almost certain to be poor. Nor is it synonymous with *lower class,* a term which, as I use it, refers to those whose style of life reflects unwillingness, or inability, to take account of the future. Some disadvantaged persons are ambitious, hard working, frugal, mindful of their obligations to family. Others more or less lack these qualities. Those who lack them entirely are the extreme case of the lower class.

THE ADMINISTRATION'S PROPOSAL

In March 1982 the president announced his long-awaited Enterprise Zone proposal. The main idea is to offer tax credits that will encourage small businesses to locate in seventy-five yet-to-be-chosen depressed urban areas. There would no federal grants or other direct intervention. In choosing among applications made by cities and states, the secretary of HUD would take into account commitments to afford tax and regulatory relief, improve public services, involve neighborhood organizations, give job training, and offer other incentives.[1]

Judging from press accounts, no one—not the president, not those of his assistants who worked out the plan, not local officials, businessmen, or community leaders—is enthusiastic about the plan. Indeed, it is pretty clear that they do not expect it to work. The best one can say for it, those most involved seem to agree, is that it will show that the administration is "trying," and the cost (in tax collections foregone) will be small. Small firms, a Heritage Foundation Bulletin points out, are by far the most effective job creators, but the administration's proposal offers these "almost nothing." What such firms typically need, the bulletin says, is start-up capital. "The prospect of a small tax benefit at some future date, and then only if a zone business is successful, will hardly prompt investors to flock to the inner cities."[2]

Even if enterprise zones *do* attract new businesses it cannot be assumed that this will have much effect on the character of the areas. In the typical depressed area there is a concentration of lower-class persons. As Anthony Downs writes in a study for the Brookings Institution, "many residents there never have the conventions of civilized life instilled in their minds and behavior."[3] Among the "intractable aspects" of life in these areas, he says, are values grounded on "low self-esteem, feelings of personal powerlessness, hostility toward others, admiration of criminal

---

[1] The President's Message to Congress on the plan is excerpted in the *New York Times*, 24 March 1982, 8. The plan is described and analyzed in the Heritage Foundation's *Issue Bulletin* of 29 March 1982.

[2] *New York Times*, 24 March 1982, 9.

[3] Anthony Downs, *Neighborhoods and Urban Development* (Washington, D.C.: Brookings Institution, 1981), 112, 120. Downs assures the reader that his conclusions are "not based upon 'racism' or any other biased perspective but upon simple recognition of reality." He says that his remarks "should not be construed as 'blaming the victims'." 122–123.

and other antisocial behavior, lack of respect for hard work or education, and cynicism." Nearly all attempts to improve the quality of life without changing the residents' values have failed, he writes, "even when those attempts were supported by significant resources."

Some will say, perhaps, that if the enterprise zones do not attract new businesses or the new businesses do not change the character of the areas, nothing much will have been lost. This is surely too simple a view of the matter. If, being politically painless, the Enterprise Zone idea precludes action—certain to be very painful—to encourage movement from the depressed areas to places where cultural and economic opportunities are better, great harm will have been done.[4] Moreover, when the futility of the undertaking has been demonstrated, the anger and cynicism of the residents of the depressed areas and that of their sympathizers elsewhere will certainly rise. This is a cost that may prove large.

THE IBM EXPERIMENT

In 1968 the IBM Corporation, at the initiative of its chairman, Thomas J. Watson, Jr., established a small manufacturing plant in the center of one of the country's most depressed areas—the Bedford-Stuyvesant section of Brooklyn.[5] Watson wanted to demonstrate that a well-managed company could successfully employ and train for promotion the so-called hard-core unemployed. Because of high land costs, stringent building codes, and other municipal regulations, and poor access to major transportation, the inner-city location was costly as compared with a suburban one and still more as compared with adding workers to existing plants. Other costs were also high. The workers—there were to be about 400 of them—were not only unskilled but unused to the discipline of a workplace. IBM's purpose being to demonstrate what could be done with the hard core, it intended to hire only men who had been unemployed for three months or more—only men because it was widely supposed that black males were victims of "matriarchy." In the planning stage no one

[4]This point has been made by Donald A. Hicks.
[5]The account of the IBM plant is based on Edward C. Banfield, "An Act of Corporate Citizenship," ed., Peter B. Doeringer, *Programs to Employ the Disadvantaged* (Englewood Cliffs, N.J.: Prentice-Hall, 1969), 26–57.

doubted that these conditions could easily be met; the unemployment rate in Bedford-Stuyvesant was notoriously high.

After about a year the plant manager found it necessary to relax the rule against hiring women and employed persons. Although he was willing to take boys as young as sixteen, to be patient with workers who did not get to work on time (a foreman would go to a man's home to get him out of bed if necessary), and to overlook arrest records for all but very serious offenses, and although the plant's mission was a simple one (assembling cables for computers), it did not seem possible to get the work done solely with recruits from among unemployed males. The awkward fact was that few of these wanted to work for IBM despite its offer of job security, high wages, training, clean toilets, patient foremen, and an all-black environment.

The lesson to be learned from IBM's experience is that many unemployed males in a depressed area do not value these benefits as much as one might expect. If he must pay for them by accepting the discipline of a well-run workplace—if he must work whether he feels like it or not, learn new skills, take responsibility, and so on—a worker who is accustomed to the lower-class style of life, or to one approximating it, may not accept opportunities of the kind that IBM presents.

The kind of firm that could succeed in a place like Bedford-Stuyvesant and in succeeding, offer some opportunities—very limited ones, to be sure—to persons who are out of, or almost out of, the labor force is in almost all respects the opposite of IBM. Such a firm pays low wages (below the minimum when possible), offers no job security (like the workers it employs, it is here today and gone tomorrow), its rest rooms are dirty, its foremen are rough, it does not trouble itself about the health and safety of its workers (they can take their chances or get out), and it does not ask them to learn skills, take responsibility, or contribute to factory morale (its investment is a short-term one). This is the only kind of firm that can profitably hire the lower-class worker. It is also the only kind of firm that the typical unemployed male will work for. If he comes to work one day and not the next, nobody cares. If he comes late and half drunk, nobody cares (although he may be told to stay away until he is sober).

Firms of this sort were once common in the cities. They were driven out by laws and ordinances intended to improve working conditions (or, if one prefers, to eliminate "unfair" competition with firms operating more nearly in the IBM manner). Driving such firms out did not improve working conditions for most workers: it merely put them out of work or moved them into illegal activities.

REMOVING GOVERNMENT BARRIERS

What is needed to improve job and other opportunities for disadvantaged workers is not tax incentives for employers but removal of a variety of barriers that have been placed in their way by government in recent decades.

There follows a list of measures that should be taken to open opportunities for the disadvantaged.

1. Repeal minimum wage laws, which cause withdrawal from the labor force and unemployment, especially among males in their early teens and twenties, and probably also an increase in felony crime.[6]

2. Remove licensing and other impediments, notably the union shop, to entry into low- and semiskilled occupations such as barber, taxi driver, and practical nurse. In New York City, Thomas Sowell has observed, where a license to drive a taxi costs $60,000, cab drivers are mostly white, whereas in Washington, where the license costs $200, they are mostly black.[7]

3. Repeal (or relax enforcement of) laws and ordinances that set unreasonably high standards for "conditions of work" and for housing construction and maintenance. Most states, Anthony Downs writes, do not prevent localities from requiring housing standards "far surpassing any required to protect the health and safety of the occupants."[8] Housing code administrators, he says, should recognize that "housing in low-income

[6]For a wide-ranging collection of essays on the minimum wage, see Simon Rottenberg, *The Economics of Legal Minimum Wages* (Washington, D.C.: American Enterprise Institute, 1981). See also the paper by John Cogan, who estimates that the minimum wage is responsible for about 40 percent of the decline in black teenage employment between 1959 and 1978, mainly by preventing those displaced from agricultural work from finding jobs in industry. John Cogan, "Black Teenage Employment and the Minimum Wage: A Time Series Analysis," *Working Papers in Economics, No. E-81-11* (Stanford, Calif.: Hoover Institution, September 1981). See also his *Working Paper No. 683*, National Bureau of Economic Research, 1982.

[7]*Meet the Press*, September 29, 1981 (Washington D.C.: Kelly Press), 6.

[8]Downs, *Neighborhoods*, 127, 164.

neighborhoods cannot be kept to the standards of housing in high-income neighborhoods."[9]

4. Reduce the flow of low-skilled and unskilled immigrants. Legal immigration to the United States has increased greatly in recent decades: from 2.5 million in the 1950s to an estimated 4.3 million in the 1970s. Until 1960, the overwhelming majority came from Europe; in the 1970s only 18 percent came from Europe. Immigration now is mainly from Latin America and Asia. The present law tends to favor the unskilled.[10] Estimates vary widely, but illegal immigration may equal or more than equal legal immigration. About half the illegal immigrants are probably Mexican nationals. The number of these will doubtless increase: the Mexican labor force, 30–40 percent of which is unemployed, is exected to double in the next twenty years.

5. Repeal laws and change policies the effect of which is to reduce the supply and raise the price of low-cost housing. This includes repealing rent control laws which reduce investment in the maintenance of housing; ending subsidization of urban renewal and other projects which displace low-income people; revising zoning laws which tend to prevent housing from being passed on to the poor more or less as secondhand automobiles are; and repealing "tenant protection" laws which make eviction of destructive occupants all but impossible, thus precipitating the rapid decline of good neighborhoods.

6. With respect to public housing, change laws and regulations so as not to discourage upward mobility on the part of tenants. Under present law no more than 10 percent of the national supply may be rented to families with incomes 50 percent or more above the median family income. This gives some families an incentive to keep their income down; without a substantial percentage of upwardly mobile families, projects are apt to be "taken over" by the disreputable poor and eventually destroyed.

7. The quality of schooling should be improved by removing students who are disruptive (Downs says that the "state should permit public

[9]Ibid., 164

[10]Barry R. Chiswick, "Guidelines for the Reform of Immigration Policy," in ed., William Fellner, *Essays in Contemporary Economic Problems* (Washington, D.C.: American Enterprise Institute, 1981), 309–47. "A cohort of unskilled immigrants," Chiswick says, "depresses the earnings of low-skilled American workers but raises the earnings of high-skilled workers and the owners of capital," 309–310.

schools to suspend or expel disruptive students"[11]), and by introducing competition among schools by means of the voucher plan or otherwise. ("Using public funds to provide at least some support to privately run schools," Downs writes, "would enable some students now attending public schools to switch to private ones."[12]) Efforts at racial integration should be dropped where it is clear that their incidental effect is to impair learning, and bilingual teaching should be stopped except as it is useful for the teaching of English.[13] The performance of pupils should be carefully monitored with standard tests, and only those able to perform at grade level should be promoted. School administrators should have authority to discharge incompetent teachers.

Whereas the barriers that have been listed make it difficult or impossible for the disadvantaged person to get his foot on the bottom rung of the ladder, welfare programs discourage him from trying to climb the ladder. The incentive effects of these programs reduce mobility in ways that are both direct and indirect.

It suffices here to refer to the striking changes in family structure that have occurred and are occurring, especially among the disadvantaged and most especially among disadvantaged blacks: the tripling of the number of unmarried couples in the past decade; the increase of 50 percent in the number of births by unwed mothers in those years; the increase in female-headed households from about 10 percent of all in the early 1960s to about 15 percent now [1982] (among blacks from about 23 percent to about 41 percent). George Gilder is doubtlessly guilty of rhetorical extravagance in attributing these changes to the growth of the welfare rolls.[14] Other forces have also been at work. But no one doubts that the poor, like other people, respond to incentives or that the tendency of

---

[11] Downs, *Neighborhoods*, 132.

[12] Ibid.

[13] In his account of his service as secretary of HEW, Joseph A. Califano, Jr., remarks that the bilingual program "had become captive of the professional Hispanic and other ethnic groups . . ." and that as a result "too little attention was paid to teaching children English, and far too many children were kept in bilingual classes long after they acquired the necessary proficiency to be taught in English." In part because of this, 40 percent of the Hispanic children dropped out of high school. *Governing America* (New York: Simon and Schuster, 1981), 313. See also the criticisms of bilingual (including black English) teaching in Richard Rodriguez, *Hunger of Memory* (Boston: David R. Godine, 1981), 34–35.

[14] George Gilder, *Wealth and Poverty* (New York: Basic Books, 1981), 12.

welfare incentives is to make it easier to evade family and other responsibilities. That the great majority of welfare recipients are old, disabled,
or children, and that no one becomes old, disabled, or a child in order
to get welfare payments, is of course true. There is no denying, however,
that welfare has contributed significantly to these changes. For many
years a mother was ineligible for Aid to Families with Dependent Children
if there was a man in the house, and even now in about half the states
only single mothers may receive AFDC and Medicaid. These rules account
in large part for the increase in the female-headed households.[15] And
Martin Kilson is right when he writes that, regardless of race, female-
headed households display an "incapacity to foster social mobility comparable to husband-wife and male-headed families."[16]

FACING POLITICAL REALITY

No one can suppose that the barriers and disincentives government
has placed in the way of mobility on the part of the disadvantaged will
be removed or even lowered. The Urban Enterprise Zone proposal is
illustrative of the general situation. The president is said to have thought
reduction in the minimum wage and relaxation of certain safety regulations
should be key elements of the plan, but he found that features of that
sort would kill any chance that the plan might have in Congress. The
poltical reality became evident when the bill sponsored by Representatives Jack Kemp (R-N.Y.) and Robert Garcia (D-N.Y.) was introduced.
"Despite any false rumors to the contrary," Representative Garcia said
then,

> the Enterprise Zone bill does not create a sub-minimum wage, elim-
> inate the OSHA or other vitally needed government bodies, nor in
> any way reduce the presence in the inner cities of health or safety
> programs. I have devoted my entire legislative life in Washington to

---

[15]For evidence that welfare assistance influences female family headship rates,
   see Marjorie Honig, "AFDC Income, Recipient Rates, and Family Dissolution,"
   Journal of Human Resources IX, no. 3: 303–22.
[16]Martin Kilson, "Black Social Classes and Intergenerational Poverty," *Public
   Interest*, 64 (Summer 1981): 61.

putting those laws on the books and I would never participate in
their dismantling.[17]

As with the minimum wage, so with the other barriers listed. Each
has interest groups that will fight for it. Organized labor will not permit
the relaxation of health and safety laws, however unreasonable. The building trades, in concert with the manufacturers of building materials, will
not stand for any tampering with housing and building codes: the more
extravagant the standard, the more material to sell and the more to install.
Changes in immigration laws and regulations to reduce the number of
unskilled immigrants are about as likely to be approved by the powerful
Hispanic organizations as changes in the schools (easy procedures for
dismissal of incompetent teachers, for example) by the organized teachers.

Beyond the opposition of special interests there lies another obstacle
to the elimination of the barriers and the removal of the disincentives.
This is the view—taken for granted by the college-educated middle class—
that these measures benefit "the poor." No doubt self-interest supports
this view: the minimum wage, for example, benefits skilled workers, and
a policy favoring unskilled immigrants benefits skilled workers, among
others. It seems safe to say, however, that it is mainly out of concern for
the less well-off that middle-class opinion supports the minimum wage,
the immigration policy, and the other obstacles to upward mobility.

Since World War II the process of "middle-classification" has gone
on at a rapid and accelerating rate. (In 1980, 16 percent of persons aged
twenty-five or over had graduated from a four-year college, double the
percentage of 1960; in 1980 nearly 40 percent of families had incomes of
$25,000 or more, twice as many [in 1980 dollars] as in 1960.) For the
same reasons that effective demand for wine, gourmet foods, and opera
has increased, the demand for "alleviation of social injustices" has also
increased. Although the rate of change may be slower, there is good
reason to expect that in the next several decades the proportion of the
population having the tastes and standards of the middle and upper-
middle classes will continue to increase.

[17]*Congressional Record*, June 3, 1981, p. E2714. This year Representative Kemp
said he would oppose the administration if there were a minimum wage waiver
because it might "turn off the whole coalition that we have put together—liberals,
Northeastern Democrats, civil rights groups and conservatives." *New York Times*,
30 January, 1982, 19.

One must expect, then, that the old barriers, instead of being torn down, will be built higher, and that new ones will be added. Consider, for example, the findings of a [1982] study by the New York State Industrial Commission of working conditions in the garment manufacturing industry.[18] It has always been easy, the report says, for contractors to set up shop if they could find employees willing to work for less than a lawful wage; increasingly, such workers have become available in New York City.

> Just as in the early 1900s they are poor immigrant women. Often undocumented, unassimilated and equipped with only general skills, they work for whatever is paid, in the shops and in their homes. While learning specialized skills they work long hours to increase their earnings. Payment at the piece rate makes the situation attractive to noncomplying employers who ignore hourly minimum wage requirements.

That the women are learning specialized skills, that it may be impossible for some to work away from home, and that they evidently deem low pay better than no pay—these considerations apparently count for nothing with the commissioner, who has proposed new regulations for the industry, violation of some of which would carry criminal penalties.

As this example suggests, the middle-class reformer tends to see as "exploitation" what to the disadvantaged worker is "opportunity." He is likely, too, to find it hard to bring himself to make some sacrifice of a principle or symbolic value for the sake of some concrete return. Dean Derrick A. Bell, Jr., a civil rights activist, points to a bias of this sort on the part of civil rights organizations. These, he writes, are

> supported by middle class blacks and whites who believe fervently in integration. At their socioeconomic level, integration has worked well, and they are certain that once whites and blacks at lower economic levels are successfully mixed in the schools, integration also will work well at those levels. Many of these supporters either reject or fail to understand suggestions that alternatives to integrated schools should be considered, particularly in majority-black districts. They will be understandably reluctant to provide financial support for policies which they think unsound, possibly illegal, and certainly disquieting.[19]

---

[18]*New York Times*, 14 March 1982, 51.
[19]Derrick A. Bell, Jr., *Yale Law Journal*, 85, no. 4 (1976): 470–516.

A LARGER CLASS OF DISADVANTAGED

I have contended that the disadvantaged are in large measure a by-product of the continuing "middle classification" of American society. Those who have climbed the ladder have, most often with kindly intentions, knocked out its lowest rungs, thereby making it difficult or impossible for those following them to get a foothold by which to rise. Also, especially in the last two decades, government has provided a set of welfare programs intended to relieve the distresses of the poor but having the effect of inducing many people to discard traditional values emphasizing work, self-sufficiency, and responsibility and to accept a life of dependence for themselves and their children. As the size and affluence of the advantaged increase, it is to be expected that the society will invest more heavily in efforts to reduce real and imagined injustices and that the outcome of these efforts will be a larger and less mobile class of the disadvantaged. It would not be surprising if in another two or three decades a considerable part of the whole population, perhaps as much as 10 or 20 percent, were to be in the society but not of it: unskilled and unschooled, unable or unwilling to work, accepting as a matter of right a level of material living—perhaps about half the median in the society— far higher than most people enjoyed a generation or two ago.

From an economic standpoint this outcome will present no great problem. Even if real income grows slowly, the compounding effect will enable the society to provide for many more dependents without taking for the purpose a larger share of GNP. Indeed, in the superaffluent society most people will have to work less if they are not to be buried under an avalanche of consumer goods.

Even at present the problem that the disadvantaged represent is not essentially an economic one. To be sure, there may be unfairness in taking from some in order to give to others, but the unfairness is not less when, as is mostly the case, the recipients are rich rather than poor. The real problem is the exclusion of the disadvantaged from a life that is truly human—from one that, as Aristotle taught, can be made so only by participation in the moral life of a community. It is one thing for persons who have been introduced, however imperfectly, to the elements of civilization to live without working or even at the expense of others. It is quite a different thing for those to live this way who (to quote Downs one last time) "never have the conventions of civilized life instilled in their minds and behavior."[20] In this latter case, life must be at a less than fully human level no matter how adequate the supply of consumer goods. What fate could be worse?

From the standpoint of the society, the presence of a large class of the permanently disadvantaged will constitute a serious political problem. As the gap between the disadvantaged and the rest of the society widens, more and more of the disadvantaged will drop into the lower class. Insofar as it tries to protect itself against the violence and irresponsibility of this class, the society will have to adopt measures which it finds obnoxious. A city, for example, must infringe upon the rights of its citizens if it is to prevent a few "problem families" from destroying good neighborhoods. It must infringe upon them still further if it is to drastically reduce street crime. If it chooses to put up with these aggressions, it thereby invites further aggressions. Either way it loses.

As "middle classification" proceeds, the goodness of the society— *compassion* is the cant word—increases. The public has more inclination (as well as more means) to be kind, generous, forgiving, and so on. It is more generous to the truly needy and, perforce, to the much larger number of those who are not truly needy but who pretend to be in order to avoid work or to gain some other benefit. But goodness, admirable as it is in private affairs, may be disastrous in public ones. What is required for the protection and good order of a public is not goodness but virtue. This is the quality of the statesman, and it often necessitates actions that are harsh or even cruel. American opinion, it is to be feared, will more and more insist upon the rule of goodness, for it is in the nature of goodness not to recognize its limitations. Carried too far, the rule of goodness will produce consequences which threaten the welfare—and the goodness— of the society.

---

[20] Downs, *Neighborhoods*, 112.

PART V

Three Problems for Democracy

Here as elsewhere in this book democracy *refers to rule by the people—that is, to direct democracy. No doubt the problems under discussion exist in some degree in all democracies and perhaps even in all regimes. But where government is most closely dependent upon public opinion they seem to be most serious.*

A century and a half ago Alexis de Tocqueville warned that where irreligion and democracy coexist, philosophers and those in power ought to strive to place the objects of human desire far beyond man's immediate range. Ordinary people's preoccupation with wealth getting, he feared, would otherwise cause them to degrade themselves by putting the gratification of their smallest desires before their concern for freedom and the public good. The tendency of modernity, the first essay of this section maintains, is indeed to encourage popular culture to enjoy the present at the expense of the future. Particular note is made of one consequence of widespread present orientedness: crime. But other consequences that may in the long run be of equal or greater public concern—family instability, failure to acquire work skills, dependency, unwillingness to accept public responsibilities—are also emphasized. As the first footnote of the essay points out, philosophers long before Tocqueville agreed that a present-oriented society cannot be a free one.

The second essay speculates upon a closely related matter: the tension, or antagonism, between individual freedom and social order. It was written to dispute the view that the big-city riots of the late 1960s were caused by "white racism," unemployment, bad housing, and poverty. Occasional outbreaks of violence, it asserts, occur naturally, especially where there are large numbers of young males. To attribute the riots to the prejudice, indifference, or callousness of "society" misrepresents the situation in a way that is likely to make matters worse by generating a diffuse hostility to institutions that need and deserve the respect of the whole public. The readiness of many reformers to derogate social institutions and to exalt the radical individualism that claims moral autonomy for all raises the question: How far can this liberty be extended without undermining the foundations of the society? The implication is that rule by the people will succeed only if the people accept and act upon some common moral principles.

"The Dangerous Goodness of Democracy" develops a theme that was implicit in several essays and explicit in the last paragraph of "The Zoning of Enterprise." This piece is the conclusion of a long essay on American foreign aid doctrines. When the article was written, some twenty years ago, the opinion was widely held that foreign aid would bring about rapid economic development in what were then called the backward *areas of the world and that economic development would bring democracy and peace. These were the "millennial ideas" referred to in the first sentence of the excerpt. Today they have lost much of their appeal, but the moral problem that they raised is as real as ever. When should policy turn on "goodness" as opposed to "virtue"? Goodness, the kindness and liberality that a democratic public characteristically displays, is very different from virtue (the original meaning of which was courage, power, efficacy, and skill), which is the morality appropriate to those whose responsibility it is to protect and serve a public—that is, to statesmen. Participatory, as opposed to parliamentary, democracy, is liable to display goodness where virtue is called for: thus, for example, well-intentioned efforts to improve the lot of the disadvantaged may actually worsen it, thereby creating a division within the society that impairs its confidence in its worth.*

At the end of this section, as at the beginning, the problem under discussion is one that is built into the logic of popular government. For goodness by its very nature is incapable of seeing that there are occasions when virtue, which may sometimes be cruel and unfair, ought to supplant it.

CHAPTER 15

Present Orientedness and Crime

Since the seventeenth century, political philosophers have maintained that an irrational bias toward present as opposed to future satisfactions is natural to both men and animals and is a principal cause of crime and, more generally, of threats to the peace and order of society.[1] It is to

---

From *Assessing the Criminal*, eds. R. E. Barnett and J. Hagel (Cambridge, Mass.: Ballinger, 1977), 133–142. Copyright 1977 by Ballinger Publishing Company. Reprinted by permission.

[1]Most men, Hobbes wrote in *The Citizen* (ch. 2, paragraphs 27 and 32), "by reason of their perverse desire of present profit" are very unapt to observe the dictates of reasons (which are also the laws of nature). If they did observe them, he said in *Leviathan* (pt. 2, ch. XVII) there would be no need for civil government "because there would be peace without subjection."

Spinoza agreed: "in their desires and judgments of what is beneficial they are carried away by their passions, which take no account of the future or anything else. The result is that no society can exist without government and force, and hence without laws to control and restrain the unruly appetites and impulses of men (*Tractatus Theologico Politicus*, ch. V).

For Locke, the "great principle and foundation of all virtue and worth" is placed in the ability of a man "to *deny himself* his own Desires, cross his own Inclinations, and purely follow what Reason directs as best, tho' the Appetite lean the other Way." One who does not know how to resist the importunity of present pleasure or pain for the sake of what reason tells him is fit to be done "is in danger never to be good for any Thing" (*Some Thoughts Concerning Education*, paragraphs 33 and 45).

According to Rousseau, the passage from the state of nature to the civil state forces man "to consult his reason before listening to his inclinations"; in the civil state man acquires (*inter alia*) "moral liberty, which alone makes him truly master

protect men against this irrationality that civil government exists. Hume makes the fullest statement of the case. All men, he says, have a "natural infirmity"—indeed a "violent propension"—that causes them to be unduly affected by stimuli near to them in time or space; this is the "source of all dissoluteness and disorder, repentence and misery," and because it prompts men to prefer any trivial present advantage to the maintenance of order, it is "very dangerous to society." Government is the means by which men cope with this defect of their nature.

> Here, then, is the origin of civil government and society. Men are not able radically to cure, either in themselves or others, that narrowness of soul which makes them prefer the present to the remote. They cannot change their natures. All they can do is to change their situation, and render the observance of justice the immediate interest of some particular persons, and its violation their more remote. These persons, then, are not only induced to observe those rules in their own conduct, but also to constrain others to a like regularity, and, enforce the dictates of equity through the whole society.[2]

The philosophers' perspective is useful for the present purposes for at least three reasons:

---

of himself; for the mere impulse of appetite is slavery." The judgment that guides the general will must be "taught to see times and spaces as a series, and made to weight the attractions of present and sensible advantages against the danger of distant and hidden evils." This makes a legislator necessary. The legislator ought "to look forward to a distant glory, and, working in one century, to be able to enjoy the next" (*The Social Contract*, bk. I, ch. VIII; bk. II, chs. VI and VII). In *A Discourse on the Origin of Inequality* he explains (in Part One) that the savage is "without any idea of the future, however near at hand" and (in Part Two) that "as men began to look forward to the future, all had something to lose, everyone had reason to apprehend that reprisals would follow from any injury he might do to another." In this situation men "had just wit enough to perceive the advantages of political institutions, without experience enough to enable them to foresee the dangers. The most capable of foreseeing the dangers were the very persons who expected to benefit by them." Law and property, Bentham maintained, exist to restrain and protect "the man who lives only from day to day . . . precisely the man in a state of nature." "To enjoy quickly—to enjoy without punishment—this is the universal desire of man; this is the desire which is terrible, since it arms all those who possess nothing, against those who possess anything. But the law, which restrains their desire, is the most splendid triumph of humanity over itself" (*Principles of the Civil Code*, ch. 9 [*Works*, vol. 2]).

[2]David Hume, *An Enquiry Concerning the Principles of Morals*, 1777 ed., sec. 6, pt. 1, paragraph 196. Other quotations are from the *Treatise of Human Nature*, 1740, bk. 3, pt. 2.

1. It emphasizes a fact—now well established by experimental psychology—that there is an innate (i.e., biologically given) tendency to choose, as between rewards that are otherwise the same, the one that is nearer in time.[3] Although not of equal strength in all organisms of the same species, some degree of present orientedness is apparently present in all. That the tendency is innate does not, of course, prevent it from being greatly affected by cultural or other nonbiological forces.

2. It calls attention to the diversity of the ways in which present orientedness may injure the society. Crime is perhaps the most conspicuous of these, but behavior that is antisocial without being illegal, or that is merely unsocial, also arises from present orientedness and may represent a greater threat to the "quality of life" or, as it used to be called, *civilization*. However much the social bond is harmed by assaults, robberies, rapes, and the like, it may be even worse harmed by behavior that is merely regardless of others' wishes, needs, interests, and rights.

3. It raises the question of how society may be protected against the consequences of present-oriented behavior and, especially, of the role of government in that connection.

PRESENT ORIENTEDNESS AS PSYCHOPATHY

For at least three-quarters of a century, psychiatric literature has discussed "the kind of person who seems insensitive to social demands, who refuses to or cannot cooperate, who is untrustworthy, impulsive and improvident, who shows poor judgment and shallow emotionality, and who seems unable to appreciate the relation of others to his behavior. Such persons are commonly called 'psychopaths.' "[4] That extreme present

[3]George Ainslie, "Specious Reward: A Behavioral Theory of Impulsiveness and Impulse Control," *Psychological Bulletin* 82 (1975): 463–96.

[4]Harrison G. Gough, "A Sociological Theory of Psychopathy," *American Journal of Sociology* 53 (March 1948): 365. Gough's theory is strikingly similar to that of Adam Smith in *The Theory of Moral Sentiments*. George Herbert Mead, Gough says, gave what is probably the most acceptable account of the "self" as a link between the individual and the social community, his view being that the self has its origin in communication and the individual's taking the role of the other. Smith's "abstract spectator" comes into being and functions exactly in the manner of Mead's "generalized other." For Smith, as for Mead, it is the internalization of the group's standards that mainly checks impulse. "The pleasure which we are to enjoy ten years hence," Smith writes, "interests us so little in comparison

orientedness is conspicuous among the traits of the psychopath is evident from the following:

> Psychopaths are characterized by an over-evaluation of the immediate goals as opposed to remote or deferred ones; unconcern over the rights and privileges of others when recognizing that they could inter-fere with personal satisfaction in any way; impulsive behavior, or apparent incongruity between the strength of the stimulus and the magnitude of the behavioural response; inability to form deep or personal attachments to other persons or to identify in inter-personal relationships; poor judgment and planning in attaining defined goals; apparent lack of anxiety and distress over social maladjustment as such; a tendency to project blame onto others and to take no respon-sibility for failures; meaningless prevarication, often about trivial mat-ters in situations where detection is inevitable; almost complete lack of dependability and of willingness to assume responsibility; and finally, emotional poverty.[5]

What David Shapiro calls "neurotically impulsive styles" involve many of the same traits.[6] Neurotically impulsive people are remarkably lacking in active interests, aims, values, or goals much beyond the imme-diate concerns of their own lives. Neurotically impulsive people usually do not have abiding, long-range personal plans or ambitions. Durable emotional involvements—deep friendships or love—are not much in evi-dence. Family interests or even personal career goals are usually not strong. When frustrated they show lack of forebearance or tolerance. Their interests tend to shift erratically in accordance with mood or opportunities of the moment and without being subjected to the critical, searching process that is called *judgment*. One whose style is impulsive tends to be

---

with that which we may enjoy today; the passion which the first excites is naturally so weak in comparison with that violent emotion which the second is apt to give occasion to, that one would never be any balance to the other, unless it was supported by the sense of propriety [which is the advice, or command, of the abstract spectator]" (pt. IV, ch. II). The psychopath, Gough writes, is unable to foresee the consequences of his own acts, especially their social implications, because he is "deficient in the very capacity to evaluate objectively his own behavior against the group's standards" (Gough, pp. 364–65).

[5]This is H. J. Eysenck's summary of the cited article by Gough. It appears in *Crime and Personality* (London: Paladin, 1971), 54.

[6]David Shapiro, *Neurotic Styles* (New York: Harper Torchbook, 1965). The author says those exhibiting impulsive styles include (among others) most persons usually diagnosed as psychopathic and certain kinds of male homosexuals, alcoholics, and probably addicts (p. 134).

without moral scruples. He is given to what in others would be called insincerity and lying but in him may be better described as a kind of glibness. He seems free of inhibitions and anxieties. Because his awareness is dominated by what is immediately striking and relevant to his immediate need or impulse, the world of the neurotically impulsive person is seen as discontinuous and inconstant—a series of opportunities, temptations, frustrations, sensuous experiences, and fragmented impressions. This style does not necessarily involve lack of intelligence—it *does* involve lack of concentration and of logical objectivity—but intelligence in the subjective world of the neurotically impulsive can function only to arrange speedy action.

THE PRESENT-ORIENTED CULTURE

That cultures (and subcultures) differ greatly in their tendency to reinforce or weaken the natural disposition of the individual to prefer present to future rewards has long been noted. Early in the last century, for example, John Rae recorded, albeit impressionistically, a great many evidences of differences in the time preferences of cultures.[7]

The normal or typical individual in some cultures exhibits a set of traits remarkably like those of the psychopaths of our culture. For example, Mayhew in his *The Life and Labour of the London Poor* (1851) notes of the "vagabond":

> his repugnance to regular and continuous labor—his want of providence in laying up a store for the future—his inability to perceive consequences ever so slightly removed from immediate apprehension—his passion for stupefying herbs and roots, and, when possible, for intoxicating fermented liquors—his extraordinary powers of enduring privation—his comparative insensibility to pain—[his] immoderate love of gaming, frequently risking his own personal liberty upon a single cast—his love of libidinous dances—the pleasures he experiences in witnessing the suffering of sentient creatures—his delight in warfare and all perilous sports—his desire for vengeance—the looseness of his notions as to property—the absence of chastity among his women, and his disregard of female honor—and lastly,—

---

[7] "State of Some New Principles on the Subject of Political Economy," first published Boston 1834, reprinted in R. Warren Jones, John Rae, *Political Economy*, vol. 2 (Toronto: University of Toronto Press, 1965).

his vague sense of religion—his rude idea of a Creator, and utter absence of all appreciation of the mercy of the Divine Spirit.[8]

The Appalachian mountaineer as described by Weller (1965) has a cultural style in many respects similar to the neurotically impulsive one.[9] The mountaineer, Weller writes, does not think ahead or plan; disregard of time is part of his makeup. As a child he learns the *feeling* of words and to grasp nuances of personal relations, but he does not learn to grasp ideas, concepts, or abstractions. He is reared impulsively, permissively, and indulgently, seldom being required to do what he does not want to do. As a youth, he holds few realistic hopes or ambitions, is seldom able to articulate goals, and is even reluctant to talk about the future. As an adult he tends to be capricious, vacillating, and volatile. He tends also to lack a sense of who he is and where he is going—of being a person in his own right. He is self-centered; all that he does has the self at heart. He does not conceive of a "public good" except as it coincides with his "private good." He sees the government as "they" and expects it to care for him. A fatalist, he has no feeling that he himself is to blame for his lot. His life is pervaded by apprehensions and anxieties, however, arising from a lack of self-confidence. His relations with others, even with members of his family, are difficult and uneasy. Married persons tend to lead separate lives and to have little in common. For the mountaineer work is a necessary evil, not an outlet for creativity or a means of fulfillment.

EFFECTS OF PRESENT ORIENTEDNESS

Insofar as they are expressed in action, the traits associated with present orientedness (both psychopathic and culturally given) tend to give rise to a characteristic set of social conditions. These in turn support and

[8] Henry Mayhew, *London Labour and the London Poor* vol. 1 (London: Griffin and Co., 1851), 4. Frederick Engels, writing at almost the same time (although his book was not published in English until 1887) remarked in *The Condition of the English Working Class*: "The failing of the workers in general may be traced to an unbridled thirst for pleasure, to want of providence, and of flexibility in fitting into the social order, to the general inability to sacrifice the pleasure of the moment to a remoter advantage."

[9] Jack E. Weller, *Yesterday's People, Life in Contemporary Appalachia* (Lexington: University of Kentucky Press, 1965). Some of the sentences in the paragraph are Weller's own and others are paraphrases.

perpetuate the traits, the relation being that of a "feedback loop." The principal conditions are listed next along with some of the traits that produce them:

Condition	Traits
1. Ignorance (including lack of work skills)	Lack of goals, inability to concentrate
2. Poverty and squalor	Improvidence, untrustworthiness, inability to accept discipline of work, fatalism
3. Unplanned births, illegitimacy	Inability to think ahead or to control impulses; lack of feelings of responsibility, lack of moral scruples
4. Weak or broken family (male absent, lack of parental care of children)	Inability to form deep or durable attachments; inability to tolerate frustration
5. Dependency (welfare, borrowing, handouts, etc.)	Preoccupation with self and with immediate wants; lack of anxiety at failure to achieve
6. Poor health	Impulsiveness (fighting, reckless acceptance of risks); inability to think ahead (failure to secure preventive health care); sensual self-indulgence (abuse of alcohol, tobacco, etc.)
7. Nonparticipation	Feelings of personal inadequacy; preoccupation with self; unwillingness to accept responsibility
8. Crime and delinquency	Lack of moral scruples; inability to control impulses, to identify with others, to exercise critical faculty called judgment; freedom from inhibitions and anxieties

As one would expect, these conditions prevail in the Appalachian community described by Weller.[10] The mountaineer acquires social but not other skills. Work is for him merely a means of making a living, and he is satisfied with a very meager one—enough food, clothing, and shelter for survival (acceptance of undesirable conditions is part of his way of life). He is content to live in squalor (he has no time to exhume himself from mounting piles of trash, but he can sit on his front porch swing doing nothing). Births are unplanned, and the illegitimacy rate is high. Households are seldom female headed, but husbands and wives have little in common and tend to lead separate lives. Small children are played with, but older ones are left pretty much to themselves. There is an off-hand attitude toward money, almost as if it did not matter, and impulsive buying of household appliances is common. The mountaineer expects the government to care for him. He does not join neighborhood groups or the larger organizations of the city (he may, however, involve himself impulsively in a community group, the style of which is not impersonal). Contrary to what one might expect, there is little delinquency and hardly any serious crime.

CRIME IN PARTICULAR

It would be an error to suppose that present orientedness (whether psychopathic or culturally given) necessarily leads to crime. The psychopath who lives among normal people may be kept out of trouble by caretakers of one sort or another—relatives, friends, lawyers, and so on. In a society the culture of which is present oriented, one's knowledge that others are as hot tempered as oneself is apt to constitute (despite Hobbes and the other political philosophers) an effective social control. In such a culture, people are likely to go to great length to avoid giving even accidental offense to others, out of fear of provoking quick reprisals. (This may explain the low crime rate in the Appalachian communities. Weller stresses the unwillingness of mountaineers to do anything that neighbors might construe as interference with them or that might otherwise stir ill will.)

It is evident, however, that a cohort of present-oriented persons could as a rule be expected to commit a good many more crimes of certain

[10]Ibid. Here again some sentences are Weller's and others are paraphrases.

types than a matched cohort of persons who are not present oriented. The qualification "of certain types" is important. Present-oriented people are, of course, incapable of crimes that require them to think and plan ahead to create organization or give it leadership, or even to be dependable.

The crime proneness of the present-oriented person has obvious connections with his characteristic traits. His inability to foresee the consequences of his actions or to control his impulses tends to behavior that, without being malicious, is criminally reckless. His inability to enter into the feelings of others and his lack of moral scruples together with the traits just mentioned may prompt him to brutal acts such as assault and rape. His improvidence, together with his inability to tolerate frustration, may lead to his "taking things" for which he has a present need; if what he needs is illegal—for example, narcotics—or if he can get what he needs most easily by violence or the threat of it, his "taking" is likely to involve other crimes as well.

For most people, Eysenck has maintained, the most effective deterrents to crime are the anticipatory pangs of conscience.[11] Early conditioning, he has pointed out, produces a disincentive—namely, the autonomic anxiety and fear reaction provoked by the idea of the crime— that is felt almost simultaneously with the temptation to the criminal act and before any possible gain can be had from it. The impulsive (present-oriented) person is resistant to early conditioning and has had little of it. Therefore, although he is usually aware of what society deems "right" and "wrong," he does not experience the unpleasant subjective state ("pangs of conscience") that for the person who has had early conditioning is usually a sufficient deterrent. It is possible, Eysenck acknowledges, that a child may be conditioned in the "wrong" direction—that is, toward behavior that society wants suppressed.[12] This, presumably, is common in present-oriented cultures.

The threat of punishment at the hands of the law is unlikely to deter the present-oriented person. The gains that he expects from his illegal act are very near to the present, whereas the punishment that he would suffer—in the unlikely event of his being both caught and punished—lies in a future too distant for him to take into account. For the normal person there are of course risks other than the legal penalty that are strong deterrents: disgrace, loss of job, hardship for wife and children if one is

[11] Eysenck, *Crime and Personality*, 120–23.
[12] Ibid., 146

sént to prison, and so on. The present-oriented person does not run such risks. In his circle it is taken for granted that one gets "in trouble" with the police now and then; he need not fear losing his job since he works intermittently or not at all, and as for his wife and children, he contributes little or nothing to their support, and they may well be better off without him.

THE PROSPECT

In countries in which irreligion and democracy coexist, de Tocqueville wrote, the instability of society fosters the instability of man's desires, hiding the future and disposing men to think only of tomorrow.[13] Moralists ought therefore to teach their contemporaries that it is only by resisting a thousand petty passions of the hour that the general and unquenchable passion for happiness can be satisfied. Men in power ought to strive to place the objects of human actions far beyond man's immediate range and, above all, to make it appear that wealth, fame, and power are the rewards of labor, not chance.

Since de Tocqueville's day, moralists have largely succeeded in persuading their contemporaries that, God being dead and existence absurd, what matters is the full and unfettered expression of self. Men in power, meanwhile, have responded to the growth and spread of democracy (why did de Tocqueville not anticipate this?) by placing the objects of human action so as to assure their reelection.

Other forces have combined with irreligion and democracy to make the predominant style of modern culture ever more present oriented or, if the reader prefers, less future oriented. The rapid growth and spread of affluence, the transfer to the state of most responsibility for providing for the individual's future, the extension of higher education to the masses (education that exalts self, sentiment, and expression while deriding institutions, reason, and subordination to a common good)—these influences have been powerful in recent decades, and there is every reason to expect them to be so for a long time to come. The sudden and tremendous increase in the number and proportion of young people in the 1960s—young people who had money in their pockets and so were free of all constraints—dramatically strengthened these forces.

[13]Alexis de Tocqueville, *Democracy in America*, vol. 2 (New York: Knopf, 1948), Chapter 17.

As the predominant cultural style becomes ever more hostile toward authority, discipline, and all constriction of individuality, and ever more indulgent toward self-expression, one must expect to see more frequently displayed the traits and conditions associated with present orientedness. Except as children internalize "a stringent morality based on fear and trembling," Bettleheim warns, they will live out their lives on a primitive ego, one which prefers the experience that gives immediate pleasure, and, although they may acquire bits of knowledge and skill, they will remain essentially uneducated and uneducable.[14] The more present oriented the culture, it seems safe to say, the less stringent will be its morality and the less that morality will be based on fear and trembling. In the more relaxed and permissive culture that is coming (if it is not already here), personalities that are now judged psychopathic or neurotically impulsive will be considered normal or, at any rate, not remarkable. As those who have not learned in childhood to control their impulses become more numerous, more caretakers will be required to guide and check their conduct. If a considerable degree of present orientedness is the norm of the culture, where are these caretakers to be found?

In the society that has overcome all concern for the future, the voice of conscience will be so still and soft as to be nearly inaudible. What for most people is still by far the most important deterrent to crime and, more generally, to socially undesirable behavior will be weakened accordingly. As for the deterrent effect of law and the machinery of law enforcement, that, even in the present state of the public mind, is generally of very little effect. In the society that does not concern itself with the future, even the pretense of such deterrence may be given up. "It is possible to imagine," wrote a prophet of the self-expressive culture,

> a society flushed with such a sense of power that it could afford to let its offenders go unpunished. What greater luxury is there for a society to indulge in? "Why should I bother about these parasites of mine?" such a society might ask. "Let them take all they want. I have plenty." Justice, which began by setting a price on everything and making everyone strictly accountable, ends by blinking at the defaulter and letting him go scot free.[15]

---

[14]Bettleheim, in eds. Nancy F. and Theodore R. Sizer, *Moral Education* (Cambridge: Harvard University Press, 1970), 90.

[15]Friedrich Nietzsche, *The Genealogy of Morals* (New York: Macmillan, 1897), essay 2, sec. 10.

CHAPTER 16

How Many, and Who, Should Be Set at Liberty?

Robberies, murders, rapes, are the sports of men set at liberty from punishment and censure.

JOHN LOCKE, *An Essay Concerning Human Understanding*

I

It is now widely held, one might almost say officially held, that not only robberies, murders, and rapes but civil disorder in general arise from society's neglect of and injustice toward the poor and the black. The Kerner Commission Report emphasized both "white racism" and the failure of society to create adequate material and social conditions as causes of riots as well as poverty and social disorganization generally. Essentially the same view was taken by the President's Commission on Crime and Violence and more recently by the President's Commission on Campus Unrest. Former Attorney General Ramsey Clark summed up the prevailing view in his best-selling book, *Crime in America*:

> The solutions for our slums, for racism and crime itself in mass society, are basically economic. We must make an economic commitment to end them. If we are to control crime, we must undertake a massive effort to rebuild our cities and ourselves, to improve the human condition, to educate, employ, house, and make healthy.[1]

The view to be taken here is that of Locke rather than of Clark and today's conventional wisdom. I trust it goes without saying that like Clark

---

From *Civil Disorder and Violence*, ed. Robert A. Goldwin (Gambier, OH: Public Affairs Conference Center, 1971), 27–45. Copyright 1971 by the Public Affairs Conference Center, Kenyon College. Reprinted by permission.

[1] Ramsey Clark, *Crime in America* (New York: Simon & Schuster, 1970), 67.

and the others I too deplore racism, slums, and poverty and favor education, employment, and the improvement of the human condition. My disagreement with them is in thinking that civil disorder and crime are not in the main caused by the material conditions of life, by social neglect, or even by racism. I do not think that the rebuilding of our cities or even the disappearance of racial prejudice would reduce the amount of crime and disorder very significantly. In my judgment the causes Locke pointed to are much more fundamental.

Ever since the Watts riot many people have assumed that the riots in the cities—and much other disorder as well—were (and are) in the nature of protest against poverty, slums, and racial injustice. I do not doubt that *some* rioters were protesting and that without them large riots could not have taken place. Locke's use of the word *men*, however, suggests that particular circumstances of time and place may be of subordinate importance in explaining behavior. Large-scale rioting is not peculiar to America, to blacks, or to cities. As I write, 80,000 Western European farmers are swarming into Brussels to demand higher prices; in the meleé one man has been killed and 140 injured. In Belfast the Irish are battling both each other and the British Army sent there to restore order. A few weeks ago 10,000 Italian policemen and army troops were trying to keep the peace in Reggio Calabria, a town that was disappointed at not having been made provincial capital. Two years ago there was mass looting in Montreal during a police strike, and the year before that the Tel Aviv city hall was stoned by rioters, injuring three policemen, in a demonstration against unemployment. At about the same time military police quelled a riot of construction workers in Amsterdam.[2] Obviously, large-scale riots are not peculiarly the product of conditions in American cities. If violence is as American as apple pie, it is also as Italian as *pasta*.

Negroes who riot in American cities do not necessarily do so *because* they are Negroes or *because* they consider themselves victims of injustice. Negroes have high arrest rates for crimes of violence, but it does not follow that they have these high rates because they are Negroes or because the police are unfair to them. A study spanning the period 1941–65 in a small industrial community in the Great Lakes region lends

---

[2]On the rioting in Reggio Calabria, *New York Times*, 1 February 1971; in Tel Aviv, *New York Times*, 15 March 1967; and Amsterdam, *New York Times*, 15 June 1966.

no credence to the explanation of the Negro–white crime rate dif-
ferential in terms of some distinctive aspect of Negro culture or in
terms of racial conflict, whether viewed as the Negro's reaction to
the frustrations resulting from racial discrimination or the expression
of racial bias by the police.

The difference in crime rates "results predominantly from the wider distribution among Negroes of lower social class characteristics associated
with crime."[3]

If, as may be, Locke meant by "men" not humankind but males as
opposed to females, his statement is no less true. Men, especially young
ones, commit far more than their proportionate share of all crimes, and
they are also mainly responsible for most civil disorders. In Belfast,
according to the *New York Times,* boys as young as six join junior branches
of I.R.A. clubs and at seventeen graduate into the I.R.A. itself. In Com-
munist China, students using spears, catapults, and iron bars battled each
other for months in the university district in 1968. Presumably the situation was no different in Locke's day. "I would there were no age between
ten and twenty-three," says a shepherd in *The Winter's Tale,* "or that
youth would sleep out the rest; for there is nothing in the between but
getting wenches with child, wronging the ancientry, stealing, fighting."
It should not be surprising or require explanation that in the so-called
ghetto riots, boys and young men did most of the burning, looting, and
sniping.

Locke's word *sport* was well chosen. It is in the nature of the human
animal—and above all of the young male human animal—to want action,
excitement, danger, and the opportunity to show off. He wants to prove
himself, to demonstrate to his own and others' satisfaction that he "can
take it" and "has heart." Generally speaking, the younger he is the less
able he is to control his impulses and to take account of the future. To
the extent that crime and civil disorder are the work of youth (and about
one-half of all serious crimes are committed by juveniles), they are only
incidentally a protest against poverty and injustice when they are that at
all. The boys in Belfast, the *Times* reporter says, "give the impression
that they are ready to die for Irish unity," and no doubt they are. But it
is significant that "the Saturday night riot has been a way of life for years."[4]

[3] Edward Green, "Race, Social Status, and Criminal Arrest," *American Sociological
Review* 35, no. 3 (June 1970): 489.
[4] *New York Times,* 14 February 1971.

As to the motivation of the student rioters brought into existence by the Maoist Cultural Revolution, we are told:

> Observers do not know what the students fight about. It is generally believed that they themselves have long since lost sight of any original political reasons for their differences and, having reached a stage of irreconcilable, irrational gang warfare, fight mostly for the fun of it.[5]

That rebuilding American cities would not make youths less unruly is suggested by the fact that Reston, Virginia, a middle-to-upper income "new town" only four years old and widely regarded as a model, has suffered seriously from teenage vandals who have smashed store windows, painted obscenities on windows and walkways, and, apparently, committed larceny.[6]

Every society has some adults who share the juvenile and adolescent restlessness, its taste for "action" and "making things happen," and its inability to control impulses and to take the future into account.[7] Such persons form a small part of the adult population today. This has not always and everywhere been the case. In the *Iliad* everyone thinks and acts like a juvenile delinquent. In Europe, Bruno Bettelheim tells us, it was not until the seventeenth century that differences began to appear between the games, manner of dress, stories, and style of life of children and adults.[8] Small as it is in proportion to the whole population in the United States today, a psychologically juvenile but chronologically adult lower class (not to be confused with the working class) is responsible for far more than its proportionate share of crime, especially violent crime. To account for the many serious problems to which this class gives rise— not only violent crime but much drug abuse, alcoholism, chronic unemployment, and functional illiteracy—on the grounds of protest against poverty and injustice is just as implausible as to account on these grounds for the problems to which boys give rise.

Although it is safe to say that there is a political element in every large-scale riot, the motives of most rioters are at most only faintly political, and assault and looting is done mostly by youths and members of

[5]Ibid., 15 July 1968.

[6]*Washington Post*, 15 August 1970.

[7]See the discussion of the lower class and of cultural and psychological factors influencing crime in E. C. Banfield, *The Unheavenly City* (Boston: Little, Brown, 1970).

[8]Bruno Bettelheim, *The Children of the Dream* (New York: Macmillan, 1969), 54–55.

the lower class whose motives are not political at all. The French Revolution of 1848, Nassau Senior wrote at the time, was partly the work of educated revolutionaries and partly also of the masses who hoped to partake of the blessings that the triumph of socialism would bring. But there was an important third class of actors "who took part in it from a mere puerile love of excitement." Senior reflected:

> It is humiliating to be forced to believe, that the secular destinies of France, and, to a considerable extent, those of the whole Continent, have been influenced for centuries to come, by a riot got up by a few hundred lads, by way of a lark. But such was the case. Boys of fifteen and sixteen, *illustres gamins* as they are seriously called by M. Caussidière, took a principal part in the little of real fighting that took place. A spectator of the Revolution told me that he saw a boy of eleven years old lurk behind a wall and fire on an officer as he rode by. The man fell, mortally wounded; the child ran away, frightened and crying.[9]

II

What Locke thought of as punishment—death, torture, mutilation, confinement in a dungeon, confiscation of property, banishment—have disappeared from our society. No doubt this is partly because of our great material wealth (the harshest punishments are the cheapest to provide!) and partly also because of our greater sensitivity to physical suffering. Increasingly we have sought to substitute the offer of reward for the threat of punishment.[10] This tendency seems to have been strengthened by experiments done twenty to thirty years ago which seemed to show that rats learn faster when rewarded with food than when punished with

[9]Nassau W. Senior, *Industrial Efficiency and Social Economy*, original manuscript arranged and edited by S. Leon Levy (New York: Henry Holt, 1928), vol. II, 295.

[10]Kenneth Boulding gives the social sciences some credit for raising doubts about the deterrent effect of punishment and for suggesting "its strong inferiority to reward as an incentive to learning." From later passages, however, it is clear that much of the credit should go to pseudoscience. Criminologists, he says, "have succeeded in achieving certain reforms" in the legal process, but "criminology has received extraordinarily little attention from the theorists and even the statisticians in the last generation, and one might also describe it as a stagnant field." *The Impact of the Social Sciences* (New Brunswick, N.J.: Rutgers University Press, 1966), 82, 107.

electric shocks. According to Dr. Barry F. Singer, a psychologist, this conclusion stood unchallenged in the literature for twenty years and was the only one to enter the criminological literature. "This is unfortunate," he writes, "because it is wrong."[11] In the past ten years, he says, a flood of experimentation has established not only that punishment is more effective than reward but also that *severe* punishment is much more effective than mild in "suppressing behavior." Holding severity constant, punishment is most effective when delivered simultaneously with or a few seconds after the behavior that is to be suppressed. Other factors being equal, the more certain the punishment the more effective. *Severity* of punishment, however, is by all odds the most important factor.

This research, like that done earlier, has been mostly on rats and other animals; only a small part of it has involved humans. Singer, however, is confident (unduly so, in my judgment) that the findings "should indicate the right direction" for research and for policy.[12] Extrapolating, he suggests that immediacy of punishment is even more important with the criminal than with the general population (this because criminals tend to be people who do not defer gratification), that delay of punishment is probably largely responsible for the apparent ineffectiveness of our current punitive system, that punishment would be more effective if it were done—even if only provisionally or symbolically (as for example by a policeman's "booking" an offender)—immediately upon apprehension of the offender, and that offenders, whether parolees or not, should be forcibly (I trust that he means forcefully) reminded after they have left prison of the possibility of punishment should they commit a new offense. A punished criminal, he concludes, should be fearful and uneasy in the presence of parole officers, courts, and police and anxious when he even thinks upon committing a crime. Punishment, he stresses, must be unpleasant if it is to be effective.

The general tendency of our society is, of course, the exact opposite of this. Most offenders know very well that the probability of their being apprehended (unless it be for homicide) is trivial, that the probability of

[11]Barry F. Singer, "Psychological Studies of Punishment," *California Law Review* 58, no. 2 (March 1970): 414.

[12]The difficulties, amounting to impossibilities, in the way of reaching general conclusions as to the efficacy of punishment as a deterrent to crime in modern society are evident from the review of the literature by Franklin E. Zimring, *Perspectives on Deterrence*, Public Health Service Publication No. 2056 (Washington, D.C.: Government Printing Office, January, 1971).

their being tried and convicted for the most serious offense that they have committed is very small (if they are tried at all it is likely to be for a lesser offense to which they have agreed to plead guilty), and that the probability of their being imprisoned if convicted is also very small (nonexistent for first offenders whose crime is not of utmost seriousness). Not only is the prospect of punishment small, it is also remote; persons accused of serious crimes often remain free on bail for a year or two before coming to trial, and they may have another period of freedom after conviction if they appeal. If it is true that those who commit the common crimes tend to have short time horizons, the law's delays must put them at a disadvantage by placing the possibility of punishment beyond their comprehension.

Psychologists, Singer tells us, are perfectly clear as to what constitutes severe punishment for a rat. With humans they are far from clear. Although the point of his article is to emphasize the efficacy of severe punishment promptly administered, his own standards turn out to be no more severe than those generally prevailing. Discussing "community treatment" (a technique currently favored over imprisonment by many criminologists), he remarks that it would allow for various punishments, the relative effectiveness of which it is difficult to judge.

> We could severely curtail the offender's liberties and privileges, or require him to report frequently to police and parole officers, or to work in the community during the day but return to prison at night. We might draw from "primitive" societies and customs, from times and places without prisons. Thus, we might require the offender to make a public apology to his victim who would then ceremoniously forgive him, or require him to wear a "scarlet letter." We might revive stocks and dunking.[13]

One might think that the experimental evidence he has described would imply the lash, the branding iron, and perhaps also the headsman's axe administered via television during prime time. (Television, after all, shows bloody scenes in Vietnam, especially ones that can be described as American atrocities.) It might be, of course, that having to *wear* a "scarlet letter" would be perceived by present-day criminals as every bit as severe a punishment as being branded was a century or more ago. If that were the case, presumably judges would also deem that punishment "cruel and unusual" and therefore unconstitutional. Even if judges did not reach that conclusion, juries, if they thought the punishment severe, would usually

---

[13]Singer, "Psychological Studies," 435.

find defendants innocent rather than subject them to it. My point is that any punishment that can properly be called severe is ipso facto unacceptable to public opinion.[14]

It is interesting that Singer uses Locke's word *censure*. Punishment in and by a community, he says, "should develop aversive reactions to courthouses, police, symbols of authority, *and probably also to the disapproving frowns and moral censure of his fellow citizens*" (emphasis added). With respect to this form of constraint too, it would seem, the conditions of modern life have changed matters very fundamentally. As compared to a generation or two ago, few Americans now live in anything that can properly be called a community. This being the case, they care little or not at all about their neighbor's disapproval.

Consider two paradigms. In the first, the individual knows from childhood that throughout the whole course of his life he will encounter only the same small set of persons (e.g., neighbors, storekeepers, schoolmates, employer and fellow employees, etc.) and that each of them will always remember what he does and what others say about him. If the individual is rational in the economist's sense, he will see that a good reputation is an investment that will bring him high returns over his lifetime, and he will behave accordingly. In the second paradigm, the individual knows from childhood that he will never encounter the same person twice and that no one will know anything of his past actions. In this case he will see some advantage in "projecting an image" but none in "building a reputation."

Our society has been, and is, moving from something like the first model toward something like the second. There was a time when most people expected to live out their lives in the same rural community, small town, or city neighborhood and thought it likely that their children would do so too. Under these circumstances it "paid" to abide by, or at least pretend to abide by, the generally accepted rules of decency. Today relatively few people expect to live long in the one place, and fewer still expect their children and grandchildren to continue there. Consequently it does not "pay" to invest so much in a "good name"; indeed the rational

[14]When a district court judge gave a teen-age offender the choice between twenty lashes with a belt and five years in prison, the incident was investigated by the F.B.I. The youth chose the lashes, which were administered in the judge's chambers by a relative in the presence of his mother. *New York Times*, 28 March 1969.

thing for some may be not to invest or even to disinvest by "trading on the reputation" of parents or, more likely, on the expectation, which arose under other conditions and survives as an instance of cultural lag, that most people will act on the rule that honesty pays.

Even if "disapproving frowns" of fellow citizens are taken into account, their "moral censure" is likely to have little or no force. Upper-middle and upper-class people (in general, "the educated") have always tended to be skeptics in religion, and today these classes are larger than ever. Characteristically the upper-middle or upper-class individual believes that he must be his own ultimate judge of what is right and wrong and that the "moral censure" of anyone claiming authority over him is mere opinion. This "postconventional" morality defines *right* action as that which is in accord with any universal or very general principle that the individual deems worthy of choice.[15] The principle may or may not coincide with what the law requires. When the two do not coincide, the individual considers it is his right, indeed his duty, to be "true to himself"—meaning, of course, to substitute his judgment for the political or other process by which the law was promulgated.

Lower-middle and working-class people, by contrast, tend to have a "conventional"morality—that is, one based on unquestioning adherence to rules laid down by some authority (priest, parent, teacher, etc.) or by habit and custom.[16] The morality of these classes is rapidly breaking down, however. Through the influence of the mass media and the educational establishment, the views of the "educated" classes are incessantly pressed upon the "uneducated" ones, undermining their confidence in the standards

---

[15]See the account of "levels and stages in moral development" in the essay by Lawrence Kohlberg in eds. Nancy F. Sizer and Theodore R. Sizer, *Moral Education, Five Lectures* (Cambridge: Harvard University Press, 1970), 71–72.

[16]On the basis of interviews with 3,100 men, representative of all men in civilian occupations in the U.S., two sociologists recently concluded (among other things) that "the lower the men's social class positions, the more likely they are to feel that morality is synonymous with obeying the letter of the law." Melvin L. Kohn and Carmi Schooler, "Class Occupation and Orientation," *American Sociological Review* 34 (October 1969): 667.

It is interesting to note that printers here and there refuse to set type when they are given copy that they consider obscene or revolutionary and that on occasion theatre projectionists have taken action against outside displays showing sexual activity (*Wall Street Journal*, 10 December 1970). Comic strips read by the working and lower-middle classes are undeviating in their support of conventional morality. So, in general, are soap and horse operas on radio and television. In "Gunsmoke," for example, virtue always triumphs.

that they have relied upon.[17] At the same time, millions of young people are moving up the class-cultural ladder and adopting the upper-middle-class style of life and with it "postconventional" morality.[18] "People no longer find it necessary to run to confession to tell the priest every little transgression," a priest recently explained to a *New York Times* reporter. "They use more judgment."[19]

Once it has been established that the individual has the right to decide moral questions affecting himself, he has ipso facto the right to decide those affecting others as well—whether to allow the fetus to be born, whether to abandon wife and children to public support, whether to fight for his country, whether to occupy the administration building, and so on.[20] If he has the right to decide what is moral truth, he has the right to require others to conform to it. If he judges the war immoral, he may judge me guilty if I support it. Having judged that, he may also judge how I should be punished.

III

It is interesting that Locke speaks of those "*set* at liberty," the implication being that they are more prone to the sports in question than those who have been at liberty all along. Perhaps this is because newly

---

[17]See, for example, the work cited previously, *Moral Education, Five Lectures.* In their introduction the editors (the dean of the Graduate School of Education at Harvard and his wife) find implicit in all of the essays the view that "the 'old morality' can and should be scrapped," and they accordingly urge teachers to introduce children to the "new morality." One of the contributors, however, seems to think that much "scrapping" has been premature. According to Bruno Bettleheim, "The mistake we still make is to hope that more and more citizens will have developed a mature morality, one they have critically tested against experience, without first having been subject as children to a stringent morality based on fear and trembling" (Sizer and Sizer, 1970, 87).

[18]"It is apparent that when men move into class positions different from those of their parents (and their grandparents), their values and orientation come to agree with those of their achieved class position" (Kohn and Schooler, "Class Occupation," 669).

[19]*New York Times*, 17 April 1970.

[20]"Moral judgment interviews" administered to more than 200 Berkeley students showed that 80 percent of those at the "highest" (postconventional) level favored a sit-in in the administration building. Kohlberg, who cites this study, observes that "protesting is a sure sign of being at the most mature moral level" (Kohlberg, in Sizer and Sizer, *Moral Education* 1970, 79).

acquired freedom has to be put to the test before one can be sure that it really is freedom; perhaps it is because one wants to explore the boundaries of a newly acquired territory. At any rate, it seems to be a psychological fact that the very gaining of freedom creates stresses and engenders a restless impatience for still more of it. What is more, the *anticipation* of being set at liberty may have these effects in exaggerated degree. In his life of Addison, Dr. Samuel Johnson illustrates the point in a manner that is disconcertingly topical:

> The boys, when the periodical vacations drew near, *growing petulant at the approach of liberty*, some days before the time of regular recess, took possession of the school, of which they barred the doors, and bade their masters defiance from the windows [emphasis added].

In large measure our present troubles arise from people having grown petulant not only at the approach of liberty but also at the *illusion* of its approach.

Mass petulance is more or less deliberately generated. The process is as follows. Members of the upper classes with a strong sense of social responsibility dramatize a social problem by expressing outrage at its existence and calling for its immediate solution. The problem may not have been recognized as one, or may not have been deemed "critical," by those directly affected. They soon learn, however, from the media, especially television, that they are suffering in ways, or to a degree, that they did not imagine. Poverty, racial injustice, school failure, bad housing— these are all problems that have been made subjectively worse for those who experience them by the efforts of reformers and publicists. Professor George Sternlieb has commented on the housing problem in a way that illustrates the point:

> Housing, perhaps more than any other facet of poverty, lends itself to visual representation. It is the very epitome of physical presence. As such it is perhaps overused by the mass media as a symbolic representation of the poverty of the poor. The poor are sold on it as well as the middle class. The media increasingly use housing—rather than the car—as the symbol of "making it." Every situation comedy uses housing amenities in terms of the mortgage references, the lawn mower, and the like to define status.
>
> Let me suggest that the symbolic impact of the tenement look upon the poor is as much a function of television as it is of the physical impact per se. As such, even when we completely alter the internal permutations of the housing unit through rehabilitation, we do little to change the secondary attributes—the psychic aura—of the slum.

This is coupled with a new negative *self-consciousness among slum dwellers*. Let me repeat: the reality of being poor, of being a slum dweller is not new; *the self-consciousness attached to it is*. In a sense the poor are saying, "Housing is lousy because it doesn't look like TV housing, and the people I know live in it, and the people I know are slum dwellers and, therefore, it is a slum."[21]

It might be a very good thing if all who suffer from social problems were made more aware of and angry about such problems as are real and about which something can be done, provided of course that they were also given an understanding of the difficulties in the way of solving them, the time that must reasonably be allowed for doing so, and the side effects and other uncertainties involved. But this of course is not the way of reformers, media, and politicians. They offer quick solutions for problems that will require decades or generations to solve if indeed they can be "solved" at all. Consider this further quotation from Ramsey Clark:

A million people now on welfare in New York City alone could add greatly to the wealth of the nation, provide needed goods and services and fulfill themselves if we work to find the way.[22]

That last clause—"if we work to find the way"—illustrates another element in the process by which mass petulance is created. A "crisis" having been brought into being in the public mind, the suggestion is then made that nothing but moral insensitivity stands in the way of its being solved. "How much do we care?" Clark asks. Having established that the "crisis" exists only because we do not care, a devastating conclusion follows: "What is to be said of the character of a people who, having the power to end all this, permit it to continue?"[23] The answer, of course, is that our society is "sick" and in need of "radical restructuring" if indeed it is "worth saving." (These clichés, which I have put in quotation marks, are not Clark's. He had no need to supply them.)

The process by which the mass mind is being prepared for revolution is not new, although because the "educated" classes are not so large, because of the effectiveness of television, and because of the exigencies of national politics under these conditions, it proceeds at an unprecedented pace. Nevertheless we should be able to judge from history what

---

[21] George Sternlieb, "Hawthornism and Housing" *Urban Affairs Quarterly* (September 1970): 96–97.
[22] Clark, *Crime in America*, 341–42.
[23] Ibid., 66.

its outcome is likely to be. Nassau Senior's further reflections on the
Revolution of 1848 are suggestive:

> For two or three years they had been reading and seeing represen-
> tations of the Great Revolution. Theatres were opened, in which it
> was acted in pieces that lasted, I believe, for whole weeks. The shops
> and the stalls along the Quays and Boulevards, and in the Courts of
> the Louvre, were covered with portraits of its chiefs, and with fine
> prints exhibiting its principal scenes. Thousands of copies of M. de
> Lamartine's "Girondins" were sold in cheap forms, in numbers, or
> by subscription; and probably as many thousands more were lent out
> to read at a price which the lowest workman could afford. The pic-
> turesque vividness with which that remarkable book is written, the
> dark grandeur with which its sanguinary heroes are invested, the
> success of every insurrection that is described, the irresistable power
> that is ascribed to the people, not only familiarized the populace with
> ideas of revolt and street war, but created, in young and ill-regulated
> minds, thirsting for new excitement, an intense desire to reproduce
> such scenes. They wished to see a tenth of August—and they made
> one![24]

IV

One would like to think that it is possible, in principle at least, for
a society to extend liberty ever more widely while at the same time
maintaining itself as a going concern. In this concluding section, I must
raise some doubts about this.

Philosophers have often pointed out that a system of universal liberty
is possible only if those who live under it voluntarily abide by certain
rules. The rational individual, however, will often find it profitable to
break the rules. Plato was probably not the first to point out that there
is an advantage in being an evil man in a good society. Against such a
one the good society is helpless, for if it threatens punishments severe
enough to deter him it will ipso facto cease to be a good society. (In Saudi
Arabia, where the hands of thieves are amputated and adulterers are
stoned to death, the crime rate is possibly the lowest in the world.)[25] As
the number and proportion of the "educated" increases in our society,
there will be more and more persons who see advantage to them as

[24]Senior, *Industrial Efficiency*, 295–96.
[25]*New York Times*, 29 March 1971.

individuals in breaking any rules that may be agreed upon and who are not restrained from doing so by conventional morality.

One can imagine a society in which all the members are morally autonomous and which is nevertheless a good, even a very good, society. For it to be such, however, all or almost all of its members would have to agree on at least some principles and on the necessity of these being the overriding ones in case of conflict among principles. In short, they would have to agree to support a constitution, however much they might disagree on substantive questions. Moralists, all of whom are their own *final* judges of what is right, cannot be depended on to constitute a stable society; and the more zealous they are the sooner they will be at each other's throats. But even if they agree on constitutional principles, this is not enough to insure that they will constitute a good or even a viable society. Even if every individual ardently desires the public good and defines it in exactly the same way as every other, the society may not cohere or be a good one. For the "enlightened" individual will see no reason why he should not violate any rule when his doing so will have no perceptible consequences on the welfare of society.[26] His not voting, not paying his income tax, not performing his military duty, and so forth, will not make any difference if it goes unnoticed because one such act is inconsequential in the aggregate of actions. Except as individuals feel a *duty*—a claim not dependent on each one's independent judgment—they cannot be depended upon to do what may be necessary for the maintenance of the society.

Obviously a good society of morally autonomous individuals, if it could exist at all, would have to consist almost entirely of people with at least enough intelligence and wisdom to accept and act upon certain indispensable social rules. In no actual society is this condition even approximately met. The most ardent of the classical apostles of liberty did not claim that children and adults of very low intelligence could be set at liberty from various "caretakers"—parents, clergymen, teachers, social workers, and the rest. J. S. Mill, for example, remarked that society, which is "armed not only with all the powers of education, but with the ascendancy which the authority of a received opinion always exercises over the minds who are least fitted to judge for themselves," has "only itself to blame for the consequences" if it "lets any considerable number

---

[26]Mancur Olson, Jr., *The Logic of Collective Action* (Cambridge: Harvard University Press, 1965).

of its members grow up mere children, incapable of being acted on by rational consideration of distant motives."[27]

Mill did not imagine that there would come a time when the "received opinion" would not only authorize but require those "least fitted . . . to judge for themselves." But this is the situation today. Witness, for example, a recent network television program which brought children aged four to eleven into a classroom setting with three experts on drug use for a "give-and-take" session in the hope, according to the producer-director-writer, "that the child can thereby come to his own rational decision not to abuse drugs."[28] Or consider the advice given a little while ago by the assistant secretary for health and scientific affairs in HEW to a meeting of the American Medical Association. We are entering an era, Dr. Egeberg said, in which doctors must provide their patients with more information, allowing them to use their judgment on many things; for example, patients should be given estimates of the hazards in various treatments. It would be a hoax, he said, to tell Americans that there will be three deaths among every 100,000 women taking birth control pills without also telling them of the risks of more ordinary things—for example, that ten times as many people have probably been killed by eating salt as will ever be killed by the pill.[29]

One may well agree that the freedom of those who "are capable of being acted on by rational consideration of distant motive" ought to be enlarged by giving more information and in other ways as well. There are, however, the others—"least fitted to judge for themselves"—a relatively small number to be sure, important nevertheless both for their own sakes and because of the ills that their errors and misfortunes may visit upon innocent bystanders and upon society at large. The benefits of enlarging the freedom of the competent must be balanced against the harm of enlarging that of the incompetent. How about the sinner who

[27]J. S. Mill, *On Utilitarianism, Liberty, and Representative Government* (New York: E. P. Dutton, 1951), pp. 186–87.

[28]Review by John J. O'Connor, *New York Times*, 29 March 1971. Mr. O'Connor remarks "Obviously this is a crucial area, prime for a massive tackling by television." Yet there are problems, he says, "The child is curious, no doubt, but that very curiosity can tend toward the morbid. One question cannot be ignored: In bringing drug data to the child, is the risk increased, however unwittingly, of stimulating that curiosity in more emotionally unstable children, of perhaps encouraging the youth attracted to dangerous 'kicks?' "

[29]*New York Times*, May 6, 1970.

does not know which transgressions are "little?" How about the child who misunderstands the "give-and-take" about the use of drugs? How about the woman whose confusion and fright increase the more medical information she gets? How about the student who does not know when it is "just" to occupy the administration building?[30]

Granted that writing prescriptions and saying masses in English makes these more "meaningful experiences" for the competent, may they not also have the opposite effect for the less than competent? To translate mysteries into ordinary language, whether of medicine or religion, destroys their mystery, and with it their authority. Those who want or need to rely upon judgments wiser than their own are now in effect told by the "received opinion" that there are none such or that, if there are, one must use one's own judgment to decide whose they are and what they are.

Perhaps a time will come when genetic engineering and vastly improved teaching techniques will greatly reduce the number of people who must be considered less than competent. At present, however, the trend seems to be in the opposite direction: the proportion who are not fitted to decide the questions life puts before them is rising. This is due not to biological or other changes in the population but to more taxing demands being placed upon people. It is as if the society were constantly raising its "passing grade," thus insuring a greater proportion of "failures" even if everyone's performance remains the same. Very likely this trend will not only continue but accelerate.

Ironically, it is the rapid progress of the most competent that constitutes the increasing disadvantage of the least. It is because so many go to college that the few who fail to graduate from high school are branded dropouts and considered to be practically unemployable. It is because so many succeed so well in acquiring complicated job skills that "dead end" jobs, not to mention low-paying ones, are considered unacceptable for all workers, including those whose limitations leave them no alternative but to be permanently dependent on welfare. It is because so many lead

---

[30]In the study of Berkeley students cited by Kohlberg (see footnote), about 60 percent of those at the next-to-the-lowest (conventional) stage of morality also favored the sit-in. These students, Kohlberg explains, were "in a state of confusion . . . unable to tell an autonomous morality of justice from one of egotistic relativism, exchange, and revenge." He assures us that they will "grow out of it to become principled adults," but he adds ominously, "If the pressures are greater and you are a Stokely Carmichael, things may take a different course" (Kohlberg, in Sizer and Sizer, *Moral Education*, 1970, 79).

satisfying and useful lives on the basis of postconventional morality that
those who must have some authority to regulate their lives are left to fail
so disastrously. As two writers on educational policy have recently
remarked:

> The educated unconsciously and without malice continue to recon-
> struct the society and economy for their own convenience, unaware
> and unconcerned that by doing so they make it still more difficult
> for the poor and uneducated to function. The uneducated are isolated,
> estranged, and alienated.[31]

By no means are all of the educated unaware and unconcerned at
what is happening to the poor and uneducated. Indeed, much of the effort
to reconstruct society that has made matters worse for the poor and
uneducated, far from being unconscious and merely without malice, is
the result of deliberate and generous efforts to improve their lots. The
"reconstruction" goes wrong partly because the educated are unable to
see, or face up to, the realities of the situation of the uneducated (for
example, that some of them are incapable of becoming in any sense
"educated")[32] and partly because of the changed structure of the situation
(for example, that much greater consequences now attach to small differences of intelligence because of the increasing complexity and importance of technology).

Seeing that life is being made more difficult for the poor and uneducated, the educated, especially the young among them, are apt to
declare the society "sick" and "not worth saving." If this view is widely
accepted by both "exploiters" and "victims," it may be a self-fulfilling
prophecy.

---

[31] A. Dale Tussing and Laurence B. De Witt, "The Costs of Illiteracy," *Notes on
the Future of Education*, Educational Research Center at Syracuse University,
vol. II, issue I (Fall 1970): 6.

[32] Kohlberg, for example, remarks that "it is quite easy to teach conventionally
virtuous behavior but very difficult to teach true knowledge of the good," but
the possibility that many or most children *cannot* be taught the true knowledge
of the good seems to escape him. He therefore takes no account of what the
social consequences may be of not trying to teach what is easy and not succeeding
in teaching what is difficult or (in many cases) impossible (Kohlberg, in Sizer
and Sizer, *Moral Education*, 1970, 82).

CHAPTER 17

The Dangerous Goodness of Democracy

The reason for our inveterate devotion to these millennial ideas is to be found in the nature of our kind of democracy.[1] Ours is the only country in which the public at large participates actively in the daily conduct of government; it is the only one in which the opinions of amateurs on foreign affairs are listened to by statesmen and taken seriously by them; consequently it is the only one in which the moral standards of the general public are decisive in the making of policy.[2]

The moral standards of a people are necessarily very different from those of its statesmen. A statesman learns early that it is his duty to act according to the rules of virtue, not those of goodness.[3] Goodness pertains

From *American Foreign Aid Doctrines*, by Edward C. Banfield (Washington, D.C.: American Enterprise Institute, 1963), 61–65. Copyright 1963 by the American Enterprise Institute for Public Policy Research. Reprinted by permission.

[1]*Democracy* means government by the people, that is, a political order in which power is widely distributed. In some democracies, for example, the British, the people exercise their power mainly by giving or withholding consent at infrequent intervals; in others, for example, our own, they exercise it by participating continuously and intimately in the day-to-day conduct of affairs. What is said about the dangerous goodness of "our kind of democracy" has little application to the other kind, in which the citizen is willing to leave the management of public affairs—and above all foreign affairs—to his elected rulers.

[2]Chester Bowles, justifying the "tough-minded" approach recommended by him on August 14, 1962, said: "It has been pre-tested over a period of years before many audiences in most states of the Union."

[3]This discussion of goodness and virtue draws upon Leo Strauss, *Thoughts on Machiavelli* (Glencoe: Free Press, 1958), 264–65.

to persons and is expressed in their everyday relations; it calls for (among other things) kindness, liberality, compassion, and the doing of justice. Virtue, by contrast, pertains to statesmen and is expressed in the actions by which they protect good citizens from both bad citizens and foreign enemies. Virtue has little to do with goodness, and may be entirely at odds with it in concrete cases; frequently the statesman must act unjustly or without kindness in order to protect the society—he must, in short, be virtuous but not good. As Churchill has written, "The Sermon on the Mount is the last word in Christian ethics. Everyone respects the Quakers. Still, it is not on these terms that Ministers assume their responsibilities of guiding states."[4]

Nations, the orthodoxy of political realism tells us, do what their vital interests require, however immoral those things may be. This may be true of nations that are governed by statesmen free to act as their judgment dictates. It is not, however, true of those governed, as ours is, by public opinion. A nation governed by public opinion may act contrary to its fundamental moral standards when swept by passion or when self-deceived. But it does not act so from deliberation or calculation. What is more, it is strongly impelled to express in action the *positive* principles of its morality, that is, its goodness.

American foreign policy has long been heavily tainted with goodness, and our country, consequently, has frequently acted against its own interests. Political realists, overlooking the difference between the morality of peoples and that of statesmen, have usually regarded American goodness as mere hypocrisy and have looked in the usual places for the "real" reasons of national interest that they were sure must exist.[5]

---

[4]Winston S. Churchill, *The Gathering Storm* (Boston: Houghton Mifflin, 1948), 320.

[5]Niebuhr, for example, finds that the Spanish-American War offers "some of the most striking illustrations of the hypocrisy of governments." He marvels that a man as intelligent as Walter Hines Page could speak of the war as a chance to clean out bandits, yellow fever, malaria, and hookworm and to make the country safe for life and investment and orderly self-government, and that he could write: "What we did in Cuba might thus be made the beginning of a new epoch in history, conquest for the sole benefit of the conquered." He sneers at McKinley's "hypocrisy" in claiming that it came to him while on his knees in prayer "that there was nothing left for us to do but to take them all, and to educate the Filipinos, and uplift and civilize and Christianize them, and by God's grace do the very best we could by them." *Moral Man and Immoral Society* (New York: Scribners, 1960), 98–102.

Perhaps there was an element of self-deception in what Page and McKinley said, but there is a world of difference morally between self-deception and

The optimistic, moralizing, and apolitical nature of American aid doctrine is a characteristic expression of this goodness. Goodness inclines men to have faith in each other, or at any rate to give each other the benefit of the doubt; public opinion therefore takes a compassionate and hopeful view of the prospects for growth and development however discouraging may be the underlying realities of the situation. It is by moralizing that one appeals to goodness; the discussion of the truly hard problems of choice, namely those in which the principles of goodness will not suffice as criteria, presupposes virtue rather than goodness. Action that is apolitical in the sense that it sees in the situation not the necessity of a struggle for power but rather the opportunity to cooperate in the realization of shared ends is consistent with goodness but not necessarily with virtue.

To know when and on what terms to subordinate goodness to virtue requires high intellectual and moral powers as well as much experience in making—and in taking responsibility for—decisions in important public matters. Few citizens can have all of these qualifications. The citizen, moreover, knows that his views will count only along with those of millions of other citizens,and so he may not trouble to go deeply enough into any public question to see its full moral complexity. Hence his confidence that the proper and the practical courses will coincide and that great affairs of states may be decided by the standards that apply in everyday life.

A public, moreover, cannot deliberately transgress the principles of its morality. Societies are held together by attachment to common values, especially ones that are held sacred. To call such values publicly into question, to consider openly the expediency of transgressing them, and then actually to do so (even though in order to realize other values) would profane and destroy the values and so weaken the mystic bonds that hold the society together. Such a thing could happen only if the values of the society had already lost their sacredness, and if, therefore, the society was in process of disintegration. A healthy society cannot subject its ultimate moral code to detached, rational scrutiny. If its code is to be scrutinized at all,the scrutinizing must be done by an elite set apart for the purpose—one which, like a bomb decontamination squad, possesses

---

hypocrisy, that is, the deception of others in order to gain one's ends. Moreover, it is not clear in what sense they were self-deceived. They seem to have had the same confident intention of doing good that the present-day advocates of aid have, and, as Niebuhr himself says (p. 102), the United States did in fact do much to improve education and sanitation in the places it conquered.

both a specialized skill and a willingness to expose itself to risk for the sake of the society. The professional statesman belongs to this elite.[6]

Much as we may wish it, the world cannot be ruled according to the Sermon on the Mount or the principles of the Quakers, and a determined effort to rule it so may lead to disaster. The goodness and optimism inseparable from democracy represent a great peril. The peril would be somewhat less if we gave our statesmen wide discretion in foreign affairs, as the other democracies do. Our statesmen, however, are trained to goodness, and they are selected for it rather than for virtue. Our tradition and the exigencies of our political system, moreover, tend more and more to subordinate them to public opinion. Confident that its goodness is the world's best hope, American public opinion reaches out eagerly for wider power in world affairs ("accepts responsibility for world leadership" is the cant phrase), thereby engendering—the United Nations is a case in point— ever more goodness in places where virtue is required, and thereby increasing ever more the incongruity between the reality of the situation in which we must act and the moral principles upon which our action is based. It is quite possible that the American people may be persuaded that the indiscriminate use of aid is folly (events may persuade them of this even though their leaders tell them the contrary). But this will not necessarily improve our foreign policy very greatly. The millennial and redemptionist character of that policy will not necessarily be changed thereby; if the American people cannot express their goodness through foreign aid they will doubtless find some other way of expressing it. To the extent that public opinion rules, our policy will reflect goodness. This is a cause for concern because goodness is, by its very nature, incapable of understanding its own inadequacy as a principle by which to govern relations among states.

---

[6]Saint Augustine, after remarking that a judge may torture and condemn an innocent man "not with any intention of doing harm, but because his ignorance compels him, and because human society claims him as a judge" concludes that although we may acquit the judge of malice "we must none the less condemn human life as miserable." If the judge must subordinate goodness to virtue, he ought at least to regret the necessity. "Surely it were proof of more profound considerateness and finer feeling were he to shrink from his own implication in that misery; and had he any piety about him, he would cry to God, 'From my necessities deliver Thou me.' " *The City of God* (New York: Modern Library), 682–83.

PART **VI**

Political Economizing

This final section deals with matters that lie along the borders of economics, an area that the late Lionel Robbins, an eminent economist, once said is the happy hunting ground of quacks and charlatans.

The first essay crosses the border to borrow from welfare economics an analytical framework within which to criticize the public library and to make suggestions for its improvement. Welfare economics assumes, rather arbitrarily, that the individual is better off when he gets what he wants, whatever that may be. By this standard an ideal market will produce an optimal allocation of goods and services. In the absence of any market, or if there are serious imperfections in the operation of a market, government intervention (or private philanthropy) is necessary if the consumer is to get even approximately what he wants. On this rationale, ought libraries to be supported by the taxpayer? If they are so supported, what sorts of services should they provide? There are, of course, broadly political reasons why these questions are not raised in a practical way with respect to libraries or other, similarly situated governmental undertakings.

"Economic Analysis of Political Phenomena"[1] was presented to an interdisciplinary Harvard–M.I.T. faculty seminar in 1967. It proposes lumping some of the subject matter of economics and some of that of political science in one discipline ("aggregation") and some of each in another ("politics"). The subject has also been treated by the economist Ronald Coase. No one will be surprised to learn that the borders of the disciplines remain very much as they were.

[1] Ronald Coase, "Economics and Contiguous Disciplines," *Journal of Legal Studies* 7 (June 1978): 201–11.

303

   The writing of the final essay was prompted by the publication of books by economists—notably Anthony Downs[2] and James M. Buchanan and Gordon Tullock[3]—which analyzed political phenomena such as parties and voting behavior on the assumption that the citizen, like the consumer, can best be understood as a utility maximizer. The usefulness of this assumption is shown to depend upon whether the goals of the actor (citizen, consumer, etc.) can be measured on some common, objective scale, such as money values. In the spheres of behavior generally thought of as political—and for that matter in much "economic" behavior as well— no such scale exists. One may construct a more realistic account of what it is that the individual is supposed to be maximizing, but this involves an almost complete sacrifice of analytical power.

   In the years that this article lay unpublished in the author's files, economists—most conspicuously Gary Becker of the University of Chicago—and some political scientists have made frequent and extensive use of "the economic approach," often with impressive results. How, one must wonder, is it possible to do what, if the argument of the essay is correct, is impossible?

---

[2]Anthony Downs, *An Economic Theory of Democracy* (New York: Harper & Row, 1957).

[3]James M. Buchanan and Gordon T. Tullock, *The Calculus of Consent* (Ann Arbor: University of Michigan Press, 1962).

CHAPTER 18

Some Alternatives for the Public Library

The public library has more users and more money today than ever before, but it lacks a purpose.[1] It is trying to do some things that it probably cannot do, and it is doing others that it probably should not do. At the same time, it is neglecting what may be its real opportunities. What the library needs is, first, a purpose that is both in accord with the realities of present-day city life and implied by some general principles and, second, a program that is imaginatively designed to carry its purpose into effect.

This paper will begin with a brief look at the principles justifying *public* action. (Why should a public body distribute reading matter and not, say, shoes?) In the light of these principles, it will then consider what the public library has been, what it is now, and what it ought to be.

From *The Metropolitan Library*, eds. Ralph W. Conant and Kathleen Molz (Cambridge: MIT Press, 1972), 89–100. Copyright 1972 by the authors. Reprinted by permission.

[1] For evidence, see the report of the eighty-fourth annual conference of the American Library Association, *New York Times*, 4 July 1966, 40. The theme of the conference was "Libraries for a Great Society," and the president of the association announced that an inventory of public and school library needs made by the U.S. Office of Education and the association revealed that $3.1 billion would have to be spent to bring the nation's libraries to the level of "adequacy" and operating budgets would have to be raised $1.2 billion a year to keep them there. "These are enormous figures, of course," he said, "but our wealthy nation can easily contribute all that is called for and then some." With regard to the library's purpose, he seems to have said nothing.

SOME GENERAL PRINCIPLES

Economists offer several justifications for governmental intervention to set the demand for a commodity or good (in this case library service).[2] One justification exists when the good is of such a nature that it cannot be supplied to some consumers without at the same time being supplied to all—examples are national defense and air pollution control; in such cases, it is impossible for the distributor of the good to charge a price for it, since he cannot withhold it from anyone who refuses to pay the price. Therefore (apart from philanthropists) only the government, which through its tax power can coerce everyone into paying, is in a position to offer the service. Clearly this justification has no application to libraries.

Another justification—and one that presumably *does* apply to the library—exists when the public will benefit in some way if the consumer consumes more (or less) of the good than he would if the government did not concern itself in the matter. If my consumption of a good—my immunizing myself against disease or my sending my children to school, for example—confers benefits of some kind upon the community at large, the government ought, in the community's interest if not in mine, to see to it that I consume a proper amount of it. In order to encourage consumption of such *merit goods* (to use an economist's term), the government may employ subsidies.

That consumption of certain goods confers benefits upon the community does not automatically justify government subsidies, however. No doubt, it is a good thing from a public standpoint that I eat well, have a safe roof over my head, and go to the doctor when I am sick. But if I am *compos mentis* and not indigent, the chances are that I will look after these matters without any encouragement from the government. The public does not have to pay me to eat; I will do so both because I must in order to stay alive and because I enjoy eating.

Public intervention to set the demand does not necessarily involve public production or distribution of the good. The school board sets the demand for school books, but it does not hire authors to write them, and it does not operate its own printing press. The Air Force sets the demand for planes, but it does not manufacture them.

[2]See Richard Musgrave, *The Theory of Public Finance* (New York: McGraw-Hill, 1959), Chapter 1.

By the same token, that a good is produced or distributed under public auspices does not imply the necessity of a public subsidy for the people who consume it. The function of the government may in some instances be merely to make up for a deficiency in the private market by offering consumers a good that from the standpoint of the community they ought to have and that for some reason no private enterprise offers. If no one saw fit to go into the shoe business, the government would have to. But if it went into the shoe business, it would not have to give shoes away or sell them for less than the cost of manufacture.

THE NINETEENTH-CENTURY PURPOSE

Let us now look at the public library of the past in the light of these principles. In the very beginning, libraries were private associations for the joint use of a facility that was too expensive for any but the well-off to own individually. Some state legislatures conferred on the associations certain corporate powers, including the power to tax their members provided that a two-thirds majority concurred. They did this on the grounds that benefits to the community at large would ensue—that is, that library service satisfied a "merit want." "These libraries," Franklin remarks in his autobiography,

> have improved the general conversation of Americans, made the common tradesmen and farmers as intelligent as most gentlemen from other countries, and perhaps have contributed in some degree to the stand so generally made throughout the colonies in defense of their privileges.

Early in the nineteenth century charitable societies were formed in the larger cities "to furnish wholesome religious, moral, and improving reading of all kinds to the poor, cheaper than they now get fanatical or depraved reading." There were complaints that the books circulated were not improving enough (a director of the Astor Library in New York wrote that "the young fry . . . employ all the hours they are out of school in reading the trashy, as Scott, Cooper, Dickens, Punch, and the 'Illustrated News' "), to which the reply was made that "if people will not come to your library you may as well establish none." No one, however, would have justified a charitable library on the grounds that it provided entertainment.[3]

Later on, the corporations thus created were made public and were supported in part by taxation of the whole public. This was about the middle of the last century, when bright and ambitious farm boys who had mastered the 3 R's but not much else were flocking to the cities to seek their fortunes. "Mechanics libraries" were established to afford these Alger characters opportunities to pick up by home study the small amount of technical knowledge that then existed. Such libraries were not supported in full by the public—philanthropists provided most of the support—but they were tax exempt and they enjoyed other advantages. There were good reasons for giving them these advantages: anything that encouraged self-improvement on the part of the "respectable poor" tended to increase the productivity and wealth of the community. Besides, to the Anglo-Saxon Protestant elite that ran the cities, self-improvement appeared good in and of itself.

It was not until near the turn of the century, however, that most sizable cities had public libraries in the present-day sense. There was no doubt about the public purpose of these libraries. They were to facilitate the assimilation of European immigrants into the urban, middle-class American style of life.

Many of the immigrants were highly receptive to what the library offered. Many of them came from cultures that respected books and learning; with few exceptions they were eager to learn the language and customs of their new country and to get ahead in a material way. There was, accordingly, a high degree of harmony between the public purposes being sought through the library and the motives and aspirations of its potential clientele.

TIMES HAVE CHANGED

Today the situation is entirely different. The Horatio Alger characters and the immigrants have long since passed from the scene. There are, to be sure, more poor people in the large cities than ever (they are not as poor in absolute terms, however, and they constitute a smaller

[3]David B. Tyack, *George Tichnor and the Boston Brahmins* (Cambridge, Mass.: Harvard University Press, 1967), pp. 208–211.

proportion of the metropolitan area's population), and the movement of the poor from backward rural areas of the South and Puerto Rico into the cities is likely to continue for some time to come. The present-day poor, however, represent a new and different problem. Their poverty consists not so much of a lack of income (although they lack that) as of a lack of the cultural standards and of the motivations, including the desire for self-improvement and for "getting ahead," that would make them more productive and hence better paid. "The culturally deprived of today's cities are not on the bottom of a ladder; they do not even know that one exists," the editor of a bulletin for librarians has written in an article extremely apposite to the present discussion.[4] Many of the poor are "functionally illiterate," some though they have gone to, or even graduated from, high school. Giving them access to books will not accomplish anything.

Assimilating the lower class into the working and the middle classes may be a public purpose of the highest urgency. (Some people, of course, assert that lower-class values—certain of them, at any rate—are as worthy of respect as any others.) But however compelling the case for assimilation is thought to be, the question has to be faced of whether the library is a fit instrument for the purpose.

Certainly no one believes that the library is now of any service to the lower class. By and large, libraries are of the middle class and for the middle class. With rare exceptions, librarians have the wrong skin color, the wrong style of dress and makeup, the wrong manner of speech, and the wrong values (among other things, they think that people should be quiet in the library!) to be acceptable to the lower class. The feeling is mutual, moreover, for most librarians are probably no freer of class and race prejudice than are other middle-class whites. The consequence is that the lower class is repelled by the library or would be if it ever got near it.

A few library boards have tried to change this but have not had much success. Some will say that their methods have not been sufficiently ingenious: they should establish storefront libraries and staff them with lower-class librarians, preferably radical ones; they should employ super-salesmen to go from door to door selling cheap reprints, and so on.

---

[4]Kathleen Molz, "The Public Library: The People's University?" *The American Scholar* 34, no. 1 (Winter 1964–65): 100. The author wishes to express his appreciation of Miss Molz's criticism of an earlier draft of this paper.

If one believes that lower-class adults can be enticed to read, there is much to be said for making this a primary purpose of the library and for trying any approach that offers the least promise. If may be, however, that the educational level of the lower class is so low and its demoralization so great that no efforts on the part of the library will have much effect. Something much more fundamental than library service may be needed— for example, compulsory nursery school attendance from the age of two or three.

Not being able or willing (or both) to serve the lower class, the public library has tended to make itself an adjunct of the school, especially of the middle-class school. Children have always been an important group of library users, but in recent years they have become the principal clientele of the public library in many places. Children sent by teachers to use books in connection with course assignments crowd some libraries after school hours to such an extent that adult users have to leave. (In certain Los Angeles schools, teachers require each pupil to borrow at least one book a week from the public library!) Here and there libraries have been forced by the sheer weight of the children's numbers to place limits on service to them.

One reason for this invasion is that, thanks to the "baby boom . . . ," there are more children than ever. Another is that the schools do not have adequate libraries of their own. Still another reason is that it has become fashionable among teachers to require research papers (in some places third graders swarm into the public library to do "research") and to assign, not a single textbook, but a list of readings, some in very short supply, selected by the teacher from a variety of sources.

Public libraries were not designed for large numbers of children and are usually not staffed to handle them. The wear and tear on books, librarians, and innocent bystanders is therefore very great. In Brooklyn, it was recently reported, book losses—not all of them caused by children— run to 10 percent of the library budget. In some places rowdyism is a serious problem.

In fairness to both the children and the adults, the schools ought to have adequate libraries of their own. Children should not be excluded from public libraries, however—it is a good thing for them to go now and then to a place with a decidedly adult atmosphere—but they should not be sent there to do assignments; they should go to the public library on their own initiative to find books that please them and in the expectation of entering a world that is not juvenile.

THE LIGHT READER

Apart from schoolchildren, the most numerous category of library users consists of light readers, especially middle-class housewives. The books these readers borrow are not *all* light, of course, and even the ones that are light are not the very lightest; public librarians do not buy out-and-out trash. Nevertheless, a considerable part of the circulation is of romantic novels, westerns, detective stories, and books on how to repair leaky faucets, take off excess fat, and make money playing the stock market. About one-half of the books public libraries lend to adults are fiction, and most of these are probably light fiction. (Unfortunately, libraries do not use more precise categories than "fiction" and "nonfiction" in their record keeping.)

It is hard to see how encouraging light reading can be regarded as a public purpose. That the housewife finds it convenient to get her detective story from a public, rather than a rental, library is certainly not a justification for the public library. Her neighbor, who may not care to borrow books and whose income may be less than hers, will be coerced into paying taxes to support a facility that is for her convenience. Why should he be? Whether she gets to sleep by reading a novel, by watching the late show, or by taking a sleeping pill—indeed, whether she gets to sleep at all—is a matter of indifference to him and to the community at large.

If it could be shown that light reading leads to serious reading, a justification for public action would exist. In the case of uneducated people who are introduced to books by the library, such a showing might possibly be made. But it is highly unlikely that it can be made in the case of the middle-class readers who constitute most of the adult library users. For the most part, light reading leads to nothing except more light reading.

Unless reason can be found for believing that light reading confers some benefit upon the community, the public library should leave the light reader to the rental library, the drugstore, and the supermarket. If for some reason these readers *must* be served by the public library, they should be charged the full cost of the service, including, of course, a fair share of the rental value of the library building and site. Charging the full cost of service would soon put the public library out of the light-reading business, but this would prove to be a benefit even from the standpoint of the light reader. He would find that when the public library stopped competing with rental libraries by giving its service free, they and other profit-making enterprises (the paperback counters of the drug-store and supermarket, for example) would fill the gap and give him better service than he got before. If there is a demand for thirty copies of *Peyton Place*, the rental library makes haste to put that many on its shelves. The public library, not being under the stimulus of the profit motive and (let us hope) feeling itself under some obligation to serve more important purposes, buys only one or two copies of such a book if it buys any at all. This, of course, accounts for the more than 3,500 rental libraries (not to mention the drugstore and supermarket counters) that are competing successfully with the tax-supported libraries.

THE SERIOUS READER

The proper business of the public library is with the *serious* reader and—assuming that the library cannot be an effective instrument for educating the lower class—with him alone. "Serious" reading is any that improves one's stock of knowledge, enlarges one's horizons, or improves one's values. Reasonable men will disagree as to where the boundary should be drawn between light and serious reading; that does not render the distinction invalid or useless, however, although it will lead to some practical difficulties.

The commonsense assumption is that all serious reading confers some benefit upon the community. This would be hard to demontrate in a rigorous way (imagine trying to specify the amounts and kinds of benefits conferred upon various sectors of the community by, say, so many man years of novel reading, so many of historical reading, and so on); but the difficulty, or impossibility, of demonstrating it does not mean that the assumption is wrong.

That an activity confers benefits upon the community does not, however (as was remarked before), constitute a sufficient justification for publicly supporting it. Perhaps those who read serious books would read as many of them if public libraries did not exist. (Indeed, conceivably they might read more of them, for if an existing institution did not stand in the way, a new and more effective one, public or private, might come into existence. Any foreigner who has observed the operation of the government salt and tobacco monopoly in Italy will agree that other and better ways of distributing these commodities are possible. To the Italian who has never been abroad, however, the idea of putting the government out of the salt and tobacco business might seem preposterous. "How then," he might ask, "could one possibly obtain these indispensable articles?") Most serious readers have adequate or more than adequate family incomes; it seems likely that if they had to pay the full cost of their reading they would not read less. If this is so, there is no reason for the public to subsidize their reading.

The relatively few serious readers who are poor—so poor that to pay for library service would entail a sacrifice of something else that is necessary to an adequate standard of living—present a problem. They could of course be given service at reduced rates or free. This is widely done by colleges, and there is no reason why there should not be "library scholarships" for all who need them. If such an arrangement involved use of an objectionable means test (would it be objectionable to give service free to all families with incomes of less than $5,000 if the user's statement that he belonged to that category were accepted without question?) or if the costs of record keeping were unduly high, the sensible thing would be to make the service—the standard service, not necessarily special services—free to all.

If it is decided that serious reading must be subsidized in order to secure for the community all of the benefits that it wants, it need not follow that the best thing for the library board to do is to own and circulate a collection of books. There may be much better ways of accomplishing the purpose. Perhaps, for example, those who have responsibility for allocating the library fund—let us now call it the *fund to encourage serious reading*—would get a greater return on the investment by inducing the local supermarket to display a big stock of quality paperbacks and to have one-cent sales of them now and then. Or, again, perhaps the fund would best be used to subsidize the rent of a dealer in used books who, because of the ravages of urban renewal or for other reasons, could not otherwise stay in business.

SOME ILLUSTRATIVE IDEAS

If one assumes, however, that such radical innovations are out of the question and that the practical problem is to make some minor changes in the existing institution, what might be done?

Here are a few suggestions.

1. Provide soundproofed cubicles that readers may rent by the week or month and in which they may keep under lock and key books (subject to call, of course), a typewriter (rented, if that is what they want), and manuscripts. Nowadays few people have space at home for a study. Many libraries have reading rooms, but there are no places where one can read, let alone write, in privacy and comfort. (A habitual smoker, for example, cannot read if he is not permitted to smoke.) The New York Public Library on 42nd Street is probably the only public library with cubicles (they are supported by an endowment); there is a long waiting list for them.

2. Offer the services of a "personal shopper" to take orders by phone and to arrange home deliveries and pickups. Many readers are too busy to go to the library, especially when there is no more than an off-chance that the book they want is in. The personal shopper could also arrange fast interlibrary loans and for the photocopying of hard-to-get, out-of-print books. (Publishers naturally object to the copying of copyrighted material. But perhaps they could be persuaded to give libraries general permission to make one copy per library of works that are not available for sale.) A fair number of the larger libraries have had "readers' advisers" ever since WPA days; the advisers' time is usually entirely taken up by children, however; in any case, only handicapped persons are assisted in absentia.

3. Buy a large enough stock of *serious* books so that no reader will have to wait more than, say, two weeks for a copy. Bentham's remark about justice can be paraphrased here: "Reading delayed is reading denied."

4. Display prominently and review in library newsletters those current books that are not widely reviewed by "middle-brow" journals. Many people suppose that all worthwhile books are listed, if not actually reviewed, by the better newspapers and magazines. This is not the case. Scholarly books are ignored as often as not; some of them are unknown to most serious readers. The natural tendency of the library is to make a fuss abut the very books that the ordinary reader would be most likely to hear of anyway. It should try instead to make up for the deficiencies of the commercial institutions by calling attention to the less advertised books.

5. Maintain up-to-date, annotated bibliographies of the sort that would help introduce a layman to a specialized field. A physician, let us suppose, wants to know what social science has to say that is relevant to problems of medical organization. What books and journals should he look at first? If the library had a file of reading lists, course outlines, and syllabuses used in colleges and universities, together with bibliographical notes and articles from academic journals, he could be helped to make

his way into the subject. A good many of the better libraries have materials of this sort—more materials, probably, than most of their serious readers realize. Even so, there is probably a good deal of room for improvement in both the quality of the materials that are collected and the methods by which they are made known to library users.

6. Offer tutorial service for readers who want instruction or special assistance. Perhaps the physician would like to discuss his questions with a social scientist. The library might have a social scientist on its staff, or it might bring one in as a consultant from a nearby college or university. The tutor would be available for an hour's discussion or, at the other extreme, to give a short course.

7. Have a mail-order counter supplied with a directory of all books in print, a list of available government publications, and the catalogs of some dealers in used and hard-to-find books. A librarian should be on hand to help buyers find what they want. In the many towns and small cities that are without proper bookstores, this kind of service might go a long way toward making up for the lack.

THE LIBRARY'S FAILURE IS TYPICAL

The library is by no means the only public institution that with passage of time has ceased to serve its original purpose and has not acquired a new one that can be justified on any general principles. Very likely it could be shown (1) that the professionals most involved, and a *fortiori* everyone else, have given little serious thought to the nature of the purposes that presumably justify not only public libraries but also public parks, museums, schools, and renewal projects (to mention only a few activities of the sort that are in question); (2) that such purposes as might plausibly be advanced to justify such activities are ill served, or not served at all, by the activities as presently conducted; (3) that these purposes could usually be better served by the market (rigged perhaps by public authorities) than by public ownership and operation; (4) that in most cases using the market would result in greater consumption of the good and in less waste in the supplying of it (public institutions tend to offer too much of those goods that are in light demand and not enough of those that are in heavy demand); and (5) that certain goods not offered by private institutions are not offered by public ones either, in spite of the fact that increased consumption of these goods would confer relatively large benefits upon the community at large.

To find the reasons for this state of affairs, one must look deep into the nature of our institutions and of our political culture. Organizations tend to perpetuate themselves and therefore to embrace whatever opportunities come along, however unrelated these may be to any previously stated purposes. Public organizations, moreover, often exist as much to symbolize something as to accomplish something. These are only two of many considerations that doubtless should be taken into account.

CHAPTER 19

Economic Analysis of Political Problems

I

My main contentions are (1) social choice processes differ in their logical structures; economics deals with one category ("aggregation"), political science with another ("politics"); (2) aggregation processes are analyzable in terms of a general deductive theory and equilibrium models, whereas political processes are not; (3) except for exchange in the market, concrete choice processes rarely involve aggregation; (4) the conceptual framework used in analyzing aggregation processes—a utilitarian one—is unsuitable for normative purposes and insufficient for conditionally normative ones; and (5) there is need to study various social choice processes as mechanisms for realizing various conceptions of social welfare.

Before entering upon the argument, it is necessary to offer some preliminary definitions and explanations.

A *social choice process* consists of the activities by which the *welfare* (somehow defined) of two or more individuals is made the basis of a social state (outcome, allocation, decision). Thus, the people of a small town, having a surplus in their treasury, are to distribute it among several uses— say education, health, and recreation. The procedures they use in reaching a decision constitute a concrete social choice process provided that these procedures "take into account" the needs, wishes, values, desires, tastes, and the like of the townspeople.

Paper presented at Harvard University Faculty Seminar, 1967.

Six analytically pure types of social choice processes may be distinguished. The townspeople, for example, may employ any of the following:

1. *Exchange.* The money is distributed to consumers who then bid against each other for health, education, and recreation goods and service that are offered on the market.
2. *Tabulation of Revealed Preferences.* Consumers submit complete accounts of their preferences with regard to all possible alternatives; these are fed into a computer.
3. *Voting.* The people vote on paired proposals until one proposal is "elected."
4. *Arbitration.* One person makes the allocation, taking into account the preferences (values, tastes, etc.) of all, himself included, and resolving conflicts arbitrarily (i.e., by invoking criteria that pertain to himself as arbitrator rather than to himself as one townsperson among many).
5. *Struggle.* People exert influence upon one another (by force, fraud, bribery, etc.) until one, or a coalition, dictates the allocation. The process by which influence is exerted is "bargaining," "fighting," and so forth, depending upon the nature of the rules governing the kinds of influence that may be used.
6. *Discussion.* People engage in a cooperative search for an outcome uniquely implied by an end that they share (perhaps for different reasons), by a common cultural value, by "the public interest," and so forth.

In a concrete process, two or more of these analytical types may occur together, either simultaneously or seriatim. (Discussion and struggle may go on in a mixture, for example, and may be followed by voting.) Also, the social choice may be to make the allocation by means of a procedure that is not a social choice process—for example lottery, dictatorship, or rule of tradition.

II

The six analytical types of social choice process may be grouped into two categories.

Aggregation

This category includes three kinds of processes: exchange, tabulation of revealed preferences, and voting. These have a common logical structure: namely individual preferences that are transitive, "are taken as data and are not capable of being altered by the nature of the decision process itself,"[1] and there exists a rule for passing from preferences to the outcome (a "rule" exists when an observer, knowing the preferences, could predict the outcome). Because they involve given (transitive) preferences and given rules, aggregation processes can be described in terms of general deductive theory. The aggregation problem is purely logical or mathematical.

Politics

This is a residual category, "preferences" *not* being given and there being *no* rule. Each of the three kinds of political process presents a different theoretical problem.

Struggle

The preferences of individuals are formed in or changed by the process of choice (struggle, after all, is a mutual effort to change others' preferences); usually the analyst has as given simple (but not comparative) value judgments; the simple value judgments, moreover, are often vague, mutually inconsistent, or without any implications for some choice alternatives. (The individual "hasn't fully made up his mind" or "is of two minds.") In lieu of a rule, the analyst has some knowledge of "the way institutions work" and of "the rules of the game" that usually apply.

Struggle does not allow of a general deductive theory or of analysis in terms of equilibrium models. Not being able to find "solutions," the theorist cannot use the concepts *efficiency* and *rationality* to much advantage. Accordingly his models tend to be explanatory rather than predictive.

This is the vineyard in which most political scientists labor; much of their work is purely descriptive, focusing on interest groups, influence relations, institutional rules of the game, and so on. Efforts to create a

[1]K. Arrow, *Social Choice and Individual Values,* 2nd ed. (New York: Wiley, 1963).

general or systematic theory of influence have not amounted to much. Game theory is used but not in its analytically powerful (mathematical) forms.

Discussion

Here the problem is how to move from a general statement to a particular criterion that is uniquely implied by it; except for analogical reasoning, there is no logic that can be applied to it (and analogical reasoning is, as Aristotle said, a bastard logic). There can be no general, deductive theory and, accordingly, no normative (or conditionally normative) theory that is general.

The discussion problem appears in three guises: (a) what does law (natural or positive) require in particular cases? (b) what do "cultural values" or "the public interest" imply in particular cases; and (c) what do the "purposes" or "goals" of formal organizations imply in particular cases?

Arbitration

Here the problem is to know how the arbitrator "makes up his mind"—that is, performs the subjective, psychological act of choice. Again, no general deductive theory seems to be possible.

In what follows I shall use *economist* to refer to an analyst of aggregation processes and *political scientist* to refer to one of political processes. Of course, this does some violence to ordinary usage. A few of these dealing with aggregation processes have degrees in political science and are employed by political science departments—for example, Duncan Black and W. H. Riker. By the same token, some economists (in the usual sense) deal with political processes—for example, Schumpeter. My division corresponds roughly to practice, however; aggregation processes are in general the domain of economists (in the usual sense), even when the concrete setting of the aggregation problem is (in the ordinary meaning of the term) *political*.

What *economic* and *political* mean *when used to distinguish among kinds of aggregation process* is explained by Black:[2]

---

[2]Duncan Black, "The Unity of Political and Economic Science," *Economic Journal* (1950): 513.

If the pure theory appropriate to the case makes use of preference schedules, the phenomena will be either economic or political. If the definition of equilibrium employed relates to quality of demand and supply, the phenomena will be economic in nature; if it relates to equilibrium attained by means of voting, they will be political.

In sum, then, the fundamental difference between the two categories of subject matter is that one (aggregation, the domain of the economist) has a logic that can be explicated deductively and mathematically as a general theory, whereas the other (politics, the domain of the political scientist) does not. Theories about struggle, discussion, and arbitration processes are possible, of course, but they have to be arrived at inductively. This means that they are not likely to have anything like the generality or power of theories about aggregation.

III

Before congratulating the economist and commiserating with the political scientist, we should first consider what the real world is like. If most social choices involve aggregation, the economist can tell us a great deal. But if few or none do, he is like an astronomer in a universe without astral bodies.

In our society, people may in general have stable, transitive preferences with respect to all of the alternatives before them. This is perhaps most likely to be the case in those spheres in which they can bring the alternatives "under the measuring rod of money." But although the assumption of rationality and of money maximizing describes, as Bagehot said, "a sort of human nature such as we see everywhere around," ours is—as Bagehot also said—"a very limited and peculiar world." It may even be that the size of this "peculiar world" is shrinking. Firms that maximized profit in Bagehot's day may now be "satisficing" in terms of objectives (e.g., prestige, market position, and rate of growth) that are not measured in money and from which accordingly an observer cannot infer a preference ordering in the concrete circumstances. If this is true of firms, it is apparently also true of consumers. Some economists, says Becker, "either deny that households maximize any function or that the function maximized is consistent and transitive."[3]

[3]Gary S. Becker, "Irrational Behavior and Economic Theory," *Journal of Political Economy*, 70, no. 1 (February, 1962): 1–13.

By and large, organizations—business or other—do not have preference orderings that can be taken as given; their goals are too few, too
vague, and too unstable to yield fixed and meaningful criteria of choice.
(In his account of operations research, Dorfman notes that operations
researchers sometimes impute objectives to organizations in order to have
something to go on.) According to H. Simon, organization goals are best
described as constraint sets that define roles at the upper levels of the
administrative hierarchy.[4] The notion of a public agency or a political
party—let alone a government—having fixed and transitive preferences
is ludicrously unreal.

To the political scientist, the real world is one in which most social
choices—certainly all "important" ones, for example war or peace—are
made in a mixed process of arbitration, struggle, and discussion. In the
political scientists' world, preferences—as well as the underlying values
that give rise to them—are products of social interaction. They are therefore not "fixed." Nor are they in any ultimate sense "individual." We
acquire our values as we do our language and everything else, from those
around us. Indeed, it is by virtue of sharing values that we cohere in
association and as a society.

Because people usually have (or think that they *may* have) the same,
or complementary, values or because they believe that the purposes of
some organization or other abstract entity (e.g., "the public") ought to be
the basis of choice, they are likely to engage in discussion. Discussion is,
therefore, an almost ubiquitous feature of social choice processes, at least
in those spheres of the world—constituting most of it—where social organization is fairly stable. Discussion alone can rarely lead to an outcome,
because there is no nonarbitrary procedure for passing from a general to
a particular criterion. Struggle of some sort, therefore, is almost always
associated with discussion. It is safe to say that even a Quaker meeting
never decides anything of importance simply by discussion, that is, without any exercise of influence or any arbitration.

In a process of discussion, and to a lesser extent in one of struggle,
information is exchanged, values are explicated, inconsistencies in preference orderings are pointed out, alternatives are canvassed, and consequences—including ones to third parties and abstract entities like "the
public"—are anticipated. If rationality is defined as a process in which all

[4]H. Simon, "On the Concept of Organizational Goal," *Administrative Science
Quarterly*, 9, no. 1 (June, 1964): 20.

alternatives are listed, the consequences that would follow from each identified, values clarified, and the course of action to be followed by the preferred set of consequences selected, then a political process may well be more rational, especially when effects on third parties are important, than one of aggregation.

Viewing the empirical situation as he or she does, the political scientist is not surprised to find that Bergson's social welfare function cannot be given content; that the paradox of voting, the inevitability of which under certain circumstances has been proved by Arrow, almost never appears in a "real" situation; that game theory, which has much to say about the strategy of pure conflict (the zero-sum game), has little or nothing to say in the empirically important cases (wars and threats of war, strikes, negotiations, etc.) where conflict is mixed with mutual dependence; that few of Black's theorems have any but academic interest; that the twenty-five "testable" propositions derived by Downs from *An Economic Theory of Democracy* are all trivial, meaningless, or wrong.[5]

IV

Even if aggregation theory is widely applicable to the real world, its value may be more limited than most economists suppose. It is one thing to predict how people will behave; it is another to say how they would have to behave in order to get from life what they *really* want; and it is still something else to say how they *ought* to behave. A mode of analysis well suited for positive studies may be inadequate for conditionally normative ones and pernicious for normative ones.

Except when on their guard, economists tend to ignore these things. They are prone to suppose that "individual preferences" are "real" and that value stuff otherwise conceived is "metaphysical." From this view they often slip into an unwitting utilitarianism. The modern social welfare functions, Samuelson says, is the old utilitarian formulation in modern dress (stripped of its "crude and materialistic calculus of hedonism").[6] Of this there can be no doubt; the social welfare function says that welfare consists in having what one prefers, whatever that may be. This is exactly

[5]Anthony Downs, *An Economic Theory of Democracy* (New York: Harper & Bros., 1957), Chapter 16.
[6]Paul Samuelson in an unpublished paper, 1967.

what Bentham said. Pushpin is as good as poetry—better if one prefers it.

Samuelson goes on to say that the utilitarian formulation, whether in new dress or old

> tends to subvert the older notions of "legitimacy"—that things are done as they are because they have always been done so; that people have a contractual right to no changes; that it is "unjust" to do certain things to people, even under the purpose of maximizing the social welfare.[7]

This is true (he might use a stronger word than *tend*), but it is odd that he does not point out the deficiencies of utilitarianism as an ethical doctrine. One may even understand him to approve the substitution of utilitarianism for the notions of legitimacy. Indeed, the form of the sentence quoted previously implies that he regards injustice as just another "notion," on a par with that of doing certain things in a certain way because they have always been done that way. Presumably he highlights *legitimacy*, *injustice*, and *equity* to indicate that it is problematical what the words mean, or whether they mean anything at all.

At any rate, he contrasts a perception of welfare based on a notion of "legitimacy" to one (which he calls *just* and *equitable*, without highlighting the words) based on a distribution of income allocated to "produce the greatest bliss for the whole universe—even if that means sacrificing something of one man's well-being in the good cause of adding more to the rest of mankind's well-being."

Apparently the just or equitable society (a) makes accurate interpersonal comparisons of the subjective states of individuals A and B ("well-being" and "bliss" are definable only in terms of their tastes); and (b) taking as the ultimate good the satisfying of tastes, without regard to their nature. If A's preference is for putting B into a gas chamber, then "bliss for the whole universe" is served by his putting him there, provided only that B's loss of satisfaction at being put there is less than A's at putting him there. Even if B claims that his loss of satisfaction will be at least as great as A's gain of it, the just and equitable society will tell him that he is mistaken and put him there anyway in the "good cause of adding more to the rest of mankind's [that is to say A's] well-being." If perchance A and his friends constitute 51 percent of the population and B and his friends only 49 percent, the matter will be simple indeed.

---

[7] Ibid.

Samuelson remarks that the notion of justice as inviolable rights raises the question of whether women, drunks, ants, carrots, and the like have the rights, too. He forgets, perhaps, that utilitarians have contended that in judging which allocation could produce the greatest bliss for the whole universe the well-being of all "sentient beings" must be taken into account. But a society that can make interpersonal comparisons can doubt- less make "interspecial" ones as well.

Aside perhaps from economists, no one believes that "economic welfare" (i.e., what A prefers) ought to be the basis of policy. In most matters, including all important ones, society more or less deliberately lays down rules as to whose preferences and what kinds of preferences are to be allowed, let alone encouraged. Even if there are no consequences for others or if they are all benefits (i.e., serve the preferences of others), the individual may not be allowed to do as he pleases. We do not allow a person to commit suicide even if no one stands to gain by his remaining alive. We do not permit incest, murder, or cannibalism, not even between consenting adults. It is true, as Samuelson says, that we do not appeal very successfully to the notion of inviolable rights in order to curb exten- sion of the activities of government. Public opinion has perhaps become almost as naively utilitarian as the economist. But this is not an argument in favor of naive utilitarianism.

Joseph Cropsey, a student of political philosophy who is also an economist, after a profound treatment of some of these matters, concludes that welfare economics "blinds itself to the largest bearing, which is the political and moral bearing, of institutions designed to implement the seemingly innocuous and incontrovertible maximization principle."[8]

V

Tied as they are conceptually and methodologically to the aggre- gation problem (i.e., to given preferences), economists fail to consider what other normative, or conditionally normative, formulations of welfare are possible. Musgrave, for example, regards the alternative to taking the individual's preference function (which, he says, includes the utility of

[8]Joseph Cropsey, *Political Philosophy and the Issues of Politics* (Chicago: Uni- versity of Chicago Press, 1977), 27.

defense along with that of door locks and ice cream as given), as that of taking the preference function of "some elite or central authority."[9]

Surely there are other possibilities, albeit perhaps not ones that lend themselves as readily to formal (mathematical) treatment. Conditionally, normative theories of welfare can have all sorts of starting places. Following Aristole one might say, "If you want to maximize happiness, meaning what you consider ultimately important, then you ought to . . ." Following J. S. Mill one might say that individuals have "two sets of preferences—those on private, and those on public grounds" and that "if you want to maximize the set of preferences on public grounds, then you ought to . . ." Or following Arrow one might say that there are "tastes" and "values" and that "if you want to maximize values, then you ought to . . ."

There is no *a priori* basis for regarding one conceputalization (e.g., tastes rather than values) or one mode of getting knowledge about the value stuff that is conceptualized (e.g., the market rather than voting) as uniquely, or even specially, appropriate for conditionally normative analysis. The relevant question is empirical, that is, what in the circumstances are the conceptions of welfare that people hold or that they would consider relevant?

To the political scientist, it appears obvious that the aim of much activity is not to give satisfaction to persons but rather to maintain, or enhance, abstract entities, not only formal organizations such as business firms, government agencies, and labor unions, but also, and especially, associations and institutions the function of which is largely symbolic. The state is certainly the most important representative of this last class, and it is with reference to its well-being that most matters are ultimately decided. In his account of "public policy" in English and French law, Dennis Lloyd says that in both countries a judge confronted with a question of public policy

> is expected to consult not his personal inclinations but the sense and needs of the community in a spirit of impartiality. Moreover it must be something which the very structure and character of the community demands; not merely something which people as a whole think desirable or otherwise, and which they would pretty generally agree to accept or reject.[10]

[9]Richard Musgrave in an unpublished paper, 1967.
[10]Dennis Lloyd, *Public Policy* (London: Athlone Press, 1953).

Like these judges, the rest of us must take into account more than "given preferences of individuals" if we are to think about public policy. Accordingly, we need more differentiated, as well as more sharply defined, concepts by which to analyze value stuff. Distinctions need to be drawn along the following lines at least: (1) permanence/impermanence (e.g., whim vs. value); (2) cognitive status (e.g., caprice vs. product of "reflective equilibrium"); (3) kind of legitimation (e.g., expressive, aesthetic, economic, moral, juridicial); and (4) interindividual incidence (e.g., widely shared, held in common, basis of a "fellowship relation," cultural value, etc.).

With adequate conceptual building blocks, one could systematically describe the many conceptions of welfare that are held and acted upon. One could also offer to policymakers (including perhaps, the English and French judges) novel conceptions of welfare—for example, welfare à la Rawls and à la Arrow. One could also inquire what kinds of social choice processes—for example, what mix of discussion, struggle, voting, and so forth—tend to be most productive.

CHAPTER **20**

Are Homo Economicus *and* Homo Politicus *Kin?*

Now, by virtue of its very nature, this utilitarian system is
incapable of taking account of political life and of the way in
which states, governments, parties and bureaucracies actually
work. We have seen that its fundamental preconceptions do
little harm in fields such as that part of economics where its
"logic of stable and barn" may be considered as a tolerable
expression of actual tendencies. But its application to political
fact spells unempirical and unscientific disregard of the essence—
the very logic—of political structures and mechanisms, and
cannot produce anything but wishful daydreams and not very
inspiring ones at that. The freely voting rational citizen, con-
scious of his (long-run) interests, and the representative who
acts in obedience to them, the government that expresses these
volitions—is this not the perfect example of a nursery tale.
Accordingly we shall expect no contribution to a serviceable
sociology of politics from this source. And this expectation is
almost pathetically verified.

J. A. SCHUMPETER[1]

In the ten years that have passed since Schumpeter published these words
ridiculing James Mill's *Essay on Government*, an extensive literature has
appeared that attempts to do the very thing that Schumpeter said could
not be done.[2] This literature, the work of economists for the most part,
applies an "economic approach" to the study of politics and government

---

[1]J. A. Schumpeter, *History of Economic Analysis* (New York: Oxford University
Press, 1954), 429.

[2]Curiously, Anthony Downs declares in one of the first and most important books
in this literature that Schumpeter's analysis of democracy "forms the inspiration
and foundation of our whole thesis." *An Economic Theory of Democracy* (New
York: Harper & Brothers, 1957), 29.

328

and aims to bring both "economic" and "political" phenomena within the compass of a "general equilibrium theory."[3]

Perhaps the most characteristic feature of the "economic approach" is that it assumes the actor is a rational and self-interested calculator. *Homo politicus,* as Downs calls him, "approaches every situation with one eye on the gains to be had, the other eye on costs, a delicate ability to balance them, and a strong desire to follow wherever rationality leads him."[4]

As Buchanan and Tullock put it, man is assumed to be a "utility-maximizer in both his market and his political activity."[5] Organizations (e.g., political parties and governments) as well as men are utility maximizers. Downs, for example, tries to provide "a generalized yet realistic behavior rule for a rational *government* similar to the rules traditionally used for rational consumers and producers"[6] (italics added).

The aim of the present paper is to consider the uses and limitations of the maximizing assumption as applied to politics and thus, by implication, the prospects for a general theory of politics and economics.

1

In economics proper (i.e., economics as distinguished from applications of the "economic approach" to politics), it is usually assumed that the actor tries to maximize a money quantity. Economists have of course always been well aware that nonmonetary, as well as monetary, considerations enter into action. In their purely formal analysis, this has never presented any difficulty for them. At a less-than-purely-formal level of

---

[3]Ibid., 3. The Thomas Jefferson Center for the Study of Political Economy, University of Virginia, has prepared a mimeographed bibliography.

In much of the literature, the distinction between "economic" and "political" phenomena is not clearly drawn. James Buchanan and Gordon Tullock, for example, equate politics to *"collective action,"* meaning action in the light of a common purpose. *The Calculus of Consent* (Ann Arbor: University of Michigan Press, 1962), 19. Presumably they regard *all* organizational behavior, including that of firms, as political.

[4]Downs, *An Economic Theory,* 7–8.
[5]Buchanan and Tullock, *The Calculus of Consent,* 23.
[6]Ibid., 3.

analysis, however, they have found it necessary to abstract from all non-monetary values. Ricardo, for example, although acknowledging that in choosing investments a capitalist may take into account "security, cleanliness, ease, or any other real or fancied advantage,"[7] proceeds on the assumption that only money profit will be sought. Present-day economists do the some thing: formally speaking, economic man is a utility maximizer; practically speaking, he is a money maximizer. The practice of giving utility a money content (whenever it is given any content at all) "confers great definiteness on economics."[8]

The usefulness of a theory based on such an assumption depends, of course, on how well the assumption fits the facts of life. Obviously a theory that assumes that people will maximize a money quantity has no application to a society without money. Obviously, too, it has no application to the behavior of those members of a commercial society—poets and philosophers, perhaps—who care nothing for money.

It has been the good fortune of economists that the money-maximizing assumption has fitted the facts reasonably well in those parts of the world that have been of most interest to them. As Bagehot said, the money-maximizing assumption describes "a sort of human nature such as we see everywhere around."[9] There are, to be sure, many cultures in which human nature is very different: as Bagehot went on to point out, the assumption applies only to "a very limited and peculiar world." Where it applies, however, its definiteness makes it an extraordinarily powerful tool of analysis.

At first sight it seems reasonable to suppose that an approach that works in analyzing behavior in one sphere of life will work—although perhaps not quite as well—in analyzing it in a different but related one. "Experience that goes back to antiquity shows," Schumpeter says, "that

[7]David Ricardo, *The Principles of Political Economy and Taxation,* Everyman's Library edition, p. 49.

[8]The quoted words are George Stigler's. He was referring to the profit-maximizing assumption, but it is clear from the context that he meant money profit. *The Theory of Price* (New York: Macmillan, 1952), 148–150. In his Presidential Address to the American Economics Association, H. S. Ellis said, "Only to the degree that these costs and utilities to individuals can express themselves in the market can economic analysis exist." "The Economic Way of Thinking," *American Economic Review* XL:1 (March 1950): 5.

[9]Walter Bagehot, "The Postulates of Political Economy," *Works* (Hartford, Conn.: Travellers Insurance Company, 1891), vol. 5, pp. 243–244.

by and large voters react promptly and rationally" when presented with "issues involving immediate and personal pecuniary profit."[10] It is in view of this experience, perhaps, that Buchanan and Tullock assert that "the most reasonable assumption is that the same basic values motivate individuals in the two cases, although the narrowly conceived hedonistic values seem to be more heavily weighted in economic than in political activity."[11]

On closer examination, the reasonableness of this assumption is questionable. As Schumpeter goes on to explain, the less immediate, personal, and pecuniary the voter's interest, the more irrational ("ignorant," "irresponsible," "infantile") his behavior.[12] That a man selects his groceries by rational calculation is therefore no indication that he will select his wife or his political party by rational calculation: in these other matters impulse, whim, or habit may prevail.

Even among those who do *everything* by rational calculation (assuming there are such) and who share the same basic values, a bewildering variety of ends may be sought. As Parsons has pointed out, differences of institutional role and personality produce differences in the *content* of motivations that are formally the same.[13] For example, the same goal ("success") that impels a businessman to take a narrowly self-serving course of action and to maximize profit may impel a member of a learned profession to take a self-sacrificing one and to maximize nonmonetary values. Similarly, an individual may be money minded in some matters and not money minded in others that are of less (or more!) importance to him. William Stanley Jevons, we are told by Keynes, far from despising money, "suffered severe pangs each time that a sacrifice was called for"; yet "at every critical stage of his affairs sacrificed his income relentlessly in order to secure his major purposes in life."[14]

There is, then, an important sense in which Buchanan and Tullock are wrong when they state the truism (as they suppose): namely, that the "same" individual acts in all spheres.[15] In fact, it is precisely because the

---

[10]J. A. Schumpeter, *Capitalism, Socialism, and Democracy* (New York and London: Harper, 1942), 260.

[11]Buchanan and Tullock, *The Calculus of Consent*, 19.

[12]Schumpeter, *Capitalism, Socialism, and Democracy*, 261–62.

[13]Talcott Parsons, "The Motivation of Economic Activities," *Essays in Sociological Theory Pure and Applied* (Glencoe: The Free Press, 1949), 211.

[14]J. M. Keynes, *Essays in Biography* (London: Ruper: Hart-Davis, 1951), 257.

[15]Buchanan and Tullock, *The Calculus of Consent*, 20.

individual acts "differently" that such distinctions as that between "economic" and "political" are indispensable.

Perhaps the "limited and peculiar world" in which actors seek to maximize money quantities is shrinking in extent; in time, perhaps, it may become so small that the present body of economic theory will have no application and will therefore be worthless. Some economists seem to think that this time has already come or is near at hand.

To the extent that the money-maximizing assumption is, or becomes inapplicable, economists will probably abandon the "economic approach." Up to now, at any rate, those of them who have tried to analyze phenomena lying outside the "limited and peculiar world" of money maximization have not employed a maximizing model in which nonmonetary values (e.g., security, cleanliness, ease, etc.) replace monetary ones. Instead, they have dropped the maximizing assumption altogether—which is to say that they have dropped the "economic approach." In his analysis of capitalist development, for example, Schumpeter does not assume that anyone is trying to maximize anything. Economic theory that does not use the maximizing assumption (e.g., "institutional" theory) does not differ from sociology in its approach.[16]

It is ironical, then, that although some economists are becoming more and more dubious about the possibilities of applying the "economic approach" to the central subject matter of their discipline (e.g., the behavior of the firm), others are turning optimistically to the task of applying the same approach (as they believe), or a similar one, to the subject matter of political science. That they cannot use a maximizing model to predict the behavior of the business firm does not make them less hopeful of using one to predict the behavior of governments, parties, interest groups, politicians, administrators, and voters.[17]

2

In applying the maximizing approach to political phenomena, it is rarely possible to employ the assumption that a *money quantity* will be maximized. One cannot, for example, assume that a senator will vote so

[16]Cf. Adolph Lowe, *Economics and Sociology* (London: Allen & Unwin, 1935).
[17]See, for example, Charles Lindblom's review of Downs' book in *World Politics* 9, no. 2 (January 1957): 240–253.

as to maximize his income simply because (unless he is corrupt) his income will not be affected by his vote.

Writers who take the "economic approach" to politics seem to think that if the situation does not allow of the assumption that money income (or profit) will be maximized, the next best thing is to assume that something similar (in the sense of related to the individual's material welfare or satisfaction) will be maximized. Buchanan and Tullock appear to be taking this position when they assume that the actor will choose more in preference to less, "more" and "less" being defined "in terms of measurable economic position."[18] Presumably they do not say "in terms of money" because they recognize that this criterion is seldom applicable in political situations. By "measurable economic position" they probably mean the material welfare or comfort of the actor.

Downs is very insistent on viewing actors as "selfish,"[19] and he excludes all goals except "economic and political" ones from their calculus.[20] That an individual works partly because he enjoys his work is a "psychological" fact that can be ignored; only the individual's desire to increase his purchasing power—an "economic" fact—need be taken into account.[21] But although he seems—most of the time[22]—to mean by "selfish" *narrowly* selfish, Downs does not assume, as Buchanan and Tullock do, that the content of self-interest is always material welfare or comfort. "Economic" goals seem to be most important in his eyes, but "political" ones cannot be left out of account altogether even if "psychological" ones can. Thus, in listing the goals of the party member, he puts income first and follows it with prestige, power, and the excitement of the political game.[23]

[18]Buchanan and Tullock, *The Calculus of Consent*, 29.
[19]Downs, *An Economic Theory*, 27.
[20]Ibid., 6.
[21]Ibid.
[22]Downs' definition of the voter's self-interest is (as he acknowledges on p. 36) circular: a "benefit" from government activity is whatever the voter votes in support of, and he votes in support of whatever is a benefit. Thus the rational, selfish voter may vote for a party that will tax him to send food to the starving Chinese (ibid., 37). Moreover, Downs "does not rule out the possibility" that self-interest may take such forms as competition for service or the striving for professional status by means of excellent work." Self-interest, he adds, "may be a far cry from a simple desire for a high income or sweeping power" (ibid., 292).
[23]Ibid., 291. On p. 28, however, only the first three of these goals are postulated. On p. 30 "love of conflict" is mentioned; presumably this is akin to the

Roland N. McKean recommends assuming "that each individual is concerned about himself, his immediate family circle, and a few close friends" and that the individual neglects "whatever concern he may feel for other living persons or for unborn generations."[24] In "self-interest *rather narrowly conceived*" (italics in the original), material comfort "is clearly a major item" (although by no means the only one).

Selfishness *broadly* conceived is useless for analysis. As McKean says, it "can explain any sort of action but predict none."[25] Selfishness *narrowly* conceived, however, may not be an applicable criterion in a political situation; this is especially likely to be the case when "important" questions are to be decided. Buchanan and Tullock, after explaining that most political theorists have gone wrong in assuming that the individual tries to maximize the "public interest" or "common good" rather than his own utility,[26] acknowledge that in the most important matters (namely "constitutional" choices) "*identifiable* self-interest is not present in terms of external characteristics"[27] (italics in the original). This means that in choosing the rules within which a political process is to be carried on, the rational individual must act "*as if* he were choosing the best set of rules for the social group."[28] In short, in constitutional choices (the kind with which Buchanan and Tullock are particularly concerned), it makes no difference whether the individual tries to maximize "the public interest" or his "own utility."

The self-interest axiom is not, as Downs asserts, the "cornerstone" of the "economic approach."[29] What makes behavior predictable to the economist is not that it is self-interested. The selections made by shoppers would be no less predictable if they made them with a view to maximizing the income of the grocer. Nor is it orientation toward material welfare or

---

"excitement of the political game" added on p. 291. Here, one would think, a "psychological" motive has crept back among the "economic or political" ones to which the actor was restricted on p. 6. Certainly "love of conflict" and "excitement of the political game" belong in the same category with "enjoyment of work."

[24] Roland M. McKean, "Costs and Benefits from Different Viewpoints" (Paper delivered at the Conference on Public Expenditure Decisions in the Urban Community, May 14 and 15, 1962).

[25] Ibid.

[26] Buchanan and Tullock, *The Calculus of Consent*, 20.

[27] Ibid., 96.

[28] Ibid.

[29] Downs, *An Economic Theory*, 28.

comfort that makes behavior predictable. If the money maximizer is an ascetic who accumulates capital for the purpose of founding a society for the propagation of asceticism, it makes no difference.

Nor is rational calculation the "cornerstone." It is, to be sure, a necessary condition of the application of the "economic approach," but it is far from being a sufficient one. Consider, for example, the behavior of the courtier:

> Those that look towards the court do not all fix upon the same ends; some the hope of gain and others the desire of honor leads; the ambition of rule draws not a few; and very many steer that course, merely out of a busy inclination, to the engrossing, crossing, or interposing in other men's affairs; while the number of those is very small who primitively intend the honor, safety, and advantage of the prince.
>
> But though their ends be diverse, yet the way to attain to whatever end any man there has pitched upon is but one and common to all that move in that sphere: to wit, the favor of the prince, in obtaining which, the industry and labor of all courtiers is employed.[30]

All of the courtier's behavior is oriented toward the maximization of a single goal, namely obtaining the favor of the prince and in a sense, "favor" is like "profit." (The courtier, incidentally, may be hard at work accumulating favor *for the benefit of others*.) It is obvious, however, that a theory that assumes favor will be maximized will not be analytically powerful. Even if, as may well be, the average courtier is more disposed to maximize favor than the average businessman is to maximize profit, one would still expect more from profit-maximizing than from favor-maximizing theory.

The superiority of the profit-maximizing principle arises from two circumstances. First, the observer has better information about the parameters of the maximizer's calculation. His information, which consists of market prices, is better in several respects: it is quantitative, objective (i.e., it is not perceived differently by different persons), and, if not altogether public in character, is at least more easily accessible (its availability does not depend upon a businessman's ability or willingness to furnish it). The observer of the courtier is not entirely without information; if he is familiar with courts, he has some idea of the parameters involved; but as compared to the observer of the businessman, his information is meager and imprecise.

---

[30] Gordon Tullock, ed., *A Practical Guide for Ambitious Politicians or Walsingham's Manual* (Columbia, S.C.: University of South Carolina Press, 1961), 1.

Second, the profit-maximizing principle leads to predictions that are definite. Frequently the course that the rational profit maximizer must take is uniquely determined, and there are seldom more than a few possibilities open to him. Accordingly, the observer who works out the profit-maximizer's problem independently is likely to come to the same conclusion as he. By contrast, the favor-maximizing principle leads to predictions that (when not completely trivial) are apt to be highly indefinite. The observer cannot predict very definitely how the rational courtier will behave because the nature of the courtier's problem does not allow of a single best solution, or even of a few solutions that are clearly better than all others. In order to say (except in trivial cases) which solution is best, one must be able to measure all of the values involved on a common scale, that is, to compute the relative net values of the various solutions. If the observer is to do this in the same way as the courtier (so as to predict the solution that the courtier will arrive at), the scale must be the same for both the observer and the courtier. To the extent that the courtier's relative scale of valuation is subjective or vague (i.e., to the extent that it is unlike a set of market prices, which is objective and precise), the observer cannot reach the same solution as he. Indeed, a rational courtier may reach a different but equally correct solution to his problem each time he confronts it; for, in a manner of speaking, he does his arithmetic with rubber numbers.

The money-maximizing assumption, then, differs from others in a very fundamental way: it, and it alone, measures all of the unlike goals that enter into the actor's choice on a common, impersonal (i.e., objective rather than subjective) numerical scale, thus making known to the observer the exact terms on which some amount of one goal will be traded for some amount of another. *This is the real "cornerstone" of the "economic approach"*—that it deals with motives that (to quote Marshall) "are measurable by a money price."[31]

Motives having to do with material comfort are no more "economic," in the sense of being amenable to analysis by the "economic approach," than are motives having to do with (say) spiritual exaltation *except as the former may be more likely to be measured in money terms*. Where, as is normally the case in the political sphere, material comfort (or "economic position") cannot be measured in money terms, the "economic approach" cannot be employed. One may employ a "maximizing approach"

---

[31] Alfred Marshall, *Principles of Economics*, 8th ed. (New York: Macmillan, 1961), vol. 1, p. 27.

in such situations, to be sure, but this differs profoundly from an "economic" one in that the goals sought are not expressible on a common scale in objective and numerical terms. Ricardo left "security, cleanliness, ease, or any other real or fancied advantage" out of his economic theory because they could not be measured on a common (money) scale. The goals "prestige, power, and love of conflict" are of precisely the same sort; not being expressible on a common (money) scale, they also must be left out of account in a truly "economic approach."

3

Any maximizing model must include assumptions with respect to three matters: (a) *goals,* that is, what the actor seeks to maximize—for example, profit, favor, votes, and the like; (b) *time horizon,* that is, the time period over which the maximum is to be figured—for example, present value, short run, long run, and so forth; and (c) *moral constraints,* that is, limitations on his pursuit of his maximum that the actor accepts out of a sense of obligation or duty—for example, prohibitions against murder, stealing, bribery, and so forth. (If the actor takes account of such prohibitions only as "costs" that it might be "unprofitable" to risk incurring, they are not, of course, moral constraints, but simply environmental conditions, like prices, which he must take as given.)

As has been shown previously, the money-maximizing assumption has conspicuous advantages over other goal assumptions. It is uniquely definite, and in the "limited and peculiar world" within which it has any application at all it is likely to be quite realistic. Other goal assumptions tend to be tautological (or otherwise nonoperational) or to be unrealistic. Even when they are neither, they usually lack definiteness.

When it comes to making assumptions about time horizons, the economist has no special advantage. Firms and consumers seem to be no more regular and uniform in their choice of time horizons—and therefore no more amenable to being theorized about—than are voters, politicians and parties. Indeed, the assumption that the political actor has his eye fixed upon the next election may be more realistic—as well as more definite—than any assumption that can be made about economic actors in general.

With respect to assumptions about moral constraints, one who analyzes the "limited and peculiar" world of business has an advantage again. As a general proposition, it can be assumed that actors in the business world feel morally free to do whatever the law allows and morally constrained from doing whatever it does not allow. Since the law is fairly detailed, clear, and objective, the influence of moral constraints on the businessman is reasonably predictable.

In the political sphere, it is not so plausible to assume either that the actors will be law abiding or that they will be *merely* law abiding. It is inherent in the nature of politics that in the most important situations the actors—some of them at least—may be ready to commit illegal acts or even to overturn the whole structure of the law. To assume, then, as Downs does, that an actor will not perform illegal acts (e.g., that he will not take bribes or violate the constitution) is to put all "great" political struggles "out of bounds." To assume further, as Downs does, that a party member will not try to benefit himself at the expense of any other member of his party is to put most lesser ones out of bounds as well.

If some moral constraint assumptions cannot be made about political matters without undue loss of relevance, others must be made for the sake of realism. One cannot, for example, explain why a rational calculator would trouble to vote (unless he thinks his vote, or his example in voting, would affect the outcome of the election or the maintenance of the political system) except on the assumption (not made by Downs) that he feels voting to be his "duty." If it is realistic to assume that the voter is constrained by a sense of duty to vote, it may also be realistic to assume that he is constrained by a sense of duty to vote *in a public-regarding rather than a private-regarding manner*. Realism may require assuming a large number of moral constraints, some of which (unlike those that appear in the form of law) are vague and ambiguous and depend for application upon the discretion of the individual and the particular circumstance of time and place. Except as an observer knows which moral constraints may be invoked in the circumstances and what concrete content they will have, he cannot predict the behavior of the rational calculator. When, as in the political sphere, moral constraints are not mainly a matter of law, the task of prediction is especially difficult.

4

It is evident that goal assumptions that are sufficiently definite and realistic to be of use in the study of political phenomena are likely to be few in number. Aside from the vote-maximizing assumption, most are likely to apply to a very restricted domain and therefore to be rather highly particularistic. Accordingly, it is a task for empirical investigation to formulate what may be called low-level maximization assumptions— assumptions that fit some social world far more limited (and perhaps more peculiar) than the one Bagehot had in mind. For example, where the assumption that politicians seek to maximize their votes in the next election cannot be applied or does not fit the facts, a more restricted one— for example, that senators with safe seats seek to maximize the approval of their most respected colleagues—may be employed instead.

To the social scientist, the search for such low-level assumptions will not be very attractive. Social science, after all, aims at explaining a great deal by a very little; doing the opposite—explaining little by much—is an enterprise without much point.

Furthermore, it must be acknowledged that in framing low-level assumptions and in estimating the parameters of the actors' choices, what is mainly required is good practical judgment, familiarity with the habits, customs, traditions, and outlooks of those whose behavior is to be predicted, and information about the particulars of the issues that are in dispute. This being so, social scientists as such may be no better equipped than others to make predictions. Indeed, they may be less qualified. The kind of knowledge that is distinctively theirs (e.g., that courtiers seek to maximize the favor of the prince) amounts to little as compared to the commonsense knowledge of particular circumstances of time and place (e.g., the gossip of the court) that they characteristically lack and that others characteristically have.

Index

Administrative management, 187, 188, 189. *See also* Organization science
Administrative task, 187
Aggregation theory, 317, 319, 321, 323, 325
Agricultural Adjustment Act, 18
Aid to Families with Dependent Children, 263
Air pollution, 251
American Management Association, 182
Anderson, Martin, 97
Appalachia, 276, 278
Arbitration, 318, 320–321
Arkansas, federal aid to, 114
Articles of Confederation, 7
Authority
  decline of, 254
  governmental, 235–237, 240–242
Autonomy, moral, 295

Becker, Gary, 133
Bedford-Stuyvesant, 258–259
Behavior, organizational, 186–192
Bentham, Jeremy, 272 n
Blacks
  crime rate, 282, 283–284
  family structure, 262–263
  suburban, 252
  urban housing, 250–251
  white racism toward, 251–252
Boss Tweed, 233
Britain
  constitution, 242
  educational system, 214
  governmental tradition, 237, 240–241, 242

Britain (*cont.*)
  law enforcement system, 213–214
  local government, 212–220
    American system contrasted with, 220–223, 226–229
    changing nature of, 218–220
    citizen–government relations in, 215–217, 218–219
    responsibilities, 212–215
Budget and Accounting Act of 1921, 131
Bureau of the Census, 140
Burns, Arthur, 98–99
Business
  chief executive's role in, 153–154, 155. *See also* Executive(s)
  corruption in, 152–157
    external constraints, 169
    organizational factors, 152–154
    profit motive and, 154, 156–157
  enterprise zones, 257–260, 263–264
  executive development programs, 182–209
  incentive systems, 153, 156–157

Califano, Joseph, A., 62, 64, 67–68, 89
Campaign funding, 18, 48–49
Canada, urban development in, 237–240
Carter, Jimmy, 3
Cavanagh, Jerome, 59, 70–71
Census data, inaccuracy of, 140
Chayes, Antonia, 61
Chicago Housing Authority, 123–124, 175–177
Chief executive. *See also* Executive(s); President
  in business, 153–154, 155

China, 284, 285
Churchill, Winston, 300
Cities, 230–244. *See also* Model Cities
    Program
  Canadian, 237–240
  civility of, 232–234
  credit rating, 111, 249
  crime rate, 233
  ethnic diversity, 232
  federal grants, 112, 114–115, 249
  fiscal crisis, 110–111
  government. *See* Local government
  growth
    government and, 211–229
    population, 231
    urban crisis and, 246–248
  material welfare in, 232
  revenue-sharing, 110–122
  revolutionary tradition, 240–244
  upward mobility in, 233–234
  urban crisis, 245–255
Civil disorder, social injustice and, 282–
    284
Civil rights movement, 129–130
Civil service, 128–129, 160
Civil War, party system during, 21
Civility, urban, 232–234
Clark, Ramsey, 282
Class system, 215–217, 218
Cloward, Richard, 133
Coalitions, of political parties, 26, 236,
    243
College graduates, as percentage of pop-
    ulation, 143–144
Community Action Agencies, 88
Competition, equality and, 168
Congress
  First, 14
  grants-in-aid and, 118–120
  power of, 14–15, 18
  presidential influence on, 32–35, 46–
    47
Congressional Budget Act of 1974, 132
Congressional Budget Office, 129
Congressional Research Service, 129
Consent of the governed, 7
Constitution
  American, 14, 15, 16, 18
  British, 242
Constitutional Convention, 6–9, 11, 12–
    13

Consumption, community's interest and,
    306
Conventions, presidential, 41, 48
Corrupt Practices Act, 149 n
Corruption, 124
  in business, 152–157
    external constraints, 169
    organizational factors, 152–154
    profit motive and, 154, 156–157
  conceptual scheme, 147–148
  constraints, 148–152
  governmental, 157–170
    cost of prevention, 163
    increase, 167–169
    local, 232–233, 236
    organizational factors, 157–163
    political "machines," 160–161, 166–
      167, 233, 236, 237
  monitoring systems, 148–149, 150–
    151, 152, 155–156, 163, 164–165
  petty, 165
Council of Economic Advisors, 131–132
Crime
  in Britain, 213
  conscience as deterent, 279
  as juvenile behavior, 284–286
  Locke on, 282, 284
  by lower class, 285–286
  moral constraints, 279, 289–291, 294–
    298
  organized, 213
  present orientedness and, 271–281
    deterents, 279–280, 281
    effects, 276–278
    as psychopathology, 273–275
  punishment, 286–289, 294
  racism and, 282–284
  sex factors, 284
  universal liberty and, 291–298
  urban, 233
Criminals, recidivism, 145
Cultural Revolution, 284, 285
Culture, present-oriented, 275–276

Daley, Richard, 70, 104, 105
Decision making. *See also* Planning
  opportunistic, 172, 175, 179
Declaration of Independence, 7
Democracy, 269–270
  irreligion and, 280–281
  moral "goodness" of, 299–302

Democracy (*cont.*)
  political system reform and, 36–37
  presidential election process and, 39,
      45, 47–52
Democratic National Convention, 48
Democratic party, 49–50. *See also* Party
      system
Democratic procedure, party system
      and, 23
Demonstration Cities and Metropolitan
      Development Act of 1966, 68–75.
      See also Model Cities Program
DeMuth, Christopher, 101
Department of Housing and Urban
      Development, 67, 69
Disadvantaged. *See also* Minority groups
  definition, 256
  education, 261–262
  enterprise zones and, 257–260, 263
  family structure, 262–263
  housing, 250–251, 260–261, 292–293
  job opportunities, 257–265
      governmental barriers, 260–263
  middle class and, 264–266, 267
  moral judgments by, 295–298
  public library use by, 308–310
  values of, 257–258
Douglas, Paul, 73
Drug addicts, 213
Duhl, Leonard, 61
Duverger, Maurice, 21

Economic analysis
  of social choice process, 317–327
    aggregation theory, 317, 319, 321,
        323, 325
    arbitration, 318, 320–321
    discussion, 318, 320, 322
    exchange, 318
    politics, 319–321
    struggle, 318, 319–320, 322
    tabulation of revealed preferences,
        318
    voting, 318
  social welfare and, 323–327
Economic growth, non-urban, 248. *See
      also* Urban growth
Economics, politics and, 303–304, 328–
      339
  maximization theory, 329–339

Economics and Urban Problems
      (Netzer), 113
Education
  British system, 214
  of disadvantaged, 261–262
  moral judgment and, 295–298
Eisenhower, Dwight D., 49, 53, 119–120
Elections
  campaign funding, 18, 42, 43, 48–49
  presidential
    direct democracy and, 39, 45, 47–52
    partisanship and, 50–51
    party system reform and, 38–52
    state party leadership and, 39–40,
        41, 42, 44–45, 47, 48, 50
Enterprise zones, 257–260, 263–264
Enthoven, Alain, 138–139
Equality, competition and, 168
Ethnic groups
  local government and, 224–225
  urban, 232
Exchange, as social choice process, 318
Executive(s). *See also* Organization
      science
  in business, 153–154, 155
  functions, 192–194
  governmental, 159–160, 161–162
    authority of, 240–243
  judgment making by, 192–195, 197,
      199, 200–202, 207–208
Executive branch, policy scientists in,
      128–129
Executive development programs, 182–
      209
  definition, 182
  efficacy, 184–185
  organization science and, 185–209

Faction, 9–11
Family Assistance Plan, 113
Family structure, 262–263
Farr, Walter, 76, 90, 91
Federal aid. *See* Grants-in-aid
Federal Election Commission, 42
Federalism
  Constitution and, 14, 15, 16, 18
  New, 115–117
  popular government and, 5–19
Federalist, The, 9, 10, 11, 12, 13, 15,
      16, 17
Ford Foundation, 183

Foreign policy, 300
Franklin, Benjamin, 7, 307
Friedman, Milton, 97–98

Gallatin, Albert, 16
Game theory, 134, 323
Garcia, Robert, 263–264
General Accounting Office, 129, 137
Glazer, Nathan, 59
Goodwin, Richard N., 58, 59
Gordon, Kermit, 63, 65
Government. *See also* Local government;
    Popular government
  British, 237
  corruption in. *See* Corruption
  Hume on, 272
  public confidence in, 47
Government Employees Training Act of
    1958, 183
Grants-in-aid
  to cities, 112, 114–115, 249
  Congress and, 118–120
  federal power and, 18–19
  revenue sharing and, 114–115, 116–
    119
  special interests and, 117, 118–121
Great Compromise, 9, 14, 16
Great Society, 130–131
Green belts, 212, 216–217, 220, 226

Haar, Charles M., 63
Hamilton, Alexander, 7, 15, 142, 241–
    242
Hart, Gary, 3
Hayes, Rutherford, 41
Headstart, 138
Heineman, Ben W., 63
Heller, Walter, 57, 65–66, 110, 111
Hitch, Charles, 134
Hobbes, Thomas, 1–2, 271 n
Homicide rate, 233
Hoover, Herbert, 130, 131
Hoover Commissions, 130, 131
Housing
  density per room, 248
  low-cost, 260–261, 292–293
  urban, 250–251
Housing authority, planning by, 123–
    124, 175–177
Hume, David, 272
Humphrey, Hubert, 71, 72, 110

Hyde, Floyd, 99–100, 102, 104–105

IBM, 258–260
Immigrants
  illegal, 261
  legal, 261
  in urban areas, 232
Implementation analysis, 138–139
Incentive systems, 153, 156–157, 159,
    165, 178–179
Institute for Congress, 129
Irish Republican Army, 284

Jackson, Andrew, 16
Jefferson, Thomas, 14–15, 16
Jevons, William Stanley, 331
Job opportunities, for disadvantaged,
    257–265
Johnson, Lyndon B., Model Cities Pro-
    gram and, 57, 58, 62–63, 67, 68,
    71
Johnson, Samuel, 292
Joint Federal-State Action Committee,
    119–120
Judgment
  executive, 192–195, 197, 199, 200–
    202, 207–208
  moral, 193, 291, 295–298
  psychopathology and, 274
  value, 192–193
Juvenile delinquency, 285

Kaiser, Edgar, 63, 65
Kennedy, John F., 50, 57, 58
Kennedy, Robert F., 61–62
Kerner Commission Report, 282
Klaman, Saul B., 59
Knapp Commission, 150 n, 156 n

LaGuardia, Fiorello, 34–35
Lane, Robert E., 143–145
Law enforcement, British system, 213–
    214
Legislative branch, policy scientists in,
    129
Liberty, universal, 291–298
Libraries, 305 n, 307. *See also* Public
    libraries
Licensing, occupational, 260
*Life and Labour of the London Poor,
    The* (Mayhew), 275–276

Lincoln, Abraham, 18
Lindsay, John, 70, 110
Lobbies, 47
Local government, 209–210, 234–235.
    See also Cities
    British, 212–220
    Canadian, 237–240
    changing nature of, 223–227
    corruption in, 232–233, 236
    ethnicity in, 224–225
    expenditures, 111
    federalism and, 5–19
    fragmentation of authority, 235–237
    revenue sharing, 110–122, 249
    special-function authorities, 224
    special interest groups, 235
    taxes, 112
    urban crisis and, 252–253
    urban growth and, 211–229
    working class control of, 221–222
Locke, John, 271 n, 282, 284
London County Council, 215, 217
Lower class. *See also* Disadvantaged
    criminal behavior, 285–286

McGill, Ralph, 59
McGovern, George, 3, 48, 74
McNamara, Robert, 134
Madison, James
    on corrupt government, 17
    on governmental authority, 241–242
    on popular government, 1, 7, 8, 9–12
Maine, Henry Sumner, 35
Majority faction, 10–12
Management. *See also* Organization
        science
    administrative, 187, 188, 189
    principles, 203
Manpower Training Act, 116
Maximization principle, 329–339
Medicaid, 142, 263
Medicare, 142
Menninger, Karl A., 59
Metropolitan and Urban Affairs Task
        Force, 58–60
Meyerson, Martin, 59
Middle class
    disadvantaged and, 264–266, 267
    flight to suburbs, 248–249
    party system and, 49
Minimum wage laws, 260, 263–264

Minority groups. *See also* Ethnic groups
    in local government, 227
Mitchell, John, 98
Model Cities Program, 53–55, 57–96,
        114, 136
    administration, 75–90, 105
    applications, 78–80, 82–85
    appropriations, 84–85, 90
    citizen participation, 81, 82–83
    City Demonstration Agency, 82, 85,
        86–87, 88–89
    constraints, 92–94
    Demonstration Cities and Metropoli-
        tan Development Act of 1966, 68–
        75
    evaluation, 97–109
        first-year results, 85–87
    interagency relations, 87–90
    Nixon administration and, 53, 54–55,
        90–91, 97–109
    planning, 58–68
    task forces, 58–68
Modeling, in policy science, 136
Mondale, Walter, 3
Morality
    of American democracy, 299–302
    crime and, 289–291, 294–298
    judgment in, 193, 291, 295–298
Morris, Gouverneur, 7, 15–16, 241, 242
Moses, Robert, 34–35, 225, 227
Moynihan, Patrick, 99
Muskie, Edmund, 73, 74

National Academy of Sciences, 131
Neighborhood Development Programs,
        60
New Deal, 18
New Federalism, 115–117
New towns, 175, 250
New York State
    Public Officers' Law, 149 n
    taxes vs. grants-in-aid, 114
New York State Urban Development
        Corporation, 253
New York City, youth development pro-
        grams, 61–62
Nichols, Thomas Low, 231, 234, 244
Nixon administration
    Family Assistance Plan, 113
    Model Cities Program under, 53, 54–
        55, 90–91, 97–109

Nixon administration (*cont.*)
  revenue-sharing policy, 115–117, 119
Nominating process, 39–40
Northern Ireland, 283, 284

O'Brien, Lawrence, 50, 71–72
Office of Economic Opportunity (OEO),
    57, 76, 88–89, 114
Office of Technology Assessment, 129
Ohlin, Lloyd, 133
One-Stop Welfare Centers, 60
Operations research, 134, 190 n, 203,
    204
Organization science, 124, 185–192
  case study, 197–199
  executive function and, 192–194
  executive judgment and, 192–195, 197,
      199, 200–202, 207–208
  limitations, 189–192
  subject matter, 185–188
  technical knowledge vs. art in, 194–
      197
  techniques, 203–206
  theoretical, 203–204
Organizations
  corruption, 124, 147–170
  descriptive studies, 203, 205
  goals, 322
  opportunistic decision making by, 172,
      175, 179
  planning by, 171–181

Parliamentary system, 33
Partisanship, 50–51
Party system, 1–3
  coalitions, 26, 236, 243
  criticisms of, 20–21, 27
  defense of, 20–37
  democratic procedure and, 23
  evaluation, 21–22
  ideological conflict in, 27, 29–30
  middle-class attitudes toward, 49
  model, 25–31
  presidential election process and, 38–
      52
    direct democracy and, 39, 45, 47–52
    nomination process, 39–40
    partisanship and, 50–51
    renomination process and, 41, 43,
        46–47

Party System (*cont.*)
  presidential election process (*cont.*)
    state party leadership and, 39–40,
        41, 42, 44–45, 47, 48, 50
    priorities, 23–24
    reform, 22–25, 31–37
      danger of, 36–37
      presidential elections and, 38–52
    responsibility of, 31–35
    social change planning and, 22–25
Patronage, 34
Pechman, Joseph, 110, 111
Pendleton Act of 1881, 131
Philadelphia
  government corruption, 233
  population growth, 231
Philanthropy, 232
Philippines, party system, 30–31
Pinckney, Charles Cotesworth, 14
Planning, 123–124, 171–181
  by housing authority, 123–124, 175–
      177
  organizational practice and, 174–181
  process, 173–174
  as rational decision making, 171–172
  theoretical model, 172–174
Planning-Programming-Budgeting Sys-
    tem, 79
Policy science, 124–125, 127–146
  effects, 129, 142–146
  historical background, 129–132
  implementation, 138–139
  "knowledgeable" society and, 144–145
  modeling, 136
  operations research, 134
  policymakers' response, 140–142
  program evaluation, 136–138
  role, 135–139
  Schlesinger on, 142
  social reform and, 129–132
  techniques, 132–135
    limitations, 139–142
  training in, 135
Political "machines," 160–161, 166–167,
    233, 236, 237
Political system. *See also* Party system
  historical development, 240–243
Politics
  economics, relationship with, 328–339
    maximization theory in, 329–339
  Woodrow Wilson on, 123

Pollution, 145
Popular government, federalism and, 5–19
Population
  foreign-born, 232
  urban
    decline, 247–248
    growth, 231
    projected, 246–247
Poverty, urban, 246
Prejudice, 145
Present orientedness, 271–281
  crime and, 279–280, 281
  cultural, 275–276
  effects, 276–278
  as psychopathology, 273–275
President
  Congress and, 32–35, 46–47
  election. See Elections, presidential
  sources of power, 34–35
President's Commission on Campus
    Unrest, 282
President's Commission on Crime and
    Violence, 282
President's Committee on Administrative
    Management, 131
Primaries, 40, 41–43
Probability theory, 134
Profit motive, 154, 156–157. See also
    Maximization principle
Program analysis, 79, 136–138
Prostitution, 213
Psychopathology, present orientedness
    and, 273–275
Public action principles, 306–307
Public libraries, 305–316
  public action principles and, 306–307
  service orientation, 305, 307–308
    children, 310
    disadvantaged, 308–309
    "light" readers, 311–312
    during nineteenth century, 307–308
    proposed innovations, 313–315
    "serious" readers, 312–313
Public Officers' Law, 149 n
Public opinion, 300
Punishment, for crime, 286–289, 294

Racism, 251–252, 282–284
Rafsky, William L., 63, 77
Rand Corporation, 134

Rand Graduate Institute, 135
Randolph, Edmund, 8, 12–13
Recent Social Trends, 130, 133
Renomination, 41, 43, 46–47
Republican party, 49. See also Party
    system
Research and development, 143
Research Committee on Social Trends,
    130, 133
Reuther, Walter, 62, 63, 64
Revenue sharing, 55, 110–122
  grants-in-aid and, 114–115, 116–119
  local vs. national needs, 113–115
  New Federalism and, 115–117
  rural migration and, 114–115
  special vs. general, 117
  state-local control, 119–121
Ribicoff, Abraham, 63, 66
Ricardo, David, 330
Riots, 283, 284–286
Rockefeller, Nelson, 110
Romney, George, 99
Roosevelt, Theodore, 41
Ross, William B., 59
Rousseau, Jean Jacques, 271–272 n
Rural migration, 114–115

Schlesinger, James R., 142
Schultz, George, 107
Schultze, Charles L., 65–66, 89
Schumpeter, J. A., 328, 331
Self-expression, 280, 281
Self-interest, 17–18, 334–335
Senior, Nassau, 286, 294
Seventeenth Amendment, 18
Singer, Barry F., 287, 288, 289
Sixteenth Amendment, 18
Slum clearance, 64–65. See also Model
    Cities Program
Smith, Adam, 167
Social chance
  planning, 22–25
  policy sciences and, 129–132
Social choice processes
  aggregation, 317, 319, 321, 323, 325
  arbitration, 318, 320–321
  discussion, 318, 320, 322
  economic analysis, 317–327
  exchange, 318
  politics, 319–321
  social welfare, 323–327

Social choice processes (*cont.*)
  struggle, 318, 319–320, 322
  tabulation of revealed preferences, 318
  voting, 318
Social injustice, civil disorder and, 282–
    284
Social sciences
  doctoral degrees in, 143
  federal-sponsored research, 127
  methodology, 132–135
Social welfare
  moral autonomy and, 295
  social choice processes and, 323–327
South Dakota, federal aid to, 114
Sparkman, John, 73
Special interests
  grants-in-aid, 117, 118–121
  local government, 235
  party system, 29
  presidential influence vs., 47
Spinoza, Baruch, 271 n
Standard Metropolitan Statistical Areas
    (SMSA), 247–248, 252
States. *See also* names of individual
    states
  federalism and, 6–19
  grants-in-aid, 112, 114. *See also* Reve-
    nue sharing
  taxes, 111, 112
Stevenson, Adlai, 49–50
Subsistence income, 145
Suburbia
  black population, 252
  juvenile delinquency in, 285
  middle-class flight to, 248–249

Tariff Act, 14
Task forces, 58, 108–109
Taylor, Frederick Winslow, 130
Taylor, H. Ralph, as Model Cities Pro-
    gram administrator, 75–77, 79–81,
    84, 86, 87, 89–92
Taxes, 111, 112
Tocqueville, Alexis de, 17, 168, 232, 280
Truman, Harry, 5, 49
Tweed, William Marcy, 233

Utilitarianism, 323–324, 325
Upward mobility, 233–234, 261

Urban Affairs Council, 60
Urban crisis, causes of, 245–255
  local government, 252–253
  middle-class flight, 248–249
  physical environment, 250–251
  public values, 253–255
  racism, 251–252
  urban growth, 246–248
Urban development, Canadian vs. Amer-
    ican, 237–240
Urban growth
  government and, 211–229
  population, 231
  urban crisis and, 246–248
Urban migration, 232
Urban reneweal, 60. *See also* Model
    Cities Program

Value judgments, 192–193
Vantage Point, The (Johnson), 62, 63
Vernon, Raymond, 59
Veto groups, 235
Vice, legality of, 213
Virginia Plan, 8
Voting crossover, 42

War on Poverty, 129–130
Washington, George, 7, 9, 14, 15, 242
Watts riots, 283
Weaver, Robert D., in Model Cities
    Program, 64, 68, 69–70, 77, 81,
    89
Welfare, 317
  family structure effects, 262–263
  One-Stop Welfare Centers, 60
  social, 295, 323–327
Wilson, James, 7, 8, 12–13
Wilson, Woodrow, 123, 210
Wirtz, Willard, 69, 89
Wood, Robert C., in Model Cities Pro-
    gram, 58–59, 63, 65, 67, 68
Working class
  conventional morality of, 290–291
  local government control by, 221–222
Wurster, Catherine Bauer, 59

Ylvisaker, Paul, 59
Young, Whitney, 63, 64–65
Youth development projects, 61–62